BioLuminescent OptoGenetics

Ute Hochgeschwender
Editor

BioLuminescent OptoGenetics

Molecular Control with Biological Light

Springer

Editor
Ute Hochgeschwender
College of Medicine
Central Michigan University
Mt. Pleasant, MI, USA

ISBN 978-3-032-18120-6 ISBN 978-3-032-18121-3 (eBook)
https://doi.org/10.1007/978-3-032-18121-3

This Springer imprint is published by the registered company Springer Nature Switzerland AG
The registered company address is: Gewerbestrasse 11, 6330 Cham, Switzerland

Preface

The aim of this book is to provide a primer for BioLuminescent OptoGenetics (BL-OG) by giving an overview of and introduction to the use of bioluminescence for controlling light-sensing molecules. The book covers the principle biological mechanisms of light emission and light sensing and the range of tools enabled when combining genetically encoded light production with light sensing for use within and across cells. Through examples covering the various applications, this book provides a practical guide for using this powerful platform for neuroscience and beyond. The book covers information from numerous publications across different fields of research with the goal of compiling this information in a single resource. My apologies if any publications were inadvertently not included. The various chapters contain exhaustive tables to make it easy to find tools and applications and many schematics to help convey the key concepts. The book addresses researchers at any stage interested in exploring BL-OG and hopes to inspire investigators across fields to incorporate bioluminescence into their tools and applications. Importantly, the book will be a resource for teachers interested in enriching educational material with the fascinating and exciting natural phenomenon of bioluminescence.

Acknowledgments to Justine Allen for valuable comments, which helped clarify the presentation of many chapters, and Springer Nature for enabling this book.

Mt. Pleasant, MI, USA Ute Hochgeschwender

Contents

About the Editor

Ute Hochgeschwender graduated from the Freie Universität in Berlin, Germany, with degrees in medicine (MD) and philosophy (MA). Her research focused on development and application of molecular genetic tools for basic and translational neuroscience. Over the past several years, she established a research program that advances the concept of using biological light, i.e., light generated by a protein, a luciferase, to control light-sensing photoreceptors. An important goal in addition to the continuous development of this BioLuminescent OptoGenetics technology is the application of these tools for research into the mechanisms and potential, noninvasive treatment of neurological and psychiatric diseases.

Chapter 1
Introduction to BioLuminescent OptoGenetics

Ute Hochgeschwender

Abstract BioLuminescent OptoGenetics (BL-OG) drives optogenetic elements with internal, locally generated, bioluminescent light created by a nearby luciferase in the presence of its luciferin. This introductory chapter to the primer on BL-OG provides a brief overview of the key elements involved, i.e., luciferases, luciferins, and photoreceptors, the mechanisms of bioluminescent light emission and sensing, the design principles and parameters for creating BL-OG tools, and the many opportunities to improve molecular tools employing bioluminescence. This is followed by a brief overview of the individual chapters that detail the different ways BL-OG has been utilized and applied to increasingly wider areas of research. This includes bioluminescence driving opsins as well as non-opsin photoreceptors and the various intracellular processes targeted, strategies for the use of bioluminescence for communication between cells and for integration of intracellular processes with cellular outcomes, a review of animal studies applying BL-OG tools, and a description of experiences with using BL-OG as a broad educational opportunity. A final Outlook chapter reflects briefly on where developments and improvements might go and on novel possibilities this technology platform might enable.

Keywords Bioluminescence · Optogenetics · Luciferase · Luciferin · Cell communication · Education

BioLuminescent OptoGenetics is the use of bioluminescence to activate optogenetic elements. Bioluminescence, "light emission by living organisms," is light produced during a chemical reaction where an enzyme, a luciferase, oxidizes a small chemical, a luciferin, thereby producing photons. Optogenetic elements are light-sensing proteins, or photoreceptors, that harness light to exert a wide variety of functions.

U. Hochgeschwender (✉)
College of Medicine, Central Michigan University, Mount Pleasant, MI, USA
e-mail: hochg1u@cmich.edu

U. Hochgeschwender (ed.), *BioLuminescent OptoGenetics*,
https://doi.org/10.1007/978-3-032-18121-3_1

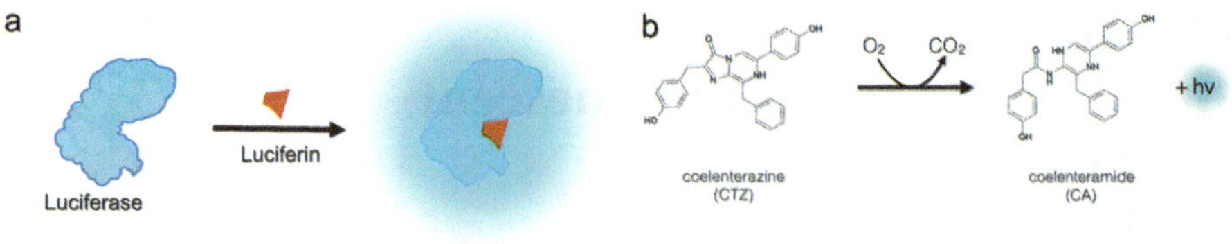

Fig. 1.1 Bioluminescence reaction. (**a**) Bioluminescent light emission requires a luciferase enzyme that interacts with a luciferin, resulting in emission of photons. (**b**) Chemical reaction of the luciferin coelenterazine (CTZ) being oxidized by a luciferase in the presence of oxygen to coelenteramide (CA) and emitting light

1.1 Bioluminescence Reaction

For bioluminescent light emission to occur, both a luciferase and a luciferin are needed. The luciferase is a genetically encoded enzyme produced in a cell while the luciferin is a small molecule externally supplied to the cell (Fig. 1.1a). The basic reaction underlying bioluminescence is shown below for the luciferin coelenterazine: in the presence of oxygen, the luciferase catalyzes the reaction of coelenterazine with molecular oxygen to produce a high-energy intermediate (oxyluciferin) that, upon returning to its stable ground state, releases the excess energy as light, ultimately forming coelenteramide and releasing carbon dioxide (Fig. 1.1b). In nature, this reaction is used by marine organisms, such as the copepod *Gaussia* and the sea pansy *Renilla*. While this basic reaction underlies all bioluminescent light production, different luciferase–luciferin pairs require different cofactors. For example, light production from the firefly luciferase–D-luciferin interaction requires adenosine triphosphate (ATP) and magnesium ions (Mg^{2+}). For BL-OG applications, luciferases that use coelenterazine and only require oxygen, readily available in all cells, provide a distinct advantage. Both pieces of the interaction, luciferases and luciferins, can be modified to change the properties of the bioluminescent light released, allowing a myriad of opportunities to create and improve molecular tools employing bioluminescence.

1.2 Luciferases

Bioluminescence is common in nature and, by recent estimates, has independently emerged at least 94 times [1] (see Chap. 14 for more background). While luciferins are generally conserved (most marine bioluminescence is accomplished by only four luciferins [2]). there are a variety of luciferases that can be exploited for BL-OG. These include natural luciferases from terrestrial and marine organisms [3] as well as, more recently, de novo synthetic luciferases [4] (Fig. 1.2). Luciferases differ in the luciferins used, the cofactors they depend on, and the wavelength their light-emitting reactions produce. Thus far, only a small number of luciferases have

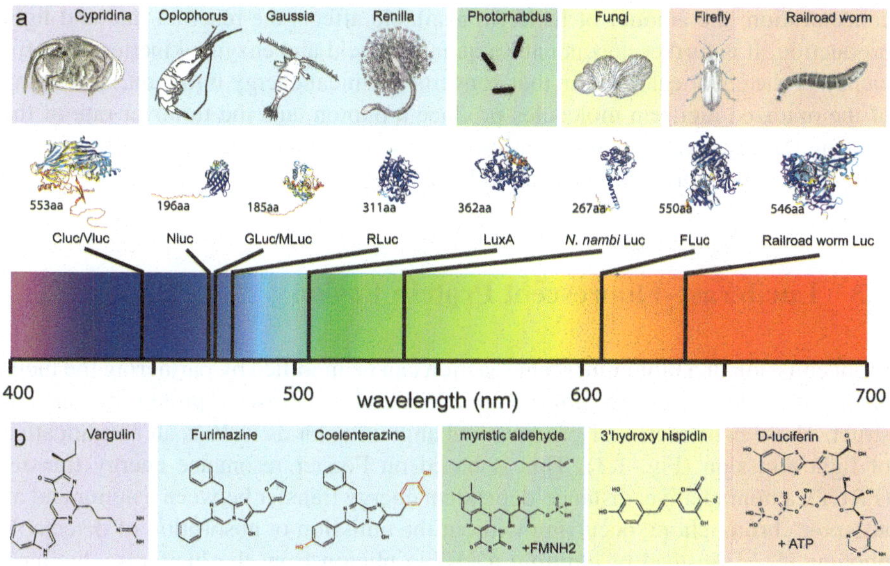

Fig. 1.2 Overview of most common luciferases and their luciferins used in research applications. (**a**) Diversity of bioluminescent organisms and their luciferases. The shading indicates the class of luciferins and their derivatives employed: purple = vargulin, blue = coelenterazine, green = FMNH2/aldehyde, yellow = hispidine, and orange = D-luciferin. The catalytic subunits of the luciferase have been modeled in AlphaFold2 and are scaled approximately to their relative size. The number of amino acids is indicated in each structure, and their peak emission wavelength is visualized below as a vertical black line within the visible spectrum. (**b**) Chemical structures of some representative luciferin substrates used by the luciferases shown in panel (**a**). [3]

been utilized in BL-OG tools. The main reason is the requirement for a substantial amount of light emission to activate nearby photoreceptors, and many natural luciferases are rather dim. The wavelength emission of natural luciferases ranges from 450–510 nm, the bioluminescence maximum of most marine species, to 490–590 nm for terrestrial organisms [5]; yet to date, most BL-OG applications utilize blue light-emitting luciferases. This reflects the combination of luciferases with the many available blue light-sensing photoreceptors. As more red-light-sensing optogenetic elements become available, these can be matched with various red-light-emitting luciferases available in the natural world or that result from engineering efforts. Most luciferases employed in BL-OG applications have been molecularly evolved to generate higher light emission compared to their native versions. Continued evolution efforts will result in improved BL-OG tools and expansion to include photoreceptors and a wider range of applications.

The specific photon flux catalyzed by a luciferase, or the number of photons emitted per second, depends on several factors. Light emission is influenced by enzyme concentration: the amount of luciferase present directly influences the number of reactions that occur and thus the light output. It is dependent on substrate concentration: the amount of luciferin available affects the reaction rate and light production. It is further dependent on quantum yield and enzyme kinetics: the efficiency of the chemical reaction that converts chemical energy into light, how many of the oxidized luciferin molecules produce a photon, and the turnover rate of the enzyme. Lastly, experimental conditions play a role. For example, different cell types or different environments can influence the reaction.

1.3 Luciferase-Fluorescent Protein Fusion

Light emission of a bioluminescent reaction can be modified by partnering the luciferase with a fluorescent protein (FP) in a luciferase-fluorescent protein fusion construct. These constructs can generate a change in color as well as an amplification of light emission (Fig. 1.3). This is based on Förster resonance energy transfer (FRET), a nonradiative, distance-dependent energy transfer between a donor and an acceptor chromophore, occurring without the emission or absorption of detectable photons [6, 7]. Instead of emitting a visible photon from the luciferase–luciferin interaction, the energy stored in the excited state of the high-energy intermediate, the oxyluciferin, is transferred to the chromophore of the interacting fluorescent protein, which then emits a photon. As light emission from the FP occurs according to the FP's own emission spectrum and quantum yield, photons emitted are tuned in color, typically of a longer wavelength, as well as amplified. Frequently, the energy transfer from luciferases to fluorescent proteins is referred to as BRET (bioluminescence resonance energy transfer); however, the underlying mechanism is FRET. BL-OG tools are increasingly taking advantage of utilizing luciferase–FP fusions as light sources, as it allows expanding the wavelengths of light emission and increasing photon output for improved activation of more diverse optogenetic receptors.

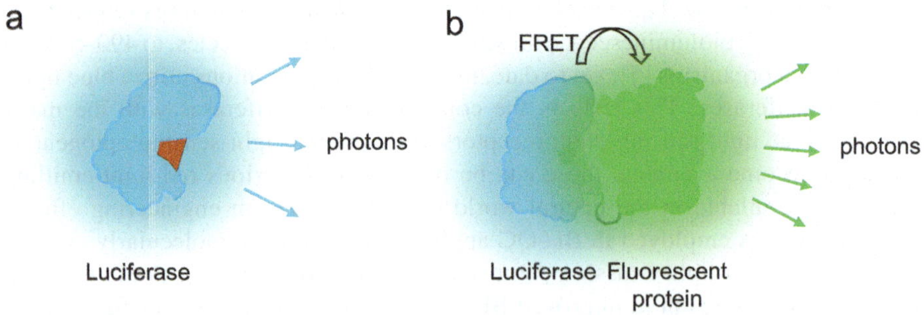

Fig. 1.3 Light emission amplified by FRET. (**a**) The luciferase–luciferin interaction generates photons of a specific wavelength and brightness. (**b**) Fusion to a fluorescent protein allows the interaction to transfer energy (Förster resonance energy transfer, FRET) to the fluorescent protein that in turn emits light with shifted wavelength and higher brightness

1.4 Luciferin

Like a lock and key system, each type of luciferase interacts with a specific luciferin. For example, the luciferin coelenterazine is used by the various luciferases from many marine organisms, including NanoLuc (NLuc) from shrimp and *Gaussia* luciferase (GLuc) from copepods. D-luciferin that is used by the firefly luciferase (FLuc) cannot be utilized by NLuc or GLuc and vice versa, providing, in principle, opportunity for multiplexing. Luciferase–luciferin interactions can be of the "flash"-type, with a high burst of light that quickly decays, or the "glow"-type, where a steady signal is maintained. The flash-decay dynamics observed in FLuc–D--luciferin reactions are based on catalytic byproduct autoinhibition, where oxyluciferin builds up in the active site of the luciferase and competes with luciferin. Due to the availability of engineered, optimized marine luciferases and their blue light emission spectra compatible with many optogenetic elements, the luciferin used in many BL-OG applications is coelenterazine (CTZ) and its variants. Chemical modifications, primarily made to its functional groups, have yielded a diverse range of CTZ analogs with enhanced stability, tuned color emissions, improved brightness, and improved functionality, such as the ability to cross cell membranes. For example, hCTZ, a 2-deoxy derivative of native CTZ, carries a hydroxyl group at the C-8 position that increases its luminescent intensity 10–20 times when used by the photoprotein aequorin. Coelenterazine 400a is a bisdeoxy derivative of CTZ that displays an emission maximum of ~400 nm when converted by *Renilla* luciferase (Rluc). Furimazine (Fz) is a CTZ derivative optimized specifically for use by NLuc. Further modifications include fluorofurimazine (FFz) and cephalofurimazine (CFz), reagents with much improved cell permeability and efficacy of crossing the blood–brain barrier. The increasing selection of luciferin variants with differences in light emission when paired with different luciferases opens possibilities for multiplexing within the CTZ-utilizing group of luciferases. It will be important to continue optimizing luciferins for improved in vivo applications of BL-OG tools. Fortunately, potential toxic effects of luciferins and their oxidation products in vivo have only been demonstrated at unusually high doses, not at effective doses.

Exogenous addition of luciferin allows the experimenter to control the timing of BL-OG reactions. However, there are instances where the continuous presence of luciferin is advantageous, for example, when using bioluminescent sensors, where light emission is dependent on an intracellular factor (e.g. Ca^{2+}) in addition to luciferin. Although the enzymes catalyzing luciferin biosynthesis are still unknown for most bioluminescent systems, the pathways of bacteria and fungi, the lux operon and the caffeic acid cycle, respectively, have been elucidated. These two fully genetically encodable systems can potentially be transferred to mammalian hosts to generate organisms with autonomous and self-sustained light emission.

1.5 Bioluminescence Imaging

Bioluminescent light created during a luciferase–luciferin chemical reaction can be imaged in real time. In fact, bioluminescence has been used in laboratory research for visualizing and quantifying physiological processes at the cellular and organismal levels for decades. BL-OG applications benefit from bioluminescence imaging because the emitted light provides optical confirmation that the luciferin reached its target, the luciferase in close proximity to the optogenetic element, removing delivery uncertainty. Photon emission from bioluminescent reactions is captured by photomultiplier tubes in plate readers and by CCD and EMCCD cameras for observations in cultured cells and in animals, including in freely behaving animals. Luciferases are employed as reporter proteins in gene expression studies and in protein complementation assays in vitro and serve to localize cells in organisms and to observe cellular function in vivo. Importantly, as light emission from bioluminescent reactions does not require external light application, in vivo imaging is carried out noninvasively. With ongoing developments improving the brightness of luciferases and the sensitivity of cameras, bioluminescent imaging will enable us to get ever more information on biological processes in awake, unencumbered, freely behaving experimental animals.

1.6 Optogenetic Elements

Photoreceptors, or photosensors, contain light-sensing molecules that detect and respond to light by initiating a biological action. These proteins contain a chromophore, a light-absorbing molecule, that changes its shape in response to light. The conformational change of the chromophore shifts the shape and/or orientation of the entire protein, creating a biological change. Examples of proteins that contain chromophores are microbial opsins, light-driven proton pumps, ion pumps and ion channels; animal opsins and G-protein-coupled receptors; phytochromes and cryptochromes in plants; and BLUF (blue light-sensing using FAD) proteins in various organisms.

Photoreceptors are involved in diverse functions including vision, plant growth, and many other cellular responses to light. In nature, photoreceptors are activated by the sun: natural light comprising the full visible spectrum. In the laboratory, researchers employ LEDs or lasers with discrete spectra to activate optogenetic elements. This was pioneered with channelrhodopsins, ion channels from algae that open in response to blue light to allow an influx of ions that results in a change in the cell's membrane potential (Fig. 1.4a). By using a laser light source (opto) to control genetically encoded opsins (genetic) expressed in mammalian neurons, optogenetics opened a wide field of experimenter control of cellular processes, including but not limited to neuronal activation. Numerous light-sensing proteins from nature have since been repurposed to interrogate and drive cellular processes

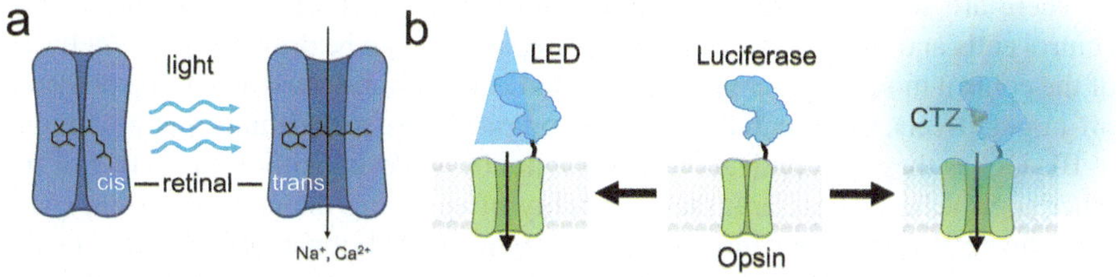

Fig. 1.4 Optogenetics and BioLuminescent OptoGenetics. (**a**) Standard optogenetics: A photoreceptor, here a channelrhodopsin (ChR), is exposed to light; the light absorbing chromophore, retinal, changes from cis to trans, opening the ion channel to allow influx of cations. In natural environments, light is supplied by the sun; in the laboratory, light is supplied by lasers or LEDs. (**b**) BioLuminescent OptoGenetics (BL-OG): a nearby luciferase, here fused to the ChR in a luminopsin (center panel), provides light upon interacting with its luciferin (right panel). The same construct can also be activated by standard optogenetics via an external light source (left panel)

with external light. Beyond the activation of proteins that are naturally sensitive to light, light-responsive domains are increasingly being incorporated into engineered proteins, enabling researchers to use light to control protein localization, interactions, and function.

1.7 BioLuminescent OptoGenetics (BL-OG)

Most optogenetic work has been accomplished by delivering light from an external source, using light fibers or small LED sources, to the target area in experimental animals via an implant. BioLuminescent Optogenetics (BL-OG) drives optogenetic elements with internal, locally generated, bioluminescent light created by a nearby luciferase in the presence of its luciferin (Fig. 1.4b). Most optogenetic systems can be activated by power densities within the mW/m^2 to W/m^2 range. This can be matched by an overall power output of bioluminescent light reactions that generate photon power outputs in the mW/m^2 range when expressed at physiologically relevant levels (μM). In principle, any photoreceptor can be activated by bioluminescence, and the chapters in this book attest to that.

BL-OG employs constructs that are combinations of luciferases and optogenetic elements. When the appropriate small molecule luciferin is added, bioluminescent light is created that activates nearby optogenetic components. The following section details the areas of research application, where BL-OG control is advantageous over standard optogenetic control or traditional pharmacological/chemical control of cellular processes. In neuroscience, these two approaches are commonly juxtaposed as optogenetic and chemogenetic control of neural activity, with optogenetics using opsins controlled by external light sources and chemogenetics using engineered metabotropic or ionotropic receptors controlled by synthetic ligands.

There are several categories that are common to each of the methods used to control cells and systems. For example, comparisons can be drawn as to the delivery of the control mode, the temporal and spatial control of targeted actuators, potential off-target effects, or potential toxicity. I contrast, there are features that are unique to BL-OG as they are applications that cannot be achieved with one or either of the other approaches. These are: the bimodality of engaging the control system (with luciferin application and bioluminescent light or traditional optogenetics in the same construct, Fig. 1.4b), the possibility of monitoring its activation (bioluminescent light as an optical reporter of the luciferin reaching its target luciferase), and applications in cell-to-cell communication (e.g., Interluminescence) and integration of intracellular processes. The comparisons between BL-OG, standard optogenetics, and chemogenetics are tabulated in Table 1.1.

Proponents of either the broad approaches of optogenetics and chemogenetics often underline the advantages of one versus the disadvantages of the other. For example, working with implanted optical fibers both causes physical damage to tissues and limits the breadth of neural structures that can be simultaneously controlled. Furthermore, light delivery is constrained by the absorption, scattering, and thermal sensitivity of biological tissues. While these disadvantages are avoided by using chemogenetic approaches, there are drawbacks inherent to chemogenetics. These include slow kinetics of activation and deactivation due to the need for the ligand to diffuse to and from the targets and its accumulation in the organism and lack of spatial precision.

It would be most productive to first identify the specific needs of a research question and then utilize the experimental approach best suited to deliver the answer. For example, if the ability to control neurons at millisecond time scales is key to the experiment, standard optogenetics is the only approach allowing this manipulation. While there are many situations where this is needed, there are equally many situations where temporal control at this level is less relevant and systemic activation is more important, making chemogenetics the best approach.

BL-OG employs a bioluminescent reaction that provides unique advantages over other chemogenetic tools. In standard chemogenetic activation, an external molecule binds to its target, but it also lingers and even accumulates, resulting in complex on/off kinetics. For BL-OG activation, the external luciferin substrate interacts with the luciferase and emits light for the duration of the activation of the target and is thereafter converted to an inert substance. This results in much tighter temporal control, even allowing for repeated administration of luciferin substrate for pulsatile activation.

Of the unique features of BL-OG tools, bimodal control provides needed flexibility: the option of standard optogenetic activation is maintained, while the combination with bioluminescence provides chemogenetic access to, in principle, any optogenetic element. This dual control with a single construct, either chemogenetically by systemic luciferin administration or locally with high spatiotemporal precision using optic fibers, allows the investigator to take advantage of the unique features of optogenetic and chemogenetic access to the same molecule in the same experimental context. This is a valuable feature when, for example, assessing the

Table 1.1 Advantages of using bioluminescence compared to external light or chemicals

	BL-OG	Optogenetics	Chemogenetics
Distinguishing features between BL-OG, optogenetics, and chemogenetics			
Actuator	Photoreceptors	Photoreceptors	Synthetic GPCRs (DREADDs) or ion channels (PSAMs)
Effector	Exogenous chemical (luciferin)	External light (LED, laser)	Exogenous chemical (CNO, DCZ, PSEMs)
Engaging actuators	Non-invasive	Invasive	Non-invasive
Kinetics of actuator activation	Seconds to minutes	Milliseconds	Hours
Option of pulsatile control	Yes	Yes	No
Reach across anatomical space	Unlimited	Limited (reach of light from fiber tip)	Unlimited
Off-target effects	None/low	None/low	Likely/possible
Toxicity	Not at effective doses	Phototoxicity	Not at effective doses
Modularity of platform	High	High	Low
Multiplexing possibility	Yes (using distinct luciferase–luciferin pairs)	Yes (using photoreceptors with distinct wavelengths)	No
Unique features of BL-OG			
Effector/actuator interaction can be tracked/monitored	Yes (bioluminescence)	No	No
Bimodal access	Yes (opto- and chemogenetic activation of same actuator)	No	No
Cell-to-cell communication	Yes (light emitter and sensor in different cells)	No	No
Integration of cellular processes	Yes (light emitter senses and drives actuator)	No	No

effect of an entire population of neurons in a behavioral sequence versus the effect of individual subcircuits.

Because activation of BL-OG constructs produces photons that trigger physiological events, control cell behavior, and can be simultaneously detected with bioluminescence imaging, users can track the time window of engagement by measuring the bioluminescent light signal. Among the various approaches used to control genetically targeted actuators, such as chemogenetics, optogenetics, sonogenetics, and magnetogenetics, BL-OG is the only approach that affords simultaneous assessment of where and when the activation occurs. The possibility of monitoring

Box 1.1 Bioluminescent Optogenetics Versus Bioluminescence + Optogenetics

Bioluminescent Ca^{2+} sensors are attractive for employment in optogenetic experiments, because the external light application required for fluorescent Ca^{2+} sensors interferes with the activation of optogenetic actuators. This parallel use of bioluminescence and optogenetics [8, 9] is not a BL-OG application because it combines optogenetics and bioluminescent imaging rather than utilizing bioluminescence to control optogenetic actuators, but it is an example of the utility of bioluminescence in diverse applications.

ligand–target interaction might not be utilized routinely in each experimental run but is useful and informative when setting up experimental systems. The visual report on successful substrate–luciferase interaction and, by inference, actuator activation can guide which route or which concentration of luciferin application to use.

Because both the light emitter and light sensor are genetically encoded, BL-OG allows the two moieties to be expressed in different cells for applications in cell-to-cell communication, a feature that allows control of neuronal transsynaptic communication. Lastly, because the light emitter (luciferase) is a protein that can be engineered and split to accommodate sensor moieties, light emission can be made dependent on the presence of various intracellular agents that report a cell's state, resulting in the integration of intracellular states with outcomes affected by optogenetic actuators. This feature has been used to link neurons' activity states with their membrane potential changes by co-expressing a Ca^{2+} sensing luciferase and optogenetic ion channels, basically allowing a neuron to self-correct its own overactivity.

1.8 Overview of This Book

The chapters in this book provide detail on the different ways BL-OG has been utilized, including the improvement, optimization, and expansion of BL-OG tools and their applications to increasingly wider areas of basic and applied research.

Light is critical for sustaining life on earth; thus, a vast variety of light-sensing molecules, broadly referred to as photoreceptors, have evolved. In principle, any photoreceptor can be combined with a luciferase for BL-OG applications. Reflecting the prominence of channelrhodopsins in neuroscience applications, initial BL-OG combinations were with opsins, ion flow-regulating channels, or pumps. These luminescent opsins, or luminopsins (LMOs), were and continue to be used to control neuronal membrane potential and thus neuron activity. At the same time, bioluminescence is starting to be utilized to drive opsins outside of neuroscience applications. Chapter 2: *Bioluminescence Driving Opsins* describes the design principles and parameters for creating efficient LMOs and showcases uses of co-expressed luciferase–opsin combinations.

Bioluminescent light is increasingly being used to activate non-opsin photosensitive proteins. These include proteins that undergo light-driven conformational changes, dimerization, enzyme activation, or generation of reactive oxygen species to manipulate the function, localization, and/or interaction of molecules in living cells. New photoactivatable proteins are being discovered, and the photophysical properties of existing photoactivatable proteins are being continuously improved, opening many opportunities for combining them with luciferases in BL-OG tools. Chapter 3: *Bioluminescence Driving Non-Opsin Photoreceptors* describes key features and design principles of the molecular tools, the intracellular processes targeted, and the applications in vitro and in vivo for proof-of-principle demonstrations and for therapeutic applications.

Bioluminescence has independently evolved many times and in many different lineages, suggesting its importance for survival (see Chap. 14 for more background). Bioluminescence allows communication via light in very dark environments, with one entity emitting and another sensing photons. Behavioral uses include attracting prey, evading predators, signaling food sources for group members, and playing a role in mating. Communication over distance via light transmission has been realized in BL-OG applications by expressing a luciferase in a presynaptic neuron and an opsin in a postsynaptic neuron, enabling synaptic communication through photons. Chapter 4: *Bioluminescence for Cell-to-Cell Communication* outlines approaches of light-induced communication and Chaps. 5, 6 and 7 detail three strategies developed for neuronal communication.

The use of locally produced light allows cells to detect their own chemical fluctuations and drive discrete outcomes. Two main pathways have been explored that couple bioluminescent light emission with downstream processes. In the first, upon administration of luciferin, an intact luciferase emits light that activates a photoprotein. This interaction between the luciferase and the photoprotein is contingent on an intracellular event. The second pathway uses a luciferase construct split by a sensing module. When an intracellular event supplying the molecule required to reconstitute the split luciferase occurs, the luciferase becomes a functional enzyme that emits bioluminescent light to activate a photoprotein. Intracellular events that have been monitored via bioluminescence include the following: intracellular Ca^{2+} increase, protein-protein interactions, GPCR activation, pH change, and protein phosphorylation. In turn, outcomes of these reactions include transcription of reporters and actuators, labeling of nearby proteins, and activation of opsins. Combining such conditional light emitting systems with light-sensing ion channels, transcription factors, and other photoreceptors opens many opportunities for self-regulation of cellular states and communication among cells. Chapter 8: *Bioluminescence for Integration of Cellular Processes* describes the concepts, approaches, and their applications and Chaps. 9 and 10 detail two strategies.

The ultimate goal of molecular tool development is application in experimental animals for basic research and preclinical studies and eventually in clinical contexts. To date, most cases of these BL-OG applications have been in the field of neuroscience, reflecting the earlier development of luminopsins. These studies demonstrate that using bioluminescence to drive light-sensing elements is working robustly in multiple scenarios in vivo and encourages the broadening of this approach to the many potential combinations with other photoreceptors. Detailed studies regarding the relationships between bioluminescent light production and neural spiking activity, luciferin concentrations, and injection parameters indicate that the dynamics of neuromodulation effects of BL-OG can be systemically controlled by the experimenter. Chapter 11: *Application of BL-OG Tools in Animal Studies* reviews these publications, Chap. 12 highlights the potential of BL-OG for therapeutic and fundamental basic science applications, and Chap. 13 details its advantages for studying neural ensembles in mammalian brains.

Seeing living matter emit light has an impact on any sighted person and especially on young learners. Further, the combination of light emitters with light sensors to affect changes in cells and animals continues to fascinate students, and the mechanisms underlying bioluminescent light emission and photoreceptor light-sensing offer countless opportunities for illustrating basic physics, chemistry, and biology concepts. Chapter 14: *Bioluminescence for Education and Outreach* describes experiences with using BL-OG as a broad educational opportunity for learners from kindergarten to graduate school.

BL-OG is a modular approach. "Off-the-shelf" luciferases and optogenetic elements can be combined, and molecular evolution can optimize partners for specific applications. The possibilities for combining light emitters with light sensors are vast, and, as a result of this, so are the possibilities for applications in basic and translational research. Chapter 15: *Outlook* reflects briefly on where developments and improvements might go and on novel possibilities this technology platform might enable.

There are many monographs and reviews available that describe the different players mentioned above: luciferases, luciferins, and photoreceptors, and the reader is referred to these extensive resources. The use of bioluminescence to activate photoreceptors (BL-OG) is a more recently realized concept. The main goal of this primer is to inform readers about the molecular tools available thus far, their design principles, and their applications in basic research and translational studies. An important objective is to convey the immense potential of this approach and to stimulate and inspire researchers to expand on the development and application of BL-OG tools to address questions made feasible by this unique approach.

References

1. Lau ES, Oakley TH. Multi-level convergence of complex traits and the evolution of bioluminescence. Biol Rev Camb Philos Soc. 2021;96(2):673–91.
2. Haddock SHD, Moline MA, Case JF. Bioluminescence in the sea. Annu Rev Mar Sci. 2010;2:443–93.

3. Porta-de-la-Riva M, Morales-Curiel LF, Carolina Gonzalez A, Krieg M. Bioluminescence as a functional tool for visualizing and controlling neuronal activity in vivo. Neurophotonics. 2024;11(2):24203.
4. Yeh AHW, Norn C, Kipnis Y, Tischer D, Pellock SJ, Evans D, et al. De novo design of luciferases using deep learning. Nature. 2023;614(7949):774–80.
5. Allen JJ, Friedman N, van der Merwe R, Bartholomew B, Buckman E, Funderbunk R, et al. BL Base. 2023. Available from: https://www.blbase.org/.
6. Förster T. Energiewanderung und Fluoreszenz. Naturwissenschaften. 1946;33(6):166–75.
7. Förster T. Zwischenmolekulare Energiewanderung Und Fluoreszenz. Ann Phys. 1948;2:55–75.
8. Yang J, Cumberbatch D, Centanni S, Shi SQ, Winder D, Webb D, et al. Coupling optogenetic stimulation with NanoLuc-based luminescence (BRET) Ca(++) sensing. Nat Commun. 2016;7:13268.
9. Narcisse D, Mustafi SM, Carlson M, Batabyal S, Kim S, Wright W, et al. Bioluminescent multi-characteristic opsin for simultaneous optical stimulation and continuous monitoring of cortical activities. Front Cell Neurosci. 2021;15:750663.

Chapter 2
Bioluminescence Driving Opsins

Ute Hochgeschwender

Abstract Opsins, light-responsive ion-conducting channels or pumps, are employed for genetically targeted modulation of a cell's membrane potential. Opsins can be activated by bioluminescent light emitted from luciferases. The most efficient way to deliver luciferase and opsin to cells at a 1:1 ratio and to ensure proximity of the light emitter and light sensor is by tethering the luciferase to the opsin in a luminescent opsin, or luminopsin (LMO). Here, we discuss basic design principles of LMOs, including the options for light-emitting and light-sensing components and strategies for combining them. With the availability of brighter luciferases and more light-sensitive opsins, it will become more feasible to drive opsins with bioluminescence generated from co-expressed light emitters, increasing flexibility and combinatorial specificity. While most applications thus far have been in neuroscience, the technology is ready to be expanded to applications in cell types other than neurons.

Keywords Optogenetic · Chemogenetic · Fusion protein · FRET · Coupling efficiency

2.1 Introduction

The initial motivation for driving optogenetic elements with bioluminescence was to modulate neural activity through light-sensitive ion channels or pumps [1–3], rather than through ligand-activated G-protein-coupled receptors [4, 5], without the need for invasive optical fibers and their limitations in spatial reach [6, 7]. Along with the rapid adoption of optogenetic applications in neuroscience and the continuous discovery and refinement of light-sensing channels and pumps, the majority of bioluminescent optogenetic (BL-OG) tools were developed for and applied to

U. Hochgeschwender (✉)
College of Medicine, Central Michigan University, Mt. Pleasant, MI, USA
e-mail: hochg1u@cmich.edu

U. Hochgeschwender (ed.), *BioLuminescent OptoGenetics*,
https://doi.org/10.1007/978-3-032-18121-3_2

neuroscience research. Now, the use of bioluminescence to drive opsins has been expanded to cell types other than neurons (see Sect. 2.6).

It became clear early on that close proximity between the light emitter and light sensor increases the likelihood of reliably activating opsins and that this proximity is best assured when fusing a luciferase and an opsin to generate a luminescent opsin, or luminopsin (LMO). With the development of brighter luciferases and more sensitive opsins, it has become feasible to co-express the two moieties. This has increased flexibility in the applications of luciferases driving opsins, including across cells (see also Chap. 4). Further, the required extent to which opsins move ions across membranes and affect changes in membrane potential is quite high for modifying neuronal firing but is potentially lower in other applications (see examples in Sect. 2.6).

This chapter details the design principles of LMOs, discusses the underlying mechanisms of opsin activation by bioluminescence, explains factors influencing the efficacy of bioluminescent activation of opsins, and provides practical considerations for choosing tools for driving opsins with bioluminescence.

2.2 LMO Design Principle

In luminescent opsins, or luminopsins (LMOs), the light source (the luciferase) is tethered to the light sensor (the opsin) via a 15-amino acid linker. Because the opsin is integrated into the cell membrane, the luciferase can be tethered to the N-terminus or the C-terminus and thus be located outside or inside the cell, respectively (Fig. 2.1).

Opsins contain a light-absorbing molecule, a chromophore, that undergoes conformational change upon absorption of light, resulting in ion flux (see also Chap. 1, Fig. 1.4). For example, channelrhodopsin, a light-gated channel, conducts cations (sodium, potassium), causing depolarization, whereas halorhodopsin, a

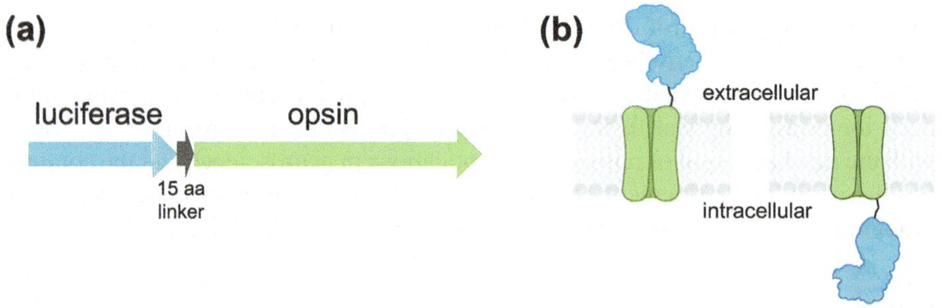

Fig. 2.1 LMO design (**a**) Components of LMO to scale (luciferase extracellular). The luciferase sequence is attached to the opsin by a 15 amino acid linker. (**b**) Configurations of LMOs. The luciferase can be attached to the N-terminus or the C-terminus of the opsin, placing the luciferase outside or inside the cell

light-activated pump, moves chloride ions into the cell, causing hyperpolarization. Light activation of the opsin from the luciferase can, in principle, occur via photon emission or, more efficiently, through radiationless Förster resonance energy transfer (FRET [8, 9]; often referred to as BRET, Bioluminescence Resonance Energy Transfer, when involving a luciferase) (Fig. 2.2). The FRET signal is dependent on the spectral properties, ratio, distance (typically within 10 nm), and relative orientation of the donor (luciferase) and acceptor (opsin) molecules, as well as on the strength and stability of the interaction between the proteins of interest.

In principle, opsin activation can be achieved by co-expressing the light emitter and light sensor, but the efficacy of this activation varies by construct. For example, bioluminescence from Nano-lantern [10] can activate halorhodopsin, NpHR, when the light-emitting and light-sensing components are co-expressed or coupled together as a single fusion protein [7]. The mean coupling efficiency was higher when Nano-lantern and NpHR were expressed together as the iLMO2 fusion protein compared to when they were co-expressed separately, but this difference was not statistically significant. In another example, no significant CTZ-induced current was observed when a step function opsin was co-transfected with sbGLuc [11]. This was in stark contrast to the scenario where the two proteins were expressed together as a fusion protein. In that case, CTZ elicited a photocurrent proportional to lamp-induced current. These latter experiments suggest that the close configuration of the two moieties in one fusion protein is necessary for optimal bioluminescence-mediated activation of opsins in this luciferase–opsin pairing.

In addition to more robust responses, the single fusion protein approach provides a practical means of delivering the light emitter and light sensor at 1:1

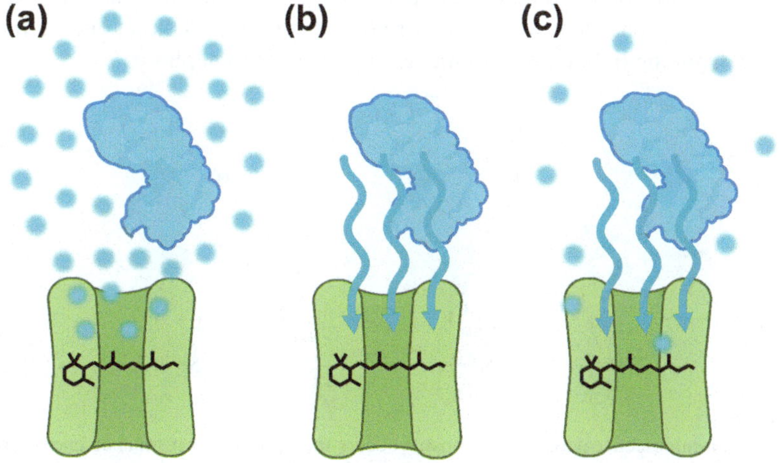

Fig. 2.2 Photon emission versus energy transfer. (**a**) Luciferases emit photons that could activate the opsin chromophore through radiative energy absorption. (**b**) Direct energy transfer through nonradiative dipole–dipole coupling from the donor luciferase to the acceptor chromophore is more efficient than radiative transfer in activating opsins. (**c**) Both mechanisms are assumed to be involved in bioluminescence-driven activation of opsins

stoichiometry. Most importantly, fusion of the two moieties guarantees proximity of the light emitter to the light sensor, placing the luciferase in the best position for allowing intramolecular resonance energy transfer from the luciferase to the channelrhodopsin chromophore versus intermolecular FRET that occurs when the co-expressed components happen to come into close proximity within a cell.

Three sets of experiments were conducted to show that BL-OG works, at least partially, through FRET from the light emitter to the chromophore of the light sensor (Fig. 2.3). All are based on the rationale that if radiationless FRET is a major component in opsin activation, then the donor, the luciferase, will recover radiation-based bioluminescent light emission when the energy acceptor, the opsin, is removed. In the first set of experiments, Berglund et al. [11] used a linker with a protease cleavage site to tether the luciferase to the opsin. Bioluminescence was measured in neurons transfected with the original LMO and the one with the modified linker, with or without a recombinant protease. Presence of the protease significantly increased bioluminescence from the LMO modified with the protease recognition site, indicating that the sbGLuc moiety recovered bioluminescence after it was detached from the acceptor (Fig. 2.3a). To further confirm FRET in LMOs, Berglund et al. [11] removed the FRET acceptor, i.e., the chromophore retinal in the opsin, by removing retinal from the culture medium. This chronic treatment of retinal deficiency also significantly increased bioluminescence from the FRET donor, sbGLuc, compared with cells treated with all-trans-retinal before the assay (Fig. 2.3b).

In the third set of experiments, Björefeldt et al. [12] compared bioluminescence output as radiance per copy of light emitter when the luciferase was tethered to an opsin versus a plain transmembrane sequence (Fig. 2.3c). Radiance per copy decreased for the opsin-tethered compared to the transmembrane-tethered light emitters. This is consistent with a loss of photon emission to direct energy transfer via FRET from the light emitter to the nearby opsin chromophore.

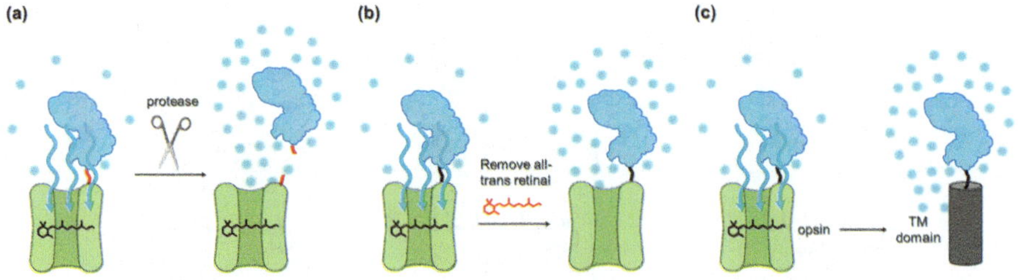

Fig. 2.3 Bioluminescence activation of opsins occurs through Förster resonance energy transfer (FRET). (**a**) Radiance increased from the luciferase after protease cleavage of the tether holding it in a configuration favoring direct energy transfer via FRET from the light emitter to the chromophore. (**b**) Radiance increased from the luciferase after removing the chromophore all-trans-retinal needed for direct energy transfer via FRET. (**c**) Radiance increased from the opsin-tethered to the transmembrane domain-tethered light emitter, as there is no more direct energy transfer via FRET from the light emitter to a chromophore

Generally, if the proximity of the two moieties causes radiationless resonance energy transfer to the opsin instead of activation of the opsin by simple photon radiation, then the donor, the light emitter, will show a reduction of photon emission due to direct energy transfer to the opsin with efficient pairing. On the other hand, the light emitter will recover emission when the acceptor, the opsin, is a less effective acceptor of FRET. This is supported by observations across a series of LMOs, where radiance increases with decreasing BL-OG efficiency and vice versa [13].

The above lines of evidence support the hypothesis of radiationless energy transfer within the luminopsin fusion protein. Assuming FRET as the mechanism underlying the most efficient bioluminescent activation of opsins explains why this class of probes works reliably, even though the absolute intensity of illumination by bioluminescent light sources is far lower than that which can be achieved using LEDs or lasers.

2.3 Comparing LMO Efficacy

The influence of LMOs on a cell's membrane potential depends on a number of factors, including the cell type and the level of expression of the LMO. A relatively straightforward way to allow comparisons between different LMOs is through whole-cell recordings in HEK293 cells. An initial measurement of the amplitude of inward current elicited with a physical light source (LED, laser, or metal halide) determines the maximum light response of this cell, thereby providing a baseline for each cell. Determining the inward current elicited by bioluminescence then allows the researcher to calculate the percentage of optogenetic activation of an opsin that can be achieved with bioluminescent stimulation, the coupling efficiency of a given LMO [6] (Fig. 2.4). This value quantifies the intrinsic coupling efficiency between luciferase and opsin and takes into account cell-to-cell variations in LMO expression. This calculation, initially introduced by Berglund et al. [6], has been used in subsequent LMO developments [11–13, 15, 16].

Figure 2.4 shows an example of calculating coupling efficiency [13]. Whole-cell recordings were obtained from cultured HEK293 cells transiently expressing LMO7 after lipofection. For each recording, opsins, here VChR1, were activated by an arc lamp to obtain the maximal photocurrent and by luciferin, here hCTZ, to determine the bioluminescence-induced current (Fig. 2.4a). We began by measuring, under voltage clamp conditions (−60 mV), the photocurrents evoked by exciting opsin actuators directly via illumination by an arc lamp, using a filter cube matching the absorption peak of the opsin (here 540 nm). As shown in the example in Fig. 2.4b, illumination evoked an inward photocurrent due to activation of VChR1 that reached a peak within a few milliseconds (ms) and exhibited a sustained plateau that persisted for as long as the excitation light was maintained (here 1 s). To determine the effectiveness of the LMO for chemogenetic control of membrane potential, we measured the ability of the luciferin to induce currents in these cells. Infusion of luciferin into the recording chamber yielded bioluminescence emission and simultaneous

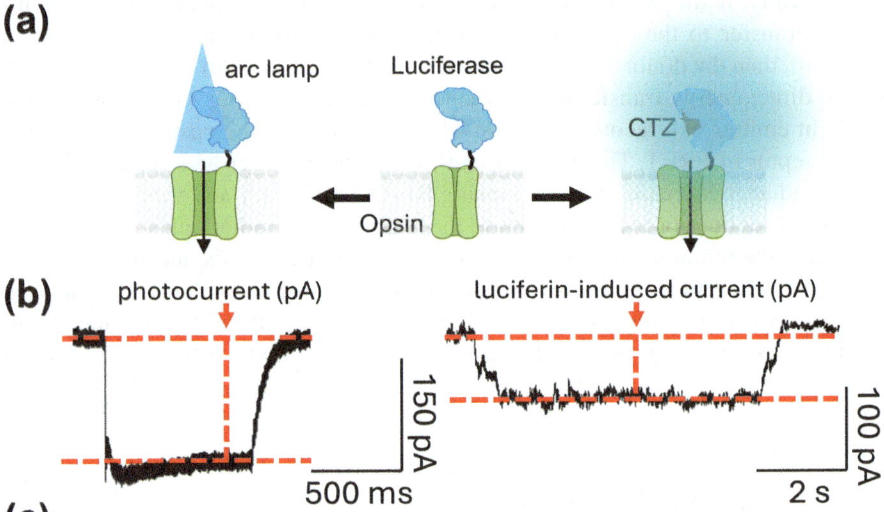

(c)

$$\text{Coupling efficiency} = \frac{\textit{luciferin-induced current}}{\text{max}\ \textit{photocurrent}}$$

Fig. 2.4 Quantifying LMO efficacy. (**a**) Membrane potential changes elicited opto (arc lamp)- or chemo (CTZ)- genetically can be measured in HEK293 cells expressing an LMO. (**b**) In whole-cell voltage patch clamp, inward current is determined in cells, here expressing LMO7, exposed to light from an arc lamp (photocurrent) and to luciferin (luciferin-induced current). (**c**) Equation used to quantify LMO efficacy using currents measured in patch clamp. (The figure was modified from reference [13])

activation of an inward current (Fig. 2.4b). This inward current generally tracked the time course of the luciferin presence in the chamber. These measurements were used to determine the efficiency of coupling between the luciferase light source and the opsin actuator. We compared the magnitude of currents evoked by maximal exposure to light from an arc lamp to those induced by luciferin by dividing the measured luciferin-induced current by the maximal photocurrent, yielding the coupling efficiency (Fig. 2.4c).

2.4 Parameters for Creating Efficient LMOs

Following the design principle of LMOs, efficiency in changing membrane potential can be expected to be influenced by the brightness of the light emitter, the sensitivity of the light sensor, the linker connecting the two moieties, the wavelength compatibility of the light emitter and sensor, and the arrangement of the two moieties with the light emitter located either extra- or intracellularly.

2.4.1 Brightness of the Light Emitter

For the initial excitatory LMOs, LMO1 and LMO2, wild-type *Gaussia* luciferase [17] (GLuc) was combined with channelrhodopsin 2 [18] (ChR2) and *Volvox* channelrhodopsin 1 [19] (VChR1), respectively [6]. While the application of luciferin induced small inward currents in HEK293 cells, bioluminescence was not able to generate action potentials in neurons. Berglund et al. determined that the minimal light intensity that induced at least one spike was 33 ± 13 μW/mm^2 (mean ± SEM) [15]. Thus, to reliably produce action potentials, it is necessary to have a luciferase that can emit sufficient bioluminescence.

We initially explored mutant variants of *Gaussia* luciferase, specifically slGLuc [20] (superluminescent, F89W/I90L), sbGLuc [21] (slowburn, M43L/M110L), and GLucM23 [22] (K25E, M35L, V88D, M102I, G159D), all of which produced more bioluminescence compared to wild-type GLuc. The combination of sbGLuc with VChR1 in LMO3 resulted in action potential firing and yielded robust and reliable activation of neurons in numerous in vivo experiments [15, 23–32]. LMO4 was created by the fusion of GLucM23 to VChR1 [16]. The GLucM23 moiety in LMO4 was comparable in bioluminescence light emission to the sbGLuc moiety in LMO3.

In the previous examples, the light emitter/energy donor in each LMO was a luciferase. Further increase in light emission/transferable energy can be achieved by using luciferase–fluorescent protein (FP) fusions. Here, the luciferase transfers energy to the chromophore of the fused FP, which then emits photons and/or functions as a donor for transferring energy to the opsin chromophore (see also Chap. 1, Fig. 1.3). Molecular evolution of luciferase–FP fusions is a promising engineering strategy for continued improvement of light emitters. A fusion of a luciferase evolved from the shrimp *Oplophorus gracilirostris* luciferase [33] (OLuc), eKL9h, and mNeonGreen, mNG [34], tethered to VChR1, resulted in LMO7 [12]. LMO7 showed a large CTZ-driven current amplitude and a coupling efficiency more than fourfold greater than the prior standard LMO3. With this improvement in activation efficiency, LMO7 more than doubled many physiological activation metrics (e.g., bioluminescence-driven depolarization of neurons ex vivo and multiunit spiking activity responses in vivo) and showed robust control of behavior [12].

This trend continued, though not as dramatically, when we utilized a new FRET construct, GeNL_SS, the brightest light emitter obtained thus far by extensive molecular evolution of a GeNL–SSLuc fusion protein [35]. Tethering GeNL_SS to VChR1 in LMO10 resulted in further improvement of coupling efficiency compared to LMO7 [13].

An illustration of the effect on coupling efficiency by improving the light emitter is the development from LMO3 to LMO7 (Fig. 2.5). The opsin in all these LMOs is VChR1. LMO3 utilizes a mutant of *Gaussia* luciferase, sbGLuc. LMO5 uses a luciferase from the shrimp *Oplophorus gracilirostris* (OLuc), NanoLuc [36], that matches sbGLuc in bioluminescence light emission. LMO6 employs a molecularly evolved version of GLuc, sbAluc1, fused to the fluorescent protein mNG to make sbAluc1-d7NmNG [12]. LMO7 is a fusion of an evolved OLuc variant with

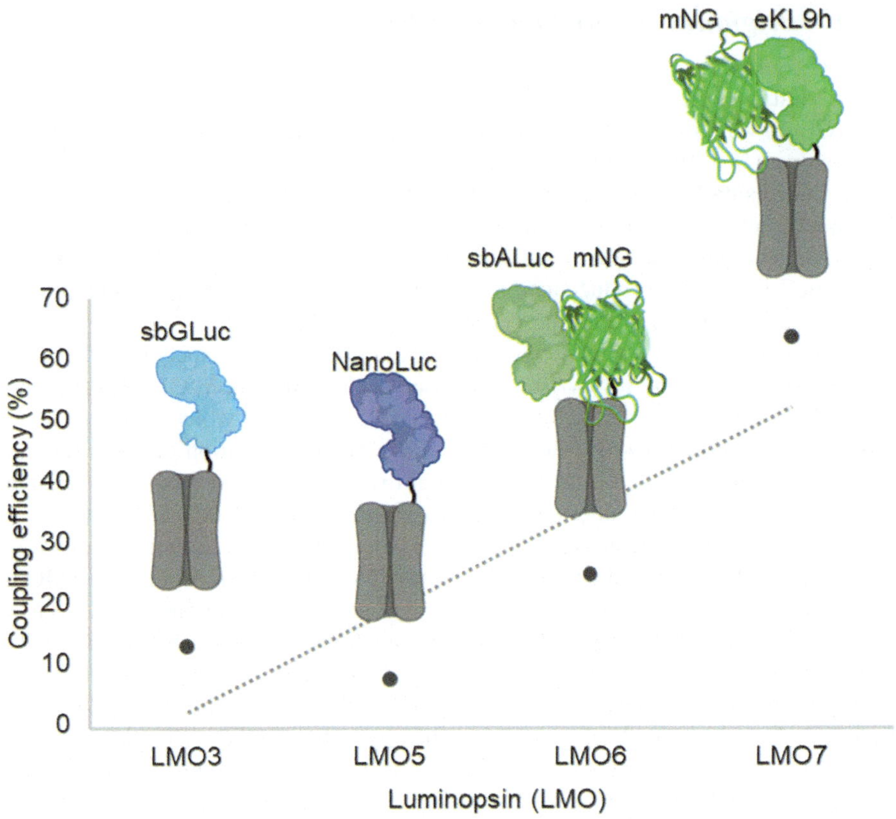

Fig. 2.5 Improving coupling efficiencies with brighter light emitters. The same channelrhodopsin, VChR1, was used to tether sbGLuc, NanoLuc, and the fusion proteins sbALuc-mNG and mNG-eKL9h to generate LMOs 3, 5, 6, and 7, respectively. The LMO schematics are shown above the plotted percentages of coupling efficiency (stippled line is the linear trendline)

improved performance compared to NanoLuc to mNG to make mNGd10C-eKL9h [12]. Reflecting the increased coupling efficiency of LMO7 compared to previous LMO versions, LMO7 is an effective tool for controlling neuronal activity in vivo and for controlling behavior [12].

By incorporating inhibitory opsins, luminopsins can also be used to inhibit neuronal activity. Figure 2.6 shows examples of determining coupling efficiencies in excitatory versus inhibitory LMOs [14]. While we did not carry out a systematic study comparing different light sources using the same inhibitory opsin as we did for the excitatory opsins (see above), the trend of improving coupling efficiency and thereby LMO performance with each iteration of LMO holds for inhibitory LMOs. For the first inhibitory LMO (iLMO [15]), we combined slGLuc with a photosensitive proton pump from the fungus *Leptosphaeria maculans* (Mac) [37]. Application of the luciferin CTZ caused bioluminescence and concomitant hyperpolarization and abolished action potential firing in cultured neurons. It also affected turning

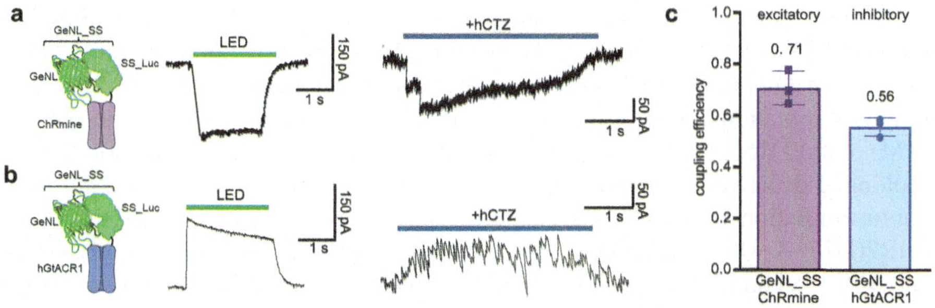

Fig. 2.6 Quantifying efficacy of inhibitory LMOs. (**a, b**) Schematics of LMOs (left: excitatory GeNL_SS-ChRmine, inhibitory GeNL_SS-hGtACR1) and examples of recordings in HEK293 cells of their photocurrent traces (middle), and luciferin-induced traces (right). showing the inward current for the excitatory and the outward current for the inhibitory LMO. (**c**) Efficiency of either LMO is quantified by calculating the percentage of chemogenetically induced versus optogenetically induced current, the coupling efficiency. (The figure was modified from reference [14])

behavior in mice [15]. By fusing GLucM23 to a hyperpolarizing anion channelrhodopsin, iChloC [38], we created the inhibitory luminopsin iLMO4, which has a high coupling efficiency of 0.36 [16]. Most recently, we tethered the FRET construct GeNL_SS to hGtACR1, a codon-optimized anion channelrhodopsin 1 from *Guillardia theta* [39] with a coupling efficiency of 0.56 [14].

Tung et al. [7] reported another form of inhibitory luminopsin based on engineered *Renilla* luciferase–fluorescent protein fusions and an inhibitory, light-driven chloride pump, *Natronomonas* halorhodopsin (NpHR) [40]. The combination of the red-shifted version TagRFP-RLuc [41], selected based on a compatible emission spectrum, with NpHR resulted in the inhibitory iLMO1 [7]. The performance was significantly improved in iLMO2 by using the Venus-RLuc fusion protein Nanolantern [10] with ~12-fold higher light emission. This probe allowed silencing of neurons via noninvasive peripheral application of the luciferase substrate in freely moving rats, resulting in observable changes in their rotational behavior [7].

Thus, in addition to excitatory LMOs, inhibitory LMOs can efficiently use bioluminescence to inhibit neuronal action potential firing.

2.4.2 Sensitivity of the Light Sensor

Higher coupling efficiencies between the light-emitting luciferase and the light-sensing opsin can be achieved, in principle, by increasing the light emission of the luciferase, the light-sensitivity of the opsin, or both. For example, we improved the performance of luminopsins utilizing GLuc by switching out ChR2 in LMO1 with the more light-sensitive VChR1 in LMO2, a change that resulted in a tenfold improvement in coupling efficiency.

With the development of step-function opsins [42, 43] (SFOs), a class of highly light-sensitive channelrhodopsins became available for incorporating into LMOs. SFOs were derived from *Chlamydomonas* channelrhodopsin 2 (ChR2) by introducing point mutations in the DC gate, namely, C128S in the transmembrane domain (TM) 3 and D156A in TM 4. These point mutations alter channel function toward prolonged deactivation kinetics, keeping the channel open even after cessation of photostimulation and making the channel more sensitive to light. ChR2(C128S) and ChR2(D156A) showed higher sensitivity to blue light compared to wild-type ChR2 and *Volvox* channelrhodopsin 1 (VChR1), which were used in the previous iterations of luminopsins (LMO1 and LMO2/3). Berglund et al. [11] generated LMOs with SFOs as the opsin moiety, named SFLMOs. There was no significant difference in coupling efficiencies between SFLMO(D156A), SFLMO(C128S), and SSFLMO (stable step function opsin D156A/C128S). However, due to the extremely high light sensitivity of the mutated channel, the coupling efficiency of SFLMOs ranged between 0.54 and 0.67, values that were significantly higher than the previous versions of luminopsins (coupling efficiencies well under 0.50). In addition, the bioluminescence-induced photocurrent lasted longer than the bioluminescence signal due to very slow deactivation of the mutated channel.

These prolonged deactivation kinetics are ideal in experiments where sustained depolarization of target neurons is desired, but this feature does not fit all applications. Taking advantage of new developments and discoveries in optogenetic tools, such as the finding of new classes of microbial opsins in a marine organism-based genomic screen [44], we next paired our brightest light emitter, GeNL_SS [35], with ChRmine from *Tiarina fusus* [44]. ChRmine was reported to have a hundred-fold improved operational light sensitivity compared with other fast, red-shifted opsin variants. The combination of GeNL_SS with ChRmine led to further improved coupling efficiency from 0.57 (LMO10, GeNL_SS-VChR1) to 0.65 (LMO11, GeNL_SS-ChRmine) [13].

Optogenetic elements used for changing neural membrane potential are defined by several features. Table 2.1 lists these parameters and examples of their actualization in previously generated LMOs.

As a general rule, any new discoveries and developments in opsins can be combined with current and future iterations of luciferases and luciferase–fluorescent protein FRET constructs toward generating powerful new LMOs (Fig. 2.7).

For selecting optogenetic elements for new LMOs, the extensive innovation applied across labs in optogenetic design provides considerable choices between light-driven pumps, channels, and pump-like channels to create a breadth of potent LMOs. A brief list of suggestions is presented in Table 2.2. We also included three opsins that, thus far, have been less widely explored, but we anticipate will prove powerful: a sodium pump for silencing (eKR2; [47]), and a cation channel that is impermeable to Ca^{2+} (virChR1, [52]) and another that mainly conducts Ca^{2+} (capChR2, [53]). Particularly as systems and behavioral studies increasingly focus on the contribution of nonneural systems, multi-option calcium control by LMOs will, we predict, be highly useful.

Table 2.1 Examples of opsins that are currently part of LMOs

Parameter	Characteristic	Opsins	Current LMOs	Reference
Molecular structure	Channel	VChR1	sbGLuc-VChR1 (LMO3)	[15]
	Pump	Mac	slGLUC-Mac (iLMO1)	[15]
	Pump-like channels	ChRmine	GeNL_SS-ChRmine (LMO11)	[13]
Ion selectivity	Cations	VChR1	GeNL_eKL9h-VChR1 (LMO7)	[12]
	Anions	hGtACR1	GeNL_SS -hGtACR1 (LMO9)	[13]
Light activation spectrum	Blue	VChR1	GeNL_eKL9h -VChR1 (LMO7)	[12]
	Red	ChRmine	GeNL_SS -ChRmine (LMO11)	[13]
Light sensitivity	Low	ChR2	sbGLuc-ChR2 (LMO2)	[6]
	High	ChRmine	GeNL_SS -ChRmine (LMO11)	[13]
Kinetics	Short	ChroME	–	
	Long	SFOs, SSFOs	SFLMO, SSFLMO	[11]
Photocurrent amplitude	Small	VChR1	sbGLuc-VChR1 (LMO3)	[15]
	Large	ChRmine	GeNL_SS -ChRmine (LMO11)	[13]

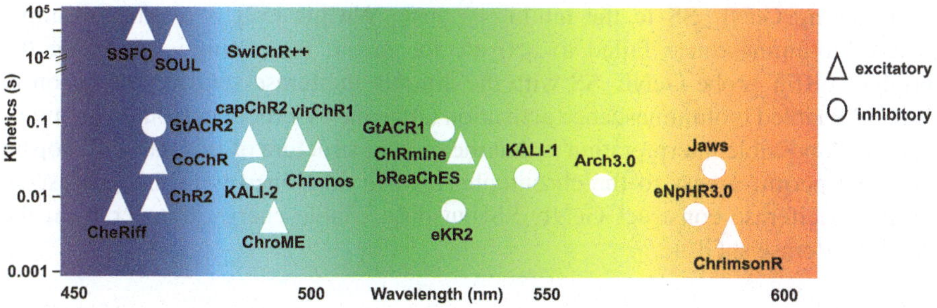

Fig. 2.7 Opsins for partnering with bioluminescent light emitters in LMOs. Excitatory and inhibitory opsins arranged along their activation wavelength and kinetics (approximations). Several have been employed in LMOs (e.g., CheRiff, ChR2, GtACR1, GtACR2, ChRmine)

There are rare instances where tethering an opsin to a bright light emitter did not result in bioluminescence-driven activation of the opsin. For example, we wanted to take advantage of the ChRger series of excitatory opsins that were optimized through a machine learning platform to yield high-photocurrent ChRs with high light sensitivity [56]. We fused mNG-eKL9h, the light emitter used successfully in LMO7 (mNG-eKL9h-VChR1), to the designer channelrhodopsin ChRger3 to create LMO8. While LMO8 emitted measurable bioluminescence and strongly responded to lamp stimulation with a change in membrane potential, it completely

Table 2.2 Opsins for new LMOs

Light-driven pumps	
H+	eArch3.0 [45]
Cl−	Jaws [46]
Na+	eKR2 [47]
Light-driven channels and pump-like channels	
Cations	
$Na^+/K^+/Ca^{2+}$	bReaChES [48], ChroME [49], Chronos [50], ChrimsonR [50], SOUL [51]
Na^+/K^+-conducting, impermeable to Ca^{2+} ions	virChR1 [52]
Ca^{2+} conducting mainly	capChR2 [53]
Anions	
Cl−	SwiChR++ [54]
K^+	*Hc*KCR1/Kali-1 [55], *Hc*KCR2/Kali-2 [55]

lacked a current response to bioluminescence [13]. Replacing the mNG-eKL9h FRET light emitter with the luciferase sbGLuc did not enable bioluminescent activation of ChRger3 either. In contrast, ChRger3 was efficiently activated by bioluminescence in an optical synapse where the luciferase is not tethered to the opsin but is instead released from a presynaptic neuron into the synaptic cleft and interacts with the opsin expressed in the postsynaptic neuron (Chap. 6). Thus, ChRger3 does not tolerate the tethering of light emitters routinely used with other opsins.

Tethering GeNL_SS to the inhibitory opsin WiChR [57] is another example where bioluminescence failed to activate the opsin. Interestingly, replacing the brighter FRET probe GeNL_SS with the dimmer luciferase sbGLuc for fusion to WiChR enabled bioluminescence activation (Johannes Vierock, personal communication). A possible interpretation is that the protein structure of this particular opsin does not permit access to the chromophore when combined with the fluorescent protein–luciferase construct GeNL_SS but does enable energy transfer from the small luciferase sbGLuc.

2.4.3 Linker Connecting Light Emitter and Light Sensor

In LMOs, the light source (the luciferase or luciferase–fluorescent protein) is tethered to the light sensor (the opsin) via a 15 amino acid linker (GDPLVQCGGIAGSAT). Thus far, the same linker has been used for all iterations of LMOs, except for iLMO1 and 2, where the opsin and the luciferase are directly fused. While this linker has been working well, it is nevertheless a potential target for improving the efficacy of LMOs.

A linker joining domains in protein fusions is generally required to maintain proper protein folding while allowing domain interactions. Linkers can vary in length, conformation, amino acid composition, and flexibility. Modifying the size and structure of linkers often allows fine-tuning of the fusion protein function. Indeed, the molecular evolution of GCaMP sensors critically involved extensive mutations of linkers that turned out to be critical for sensor function [58]. Further, the development of the first genetically encoded neurotransmitter indicator based on bioluminescent light emission was the result of optimizing linkers by high-throughput screening of linker libraries containing ~400 variants, followed by targeted linker optimization [59].

Thus, systematic molecular evolution of LMOs including extensive linker library screens will likely result in further improvements of LMO performance.

2.4.4 *Wavelength Compatibility*

Each light emitter and light sensor partnered for an LMO has unique peak emission and absorption spectra, respectively. A general assumption is that closely matching these spectra will improve the function of LMOs. While we did not systematically vary the emission spectrum with a given absorption spectrum or vice versa, there are several reports that suggest that matching the wavelength between light emitter and light sensor is less critical to efficient cellular control than the strength of light emission and/or the sensitivity of the opsin.

The first example is the tenfold improvement in coupling efficiency from LMO1 to LMO2. The light emitter for both LMOs, GLuc, has a peak emission of 480 nm. While this is a perfect match to the peak absorption of 480 nm for ChR2 in LMO1, the less favorable match to the peak absorption of 535 nm of VChR1 in LMO2 still resulted in better coupling efficiency of LMO2 due to the higher light sensitivity of VChR1 across wavelengths [6]. Another example is the fusion of NpHR to Rluc variants in iLMO1 and iLMO2. Although the red-shifted emission spectrum of TagRFP-RLuc appeared to be more preferable for activating NpHR in iLMO1 (NpHR-TagRFP-RLuc) than Nano-lantern, total luminescence from Nano-lantern was significantly higher by a factor of 12.3 and resulted in more robust performance of iLMO2 (NpHR-Nano-lantern; [7]). Most recently, we fused the bright FRET light emitter GeNL_SS (emission peak 520 nm) [35] to excitatory and inhibitory opsins with absorption peaks ranging from 515 nm (hGtACR1) to 585 nm (ChRmine), resulting in highly efficient LMOs for in vivo studies [14].

For radiationless energy transfer, a donor emission spectrum that overlaps well with the recipient excitation spectrum leads to highly efficient energy transfer. It is possible that the rather broad activation spectra of opsins provide tolerance for off-peak emission spectra. For example, activation of VChR1 with blue (480 nm) and green (540 nm) light results in similar photocurrent amplitudes [12]. Overall, wavelength compatibility alone cannot outcompete other structural requirements of LMO performance.

2.4.5 Arrangement of Moieties

The moieties within LMOs can be arranged with the light emitter on the N-terminus or the C-terminus of the opsin. In all of our excitatory LMO designs, the luciferase or luciferase–fluorescent protein FRET probe is tethered to the N-terminus of the opsin. For the inhibitory iLMO2, the RLuc-based luciferase–fluorescent fusion protein Nano-lantern was tethered to the C-terminus of the opsin NpHR [7]. There is no inherent reason for one orientation being preferable to the other for the light activation of the opsin. There might, however, be a difference with respect to accessibility to the luciferin. With intracellular luciferase placement, the luciferin has to cross the cell membrane, which it does efficiently. In a limited experiment, we directly compared the coupling efficiency of each arrangement. The light emitter mNG-eKL9h was tethered to the N- or C-terminus of VChR1, generating LMO7 and LMO7.2, respectively [13]. While the N-terminal arrangement (light emitter outside the cell) showed a luciferin-induced current response with a coupling efficiency of 0.56, the C-terminal fusion of the light emitter (light emitter inside the cell) in LMO7.2 had a diminished bioluminescent response reflected in the lower coupling efficiency of 0.30. This coupling efficiency will likely have to be experimentally tested for intra- versus extracellular arrangement in a given luciferase–opsin combination.

2.5 Practical Considerations for Using LMOs

There was a tenfold improvement in coupling efficiency from the initial combination of wild-type *Gaussia* luciferase and channelrhodopsin-2 (ChR2(H134R)) to the construct where the opsin ChR2(H134R) was swapped for VChR1 (to create LMO2) [6]. From there, we saw another tenfold improvement in coupling efficiency after exchanging the luciferase GLuc for sbGLuc (to create LMO3) [15]. LMO3 was the first excitatory LMO that drove action potential firing in cultured neurons and showed robust neural activation in vivo. LMO3 has been widely used in animal experiments, both in rats and mice (see Table 2.3), either supplied via AAV transduction or through use of the lox-stop-lox-LMO3 transgenic mouse line crossed with various Cre driver lines [29]).

Different from the LMO3 construct in the initial publication (sbGLuc-VChR1-EYFP; [15], LMO3 plasmids were subsequently modified by introducing a membrane trafficking signal from Kir2.1 (TS: KSRITSEGEYIPLDQIDINV) before and an endoplasmic reticulum export signal (ER: FCYENEV) after the EYFP sequence to optimize expression, as has been observed for other optogenetic constructs [73]. Thus, the pAAV-LMO3 plasmids distributed at Addgene (plasmids # 114099, 114105, 114104) and the conditional LMO3 knock-in mouse distributed at the Jackson Lab (strain #034853) have the sbGLuc-VChR1-(TS)EYFP(ER) sequence. Zhang et al. [61] carried out a detailed comparison of LMO3 constructs with and

Table 2.3 Bioluminescence driving opsins Tabulated are the luciferases, their partner opsins, and their arrangement (tethered in LMOs or co-expressed)

Luciferase	Opsin	Arrangement	In vitro	In vivo
Excitatory				
GLuc	ChR2(H134R)	LMO1	[6]	
GLuc	VChR1	LMO2	[6]	[60]
sbGLuc	VChR1	LMO3	[15]	[15, 23–32, 61–65]
sbGLuc	CheRiff	LMO3.2	[30]	[30]
GLucM23	VChR1	LMO4	[16]	[16]
NanoLuc	VChR1	LMO5	[12]	
sbALuc-mNG	VChR1	LMO6	[12]	
mNG-eKL9h	VChR1	LMO7	[12]	[12, 66]
GeNL_SS	VChR1	LMO10	[13]	
GeNL_SS	ChRmine	LMO11	[13]	[14]
sbGLuc	ChR2(C128S)	SFLMO.CS	[11]	[11]
sbGLuc	ChR2(D156A)	SFLMO.DA	[11]	
sbGLuc	ChR2(C128S/D156A)	SSFLMO.CS/DA	[11]	
GLuc	ChR2(H134R)	Co-expressed	[67]	
NLuc	CoChR	Co-expressed	[68]	[68]
Inhibitory				
slGLuc	Mac	iLMO	[15]	[15, 23]
TagRFP-RLuc	NpHR	iLMO1	[7]	
NanoLantern	NpHR	iLMO2	[7]	[7, 69, 70]
GLucM23	iChloC	iLMO4	[16, 71]	
sbGLuc	hGtACR2	tethered	[71]	
GeNL_SS	hGtACR1	tethered	[14]	[14]
RLuc8	NpHR	pHIL	[72]	[72]

All constructs were tested in vitro, with the best versions applied in vivo

without the Kir2.1 membrane trafficking signal and demonstrated improvements in expression, electrophysiological performance, and effect on behavior in the versions with Kir2.1. Some subsequent publications used this modified version of LMO3, named eLMO3 (sbGLuc-VChR1-(TS)EYFP).

The graph in Fig. 2.8 plots the coupling efficiency of various excitatory LMOs over time. While further improvements after LMO3 are considerable, they are no longer increasing the coupling efficiency exponentially. Rather, the high overall efficacy of a variety of LMOs allows the experimenter to choose the best tool for the intended application, for example, by selecting opsins with different photocurrent amplitudes and kinetics (VChR1 versus CheRiff versus ChRmine with LMO3, LMO3.2, and LMO11). LMOs now show robust activation properties that allow the drive and silencing of neurons in vivo and during behavior. Table 2.3 lists luciferase–opsin combinations developed and applied in in vivo studies. This list will continue to expand with new combinations of bright luciferases and opsins with evolved properties.

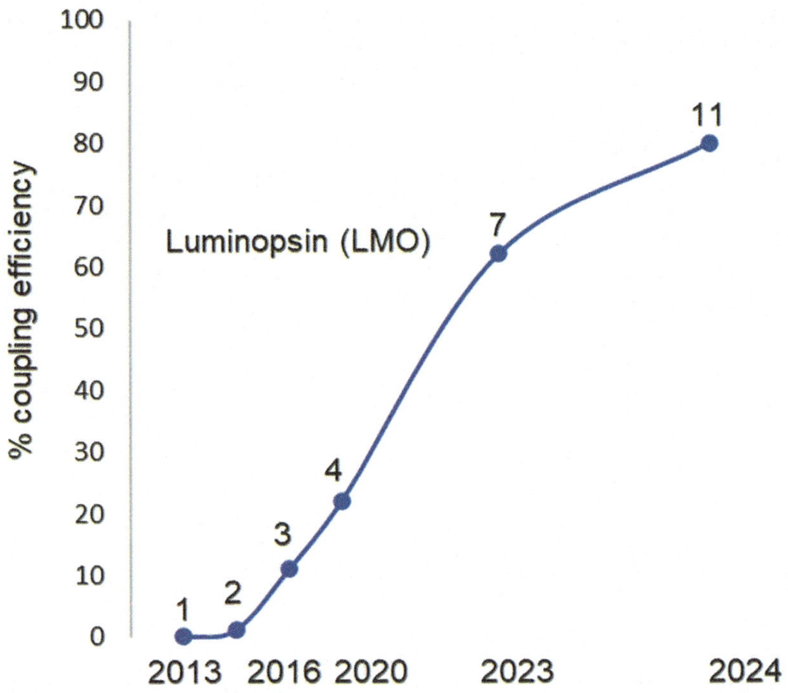

Fig. 2.8 Increasing LMO performance over time. Coupling efficiency is the ratio of bioluminescent (chemogenetic) activation proportional to maximum direct optogenetic activation. This metric is plotted across LMO versions. (Compiled from Refs. [6, 12, 13, 15, 16])

2.6 Bioluminescence Driving Opsins Beyond Neuroscience

The majority of publications on bioluminescent light emission activating opsins fall into the area of neuroscience. Recently, two reports applied BL-OG to change the membrane potential in cell types other than neurons. In both studies, the luciferase and the channelrhodopsin were co-transfected rather than tethered.

In the first study, Kim et al. [67] transduced iPSC-derived cardiomyocytes (CM) with ChR2(H134R). Light activation of ChR2(H134R) leads to depolarization and calcium influx, triggering contractions of CMs. The addition of *Gaussia* luciferase (GLuc) protein encapsulated in polymersomes as artificial organelles resulted in modulation of cardiac beating frequency upon treatment with the GLuc luciferin coelenterazine (CTZ). The intracellular bioluminescent light source was capable of a near-instantaneous increase in the CM beating rate for at least 10 s, determined by calcium transient imaging. Acceleration of the mechanical beating rate was observed for at least 5 s after CTZ addition, confirming that the CTZ-induced changes in intracellular calcium were mirrored in the mechanical contraction behavior of the ChR2-transduced iPSC-CMs. The authors also took advantage of being able to directly monitor the artificial organelle activity within live cells and over prolonged time periods (days) through bioluminescence imaging. The use of brighter luciferases and opsins with higher light sensitivity will likely further improve this application.

Chen et al. [68] used BL-OG for the manipulation of cell membrane excitability in the mitochondria of cancer cells. Targeting mitochondria in cancer cells is a promising strategy to specifically and effectively destroy cancer cells in vivo without affecting other tissues. Proper mitochondrial function relies on maintaining an electrical potential gradient across the inner mitochondrial membrane, and disruption of this potential induces cell death primarily through the intrinsic apoptotic pathway, accompanied by DNA damage and autophagy. Using a mitochondrial leading sequence, the authors targeted the blue light-gated cationic channelrhodopsin CoChR from *Chloromonas oogama* [50] to the inner mitochondrial membrane. The blue light-emitting NanoLuc was expressed in the cytosol of the same cells by proteolytic cleavage from the same propeptide as the opsin, with the construct selectively delivered to cancer cells. Application in vitro and in vivo of the luciferin ViviRen elicited robust intracellular bioluminescence, which activated the mitochondrial opsin, triggering mitochondrial depolarization and subsequent death of cancer cells. The ability of this BL-OG application to induce mitochondrial depolarization and cytotoxicity was demonstrated across multiple different cancer cell types. Importantly, it was highly efficient in reducing tumor burden and killing tumor cells in glioblastoma and triple-negative breast cancer xenograft mouse models without impairing normal organs or tissues.

2.7 Summary

In principle, any opsin can be equipped with a biological light source, enabling chemogenetic access to ion-conducting molecular tools for changing membrane potential. For LMOs, luciferase–opsin fusion proteins, efficacy in activating opsins increased by combining brighter light emitters with more sensitive light sensors. Currently available LMOs are widely usable for the modulation of membrane potential in experimental animal studies. The continued discovery and development of novel opsins and luciferases allows the engineering of new LMOs that meet experimenters' needs for fine-tuning the kinetics and amplitudes of bioluminescence-driven membrane potential changes for the desired studies. The best strategy for systematically generating new LMOs will likely be molecular evolution of the entire fusion protein, encompassing the luciferase, fluorescent protein, linker, and opsin to select for the most compatible energy transfer pairs of light emitters and light sensors. Another benefit of continued improvements in light emitters and sensors will be the option to drive opsins with bioluminescence generated from co-expressed light emitters, increasing flexibility and combinatorial specificity. Lastly, the technology, mainly applied in neuroscience research thus far, is ready to be expanded to other research areas.

References

1. Boyden ES, Zhang F, Bamberg E, Nagel G, Deisseroth K. Millisecond-timescale, genetically targeted optical control of neural activity. Nat Neurosci. 2005;8(9):1263–8.
2. Li X, Gutierrez DV, Hanson MG, Han J, Mark MD, Chiel H, et al. Fast noninvasive activation and inhibition of neural and network activity by vertebrate rhodopsin and green algae channelrhodopsin. Proc Natl Acad Sci USA. 2005;102(49):17816–21.
3. Ishizuka T, Kakuda M, Araki R, Yawo H. Kinetic evaluation of photosensitivity in genetically engineered neurons expressing green algae light-gated channels. Neurosci Res. 2006;54(2):85–94.
4. Armbruster BN, Li X, Pausch MH, Herlitze S, Roth BL. Evolving the lock to fit the key to create a family of G protein-coupled receptors potently activated by an inert ligand. Proc Natl Acad Sci USA. 2007;104(12):5163–8.
5. Zhu H, Roth BL. DREADD: a chemogenetic GPCR signaling platform. Int J Neuropsychopharmacol. 2014;18(1):1–6.
6. Berglund K, Birkner E, Augustine GJ, Hochgeschwender U. Light-emitting channelrhodopsins for combined optogenetic and chemical-genetic control of neurons. PLoS One. 2013;8(3):e59759.
7. Tung JK, Gutekunst CA, Gross RE. Inhibitory luminopsins: genetically-encoded bioluminescent opsins for versatile, scalable, and hardware-independent optogenetic inhibition. Sci Rep. 2015;5:14366.
8. Förster T. Energiewanderung und Fluoreszenz. Naturwissenschaften. 1946;33(6):166–75.
9. Förster T. Zwischenmolekulare Energiewanderung Und Fluoreszenz. Ann Phys. 1948;2:55–75.
10. Saito K, Chang YF, Horikawa K, Hatsugai N, Higuchi Y, Hashida M, et al. Luminescent proteins for high-speed single-cell and whole-body imaging. Nat Commun. 2012;3:1262.
11. Berglund K, Fernandez AM, Gutekunst CAN, Hochgeschwender U, Gross RE. Step-function luminopsins for bimodal prolonged neuromodulation. J Neurosci Res. 2020;98:422–36.
12. Björefeldt A, Murphy J, Crespo EL, Lambert GG, Prakash M, Ikefuama EC, et al. Efficient opto- and chemogenetic control in a single molecule driven by FRET-modified bioluminescence. Neurophotonics. 2024;11(2):21005.
13. Slaviero AN, Gorantla N, Simkins J, Crespo EL, Ikefuama EC, Tree MO, et al. Engineering luminopsins with improved coupling efficiencies. Neurophotonics. 2024;11(2):24208.
14. Ikefuama EC, Slaviero AN, Silvagnoli AD, Crespo EL, Schalau R, Gott M, Tree MO, Dunbar GL, Rossignol J, Hochgeschwender U. Presymptomatic targeted circuit manipulation for ameliorating Huntington's disease pathogenesis. iScience. 2025;28(3):112022.
15. Berglund K, Clissold K, Li HE, Wen L, Park SY, Gleixner J, et al. Luminopsins integrate opto- and chemogenetics by using physical and biological light sources for opsin activation. Proc Natl Acad Sci USA. 2016;113(3):E358–67.
16. Park SY, Song SH, Palmateer B, Pal A, Petersen ED, Shall GP, et al. Novel luciferase–opsin combinations for improved luminopsins. J Neurosci Res. 2020;98(3):410–21.
17. Verhaegen M, Christopoulos TK. Recombinant Gaussia luciferase. Overexpression, purification, and analytical application of a bioluminescent reporter for DNA hybridization. Anal Chem. 2002;74(17):4378–85.
18. Nagel G, Szellas T, Huhn W, Kateriya S, Adeishvili N, Berthold P, et al. Channelrhodopsin-2, a directly light-gated cation-selective membrane channel. Proc Natl Acad Sci USA. 2003;100(24):13940–5.
19. Zhang F, Prigge M, Beyrière F, Tsunoda SP, Mattis J, Yizhar O, et al. Red-shifted optogenetic excitation: a tool for fast neural control derived from Volvox carteri. Nat Neurosci. 2008;11(6):631–3.
20. Kim SB, Suzuki H, Sato M, Tao H. Superluminescent variants of marine luciferases for bioassays. Anal Chem. 2011;83(22):8732–40.
21. Welsh JP, Patel KG, Manthiram K, Swartz JR. Multiply mutated Gaussia luciferases provide prolonged and intense bioluminescence. Biochem Biophys Res Commun. 2009;389(4):563–8.

22. Lindberg E, Mizukami S, Ibata K, Miyawaki A, Kikuchi K. Development of luminescent coelenterazine derivatives Activatable by β-galactosidase for monitoring dual gene expression. Chem Eur J. 2013;19(44):14970–6.
23. Song D, Yang Q, Lang Y, Wen Z, Xie Z, Zheng D, et al. Manipulation of hippocampal CA3 firing via luminopsins modulates spatial and episodic short-term memory, especially working memory, but not long-term memory. Neurobiol Learn Mem. 2018;155:435–45.
24. Lee CH, Lin JH, Lin SH, Chang CT, Wu YW, Bewick G, et al. A role for proprioceptors in sngception. Sci Adv. 2025;11(5):eabc5219.
25. Yu SP, Tung JK, Wei ZZ, Chen D, Berglund K, Zhong W, et al. Optochemogenetic stimulation of transplanted iPS-NPCs enhances neuronal repair and functional recovery after ischemic stroke. J Neurosci. 2019;39(33):6571–94.
26. Zenchak JR, Palmateer B, Dorka N, Brown TM, Wagner LM, Medendorp WE, et al. Bioluminescence-driven optogenetic activation of transplanted neural precursor cells improves motor deficits in a Parkinson's disease mouse model. J Neurosci Res. 2020;98:458–68.
27. Gomez-Ramirez M, More AI, Friedman NG, Hochgeschwender U, Moore CI. The BioLuminescent-OptoGenetic in vivo response to coelenterazine is proportional, sensitive and specific in neocortex. J Neurosci Res. 2020;98(3):471–80.
28. English AW, Berglund K, Carrasco D, Goebel K, Gross RE, Isaacson R, et al. Bioluminescent optogenetics: a novel experimental therapy to promote axon regeneration after peripheral nerve injury. Int J Mol Sci. 2021;22(13):7217.
29. Medendorp WE, Bjorefeldt A, Crespo EL, Prakash M, Pal A, Waddell ML, et al. Selective postnatal excitation of neocortical pyramidal neurons results in distinctive behavioral and circuit deficits in adulthood. iScience. 2021;24(3):102157.
30. Ikefuama EC, Kendziorski GE, Anderson K, Shafau L, Prakash M, Hochgeschwender U, et al. Improved locomotor recovery in a rat model of spinal cord injury by BioLuminescent-OptoGenetic (BL-OG) stimulation with an enhanced luminopsin. Int J Mol Sci. 2022;23(21):12994.
31. Petersen ED, Sharkey ED, Pal A, Shafau LO, Zenchak-Petersen J, Peña AJ, et al. Restoring function after severe spinal cord injury through BioLuminescent-OptoGenetics. Front Neurol. 2022;12:792643.
32. Anderson KA, Whitehead BJ, Petersen ED, Kemme MR, Wedster A, Hochgeschwender U, et al. Behavioral context improves optogenetic stimulation of transplanted dopaminergic cells in unilateral 6-OHDA rats. Behav Brain Res. 2023;441:114279.
33. Inouye S, Watanabe K, Nakamura H, Shimomura O. Secretional luciferase of the luminous shrimp Oplophorus gracilirostris: cDNA cloning of a novel imidazopyrazinone luciferase(1). FEBS Lett. 2000;481(1):19–25.
34. Shaner NC, Lambert GG, Chammas A, Ni Y, Cranfill PJ, Baird MA, et al. A bright monomeric green fluorescent protein derived from Branchiostoma lanceolatum. Nat Methods. 2013;10(5):407–9.
35. Lambert GG, Crespo EL, Murphy J, Boassa D, Luong S, Celinskis D, et al. CaBLAM! A high-contrast bioluminescent Ca(2+) indicator derived from an engineered oplophorus gracilirostris luciferase. Nat Methods. 2026;23(1):205–15.
36. Hall MP, Unch J, Binkowski BF, Valley MP, Butler BL, Wood MG, et al. Engineered luciferase reporter from a deep sea shrimp utilizing a novel imidazopyrazinone substrate. ACS Chem Biol. 2012;7(11):1848–57.
37. Chow BY, Han X, Dobry AS, Qian X, Chuong AS, Li M, et al. High-performance genetically targetable optical neural silencing by light-driven proton pumps. Nature. 2010;463(7277):98–102.
38. Wietek J, Beltramo R, Scanziani M, Hegemann P, Oertner TG, Simon WJ. An improved chloride-conducting channelrhodopsin for light-induced inhibition of neuronal activity in vivo. Sci Rep. 2015;5:14807.
39. Govorunova EG, Sineshchekov OA, Janz R, Liu X, Spudich JL. Natural light-gated anion channels: a family of microbial rhodopsins for advanced optogenetics. Science. 2015;349:647–50.

40. Gradinaru V, Thompson KR, Deisseroth K. eNpHR: a Natronomonas halorhodopsin enhanced for optogenetic applications. Brain Cell Biol. 2008;36(1–4):129–39.
41. Dragulescu-Andrasi A, Chan CT, De A, Massoud TF, Gambhir SS. Bioluminescence resonance energy transfer (BRET) imaging of protein-protein interactions within deep tissues of living subjects. Proc Natl Acad Sci USA. 2011;108(29):12060–5.
42. Berndt A, Yizhar O, Gunaydin LA, Hegemann P, Deisseroth K. Bi-stable neural state switches. Nat Neurosci. 2009;12(2):229–34.
43. Yizhar O, Fenno LE, Prigge M, Schneider F, Davidson TJ, O'Shea DJ, et al. Neocortical excitation/inhibition balance in information processing and social dysfunction. Nature. 2011;477(7363):171–8.
44. Marshel JH, Kim YS, Machado TA, Quirin S, Benson B, Kadmon J, et al. Cortical layer-specific critical dynamics triggering perception. Science. 2019;365(6453):eaaw5202.
45. Mattis J, Tye KM, Ferenczi EA, Ramakrishnan C, O'Shea DJ, Prakash R, et al. Principles for applying optogenetic tools derived from direct comparative analysis of microbial opsins. Nat Methods. 2011;9(2):159–72.
46. Chuong AS, Miri ML, Busskamp V, Matthews GAC, Acker LC, Sørensen AT, et al. Noninvasive optical inhibition with a red-shifted microbial rhodopsin. Nat Neurosci. 2014;17(8):1123–9.
47. Grimm C, Silapetere A, Vogt A, Bernal Sierra YA, Hegemann P. Electrical properties, substrate specificity and optogenetic potential of the engineered light-driven sodium pump eKR2. Sci Rep. 2018;8(1):9316.
48. Rajasethupathy P, Sankaran S, Marshel JH, Kim CK, Ferenczi E, Lee SY, et al. Projections from neocortex mediate top-down control of memory retrieval. Nature. 2015;526(7575):653–9.
49. Mardinly AR, Oldenburg IA, Pégard NC, Sridharan S, Lyall EH, Chesnov K, et al. Precise multimodal optical control of neural ensemble activity. Nat Neurosci. 2018;21(6):881–93.
50. Klapoetke NC, Murata Y, Kim SS, Pulver SR, Birdsey-Benson A, Cho YK, et al. Independent optical excitation of distinct neural populations. Nat Methods. 2014;11(3):338–46.
51. Gong X, Mendoza-Halliday D, Ting JT, Kaiser T, Sun X, Bastos AM, et al. An ultra-sensitive step-function opsin for minimally invasive optogenetic stimulation in mice and macaques. Neuron. 2020;107(1):38–51.e8.
52. Zabelskii D, Alekseev A, Kovalev K, Rankovic V, Balandin T, Soloviov D, et al. Viral rhodopsins 1 are an unique family of light-gated cation channels. Nat Commun. 2020;11(1):5707.
53. Fernandez Lahore RG, Pampaloni NP, Schiewer E, Heim MM, Tillert L, Vierock J, et al. Calcium-permeable channelrhodopsins for the photocontrol of calcium signalling. Nat Commun. 2022;13(1):7844.
54. Berndt A, Lee SY, Wietek J, Ramakrishnan C, Steinberg EE, Rashid AJ, et al. Structural foundations of optogenetics: determinants of channelrhodopsin ion selectivity. Proc Natl Acad Sci USA. 2015;113(4):822–9.
55. Tajima S, Kim YS, Fukuda M, Jo Y, Wang PY, Paggi JM, et al. Structural basis for ion selectivity in potassium-selective channelrhodopsins. Cell. 2023;186(20):4325–4344.e26.
56. Bedbrook CN, Yang KK, Robinson JE, Mackey ED, Gradinaru V, Arnold FH. Machine learning-guided channelrhodopsin engineering enables minimally invasive optogenetics. Nat Methods. 2019;16(11):1176–84.
57. Vierock J, Peter E, Grimm C, Rozenberg A, Chen IW, Tillert L, et al. WiChR, a highly potassium-selective channelrhodopsin for low-light one- and two-photon inhibition of excitable cells. Sci Adv. 2022;8(49):eadd7729.
58. Akerboom J, Chen TW, Wardill TJ, Tian L, Marvin JS, Mutlu S, et al. Optimization of a GCaMP calcium indicator for neural activity imaging. J Neurosci. 2012;32(40):13819–40.
59. Petersen ED, Lapan AP, Castellanos Franco EA, Fillion AJ, Crespo EL, Lambert GG, et al. Bioluminescent genetically encoded glutamate indicators for molecular imaging of neuronal activity. ACS Synth Biol. 2023;12(8):2301–9.
60. Birkner E, Berglund K, Klein ME, Augustine GJ, Hochgeschwender U. Non-invasive activation of optogenetic actuators. Proc SPIE Int Soc Opt Eng. 2014;8928:89282F.

61. Zhang JY, Tung JK, Wang Z, Yu SP, Gross RE, Wei L, et al. Improved trafficking and expression of luminopsins for more efficient optical and pharmacological control of neuronal activity. J Neurosci Res. 2020;98:481–90.
62. Ecanow A, Berglund K, Carrasco D, Isaacson R, English AW. Enhancing motor and sensory axon regeneration after peripheral nerve injury using bioluminescent optogenetics. Int J Mol Sci. 2022;23(24):16084.
63. Klein E, Marsh S, Becker J, Andermann M, Lehtinen M, Moore CI. BioLuminescent OptoGenetics in the choroid plexus: integrated opto- and chemogenetic control in vivo. Neurophotonics. 2024;11(2):24210.
64. Wang Y, Liu X, Zhang Q, Zhao D, Zhou B, Pan Z, et al. Bioluminescence-optogenetics-mediated gene therapy in a sleep-disordered breathing mouse model. Biomed Pharmacother. 2024;178:117159.
65. Tian T, Kim D, Yu K, Hartzell HC, Ward PJ. Regenerative failure of sympathetic axons contributes to deficits in functional recovery after nerve injury. Neurobiol Dis. 2025;209:106893.
66. Murphy EF, Means A, Li C, Baez H, Gomez-Ramirez M. Strength of activation and temporal dynamics of bioluminescent-optogenetics in response to systemic injections of the luciferin. Neuroimage. 2024;301:120882.
67. Kim H, Yeow J, Najer A, Kit-Anan W, Wang R, Rifaie-Graham O, et al. Microliter scale synthesis of luciferase-encapsulated polymersomes as artificial organelles for optogenetic modulation of cardiomyocyte beating. Adv Sci. 2022;9(27):e2200239.
68. Chen K, Ernst P, Sarkar A, Kim S, Si Y, Varadkar T, et al. mLumiOpto is a mitochondrial-targeted gene therapy for treating cancer. Cancer Res. 2024;84(23):4049–65.
69. Jaiswal PB, Tung JK, Gross RE, English AW. Motoneuron activity is required for enhancements in functional recovery after peripheral nerve injury in exercised female mice. J Neurosci Res. 2020;98:448–57.
70. Tung JK, Shiu FH, Ding K, Gross RE. Chemically activated luminopsins allow optogenetic inhibition of distributed nodes in an epileptic network for non-invasive and multi-site suppression of seizure activity. Neurobiol Dis. 2018;109(Pt A):1–10.
71. Prakash M, Medendorp WE, Hochgeschwender U. Defining parameters of specificity for bioluminescent optogenetic activation of neurons using in vitro multi electrode arrays (MEA). J Neurosci Res. 2020;98(3):437–47.
72. Merolla A, Michetti C, Moschetta M, Vacca F, Ciano L, Emionite L, et al. A pH-sensitive closed-loop nanomachine to control hyperexcitability at the single neuron level. Nat Commun. 2024;15(1):5609.
73. Gradinaru V, Zhang F, Ramakrishnan C, Mattis J, Prakash R, Diester I, et al. Molecular and cellular approaches for diversifying and extending optogenetics. Cell. 2010;141(1):154–65.

Chapter 3
Bioluminescence Driving Non-Opsin Photoreceptors

Ute Hochgeschwender

Abstract Light sources, including bioluminescence, have been used extensively to activate opsins, light-sensing ion channels, and pumps, to change membrane potential and thereby the function of neurons. However, opsins are but a small part of the vast pool of photoreceptors, proteins that contain light-sensing domains. Recruited from across phyla, these photosensory proteins enable light-driven conformational changes, dimerization, enzyme activation, or the generation of reactive oxygen species. Extension of the optogenetic approach to these non-opsin photoreceptors has enabled light control of intracellular processes for applications in cell and developmental biology as well as in therapeutic contexts. This review details the application of bioluminescence as the light source for activating non-opsin photoreceptors for control of biological processes from transcription to cell death. Key features of the design principles of the molecular tools, the intracellular processes targeted, and the applications in vitro and in vivo for proof-of-principle demonstrations and for therapeutic applications are described.

Keywords Bioluminescent optogenetics · Luciferase · Luciferin · Transcription · Recombination · Protein localization · Protein allostery · Second messenger signaling · Cell death · Reactive oxygen species production

3.1 Introduction

Bioluminescence, light emitted by a luciferase enzyme when oxidizing a small molecule substrate, a luciferin, has been widely used as a light source to activate light-sensing ion channels, opsins, for changing membrane potentials in neurons (see

U. Hochgeschwender (✉)
College of Medicine, Central Michigan University, Mt. Pleasant, MI, USA
e-mail: hochg1u@cmich.edu

U. Hochgeschwender (ed.), *BioLuminescent OptoGenetics*,
https://doi.org/10.1007/978-3-032-18121-3_3

Chap. 2). Opsins are a small part of the many photosensory proteins provided by nature to enable living organisms, including plants, bacteria, algae, and fungi, to sense light and transduce this signal into biochemical outputs to elicit appropriate cellular responses in support of the organism's physiology, development, and behavior [1]. Including the vast pool of photoreceptors as tools for experimental light control of biological processes has expanded the scope of optogenetics beyond the original application of light-controlled ion flow [2, 3]. Increasingly, bioluminescence is being employed to activate these non-opsin photoreceptors, thereby adding another dimension to light stimulation and enabling manipulation of a multitude of light-mediated functions in cells across temporal and spatial scales.

As with luminopsins, using luciferases as a light source to activate light-sensing domains of photosensory proteins has several unique advantages (see also Table 1.1 in Chap. 1). In contrast to optical fiber light activation, bioluminescence reaches every light-sensing domain expressed in the target cell population, as the light source is genetically encoded and the light is generated locally. Using bioluminescence alleviates concerns over tissue and cell damage by fiber optics and extended physical light exposure. Yet, dual-light stimulation within the same construct allows for comparing the effects of transient with sustained activation. Different from chemical induction systems, the light is turned on with the application of the luciferase substrate, and the onset is immediate. Importantly, bioluminescent activation ceases with consumption of the luciferin in contrast to chemical inducers that continue to be present longer, as it takes time for these chemical compounds to be degraded or metabolized. There are multiple options for fine-tuning the kinetics of bioluminescent light emission, by varying the time the luciferin is made available and the concentration of the luciferin. Depending on the route of administration, the type of luciferin, and the concentration of the luciferin, effects last for minutes to hours, and phasic stimulation of light can be achieved with repeated applications of luciferin. Lastly, bioluminescent systems are heterogeneous and orthogonal in mammals, indicating the low possibility of cross talk and activation of endogenous actuators. The examples of applications of bioluminescent optogenetic (BL-OG)-driven photoreceptor activation given below will highlight these advantages.

In principle, any photosensory protein can be combined with a luciferase, with light emission wavelengths matching activation spectra of photosensory proteins from blue to red light. In practice, photoproteins used for light activation have been recruited mainly among the blue-light-sensing actuators. Accordingly, the luciferases that were combined with photoproteins are mostly blue-light-emitting. Light emitter and light sensor can be co-expressed or arranged in fusion proteins. Fusion constructs have the advantage of keeping a 1:1 molar ratio of emitter and sensor and of bringing the light emitter close to the light-sensing domain, while considerations about where to tether (N- or C-terminus) and how to link (linker length and composition) without impacting the performance of the photosensory actuator have to be tested experimentally, often extensively. In contrast, co-expression can enable robust activation across a variety of optogenetic systems without the need for additional engineering and functional validation of each new protein fusion pair. In either case, fusion or co-expression, the most likely mechanism of activation of the light-sensing domain is through energy transfer from the luciferase-luciferin reaction to the photoreceptor (Förster resonance energy transfer, FRET) [4, 5]. In fusion proteins, intramolecular FRET is clearly favored, while for co-expressed systems, intermolecular FRET requires high enough concentrations of the non-fused partners for FRET to occur. Table 3.1 lists the luciferases sourced for BL-OG of non-opsin photoreceptors, the various photoreceptors employed, and their arrangement as fused proteins or co-expressed.

From their application in nature to their exploitation as a biomolecular tool, some photoreceptors can be simply transferred essentially in their naturally occurring forms into different organisms; examples are the light-sensing bacterial transcription factor EL222 [6] and the bacterial light-activated adenylate cyclase bPAC [7]. However, most current tools have been molecularly engineered to exploit light-sensing mechanisms to outfit non-light-sensing molecules to enable light control. Under illumination, the associated chromophore isomerizes and induces an overall conformational change of the protein, translating light into manipulation of protein activity. With respect to their basic mechanisms of action, the photoproteins employed in BL-OG fall into four main groups: conformational changes, dimerization, enzyme activation, and generation of reactive oxygen species (ROS). These mechanisms have been leveraged to enable BL-OG control of biologic processes, including transcription, recombination, protein localization, protein allostery, second messenger signaling, cell death, and ROS production. Table 3.1 summarizes the key features of BL-OG-driven photoproteins, and the sections below detail the design principles of the molecular tools, the intracellular processes targeted, and the applications in vitro and in vivo for proof-of-principle demonstrations and for therapeutic applications.

3.2 BL-OG Control of Transcription

Transgene systems that are regulated by a single genetically encoded, photosensitive transactivator combine the advantages of simplicity with speed and versatility. Two such single-component systems have been used for bioluminescence-driven transcription. The first one employed the bacterial photoprotein EL222. Blue light induces dimerization of EL222 and enables its binding to the 5 x C120 promoter sequence, triggering transcription of the gene of interest [8] (Fig. 3.1a). In one application, NanoLuc (NLuc [9]) was the light source and was either co-transfected with or fused to EL222 to drive transcription of a reporter firefly (FLuc) luciferase in HEK293 cells [10]. While co-expression of light emitter and light sensor worked well, fusion proteins led to significantly higher levels of transcription compared to co-transfected components. The efficiency of transcription was comparable when using bioluminescence versus LED; if anything, bioluminescence was more efficient. Adir et al. [11] fused the bright *Gaussia* luciferase variant sbGLuc to the N-terminus of EL222 via a short linker. Synthetic cells containing the fusion protein and a target plasmid with red fluorescent protein (RFP) under control of the 5 x

Table 3.1 Bioluminescence driving non-opsin photoreceptors

Luciferase	Photoreceptor	Arrangement	Light triggers	Application—in vitro	Application—in vivo	Ref
Transcription						
NLuc	EL222	co-expression	Homodimerization	luciferase reporter expression in HEK cells		[10]
NLuc	EL223	Fusion	Homodimerization	Luciferase reporter expression in HEK cells		[10]
sbGLuc	EL222	Fusion	Homodimerization	Fluorescent reporter expression in synthetic cells		[11]
NLuc	CRY2/CIB	Co-expression	Heterodimerization	Luciferase reporter expression in HEK cells		[10]
NLuc	CRY2/CIB	Fusion	Heterodimerization	Luciferase reporter expression in HEK cells		[10]
NLuc	FKF1/GI	Co-expression	Heterodimerization	Luciferase reporter expression in HEK cells		[15]
NLuc	VVD	Fusion	Homodimerization	Luciferase reporter expression in different mammalian cell lines	Fluorescent reporter expression in liver and muscle in mice; insulin production in type 1 diabetic (T1D) mice	[13]
sbGLuc	CRY2/CIB	Co-expression	Heterodimerization	Luciferase reporter expression in HEK cells		[10]

NLuc	CRY/CIB	Co-expression	Heterodimerization	Luciferase reporter expression in HEK cells		[15]
Recombination						
NLuc	CRY/CIB	Co-expression	Heterodimerization	Luciferase reporter expression in HEK cells		[15]
NLuc	VVD	Co-expression	Heterodimerization	Luciferase reporter expression in HEK cells		[15]
sbGLuc	VVD	Co-expression	Homodimerization	Fluorescent reporter expression in HEK cells		[10]
Protein localization						
NLuc	CRY/CIB	Fusion	Heterodimerization	Recruitment of fluorescent protein from cytoplasm to cell membrane in HeLa cells		[13]
NLuc	LOV	Fusion	Conformational change	Translocation of fluorescent protein between mitochondrial membrane and cytoplasm in HeLa cells (LOVTRAP)		[13]

(continued)

Table 3.1 (continued)

Luciferase	Photoreceptor	Arrangement	Light triggers	Application—in vitro	Application—in vivo	Ref
NLuc	LOV	Co-expression	Conformational change	Recruitment of fluorescent protein from cytoplasm to mitochondrial surface in HeLa cells (iLID)		[15]
sbGLuc	LOV	Fusion	Conformational change	Recruitment of fluorescent protein from solution to outside membrane in synthetic cells (iLID)		[11]
Bacterial lux	VVD	Co-expression	Heterodimerization	Bacterial cell-cell adhesion		[22]
Protein allostery						
NLuc	LOV	Co-expression	Conformational change	Biotinylation of proteins in HEK cells (LOVTurbo)		[23]
NLuc	LOV	Fusion	Conformational change	VAMP2 cleavage in HEK cells (LATeNT)	VAMP2 cleavage in hippocampal neurons in mice	[24]
Second messenger signaling						
Nluc	bPAC	Fusion	Enzyme activation	cAMP production in thyroid cells		[27]
GLucM23	bPAC	Co-expression	Enzyme activation	cAMP production in CHO, HEK293, and vascular smooth muscle cells	cAMP production in vasculature for converting bioluminescence to hemodynamic contrast for MRI imaging	[28]

Cell death						
NLuc	CRY2	Fusion	Oligomerization	Necroptosis in HeLa, murine melanoma B16, and renal adenocarcinoma cancer cells	Necroptosis in human 786-O cell subcutaneous tumors in mice	[30]
NLuc	CRY/CIB	Co-expression	Heterodimerization	Apoptosis in Y79 retinoblastoma cells	Apoptosis of Y79 retinoblastoma cells in mouse eyes	[31]
ROS production						
NLuc	miniSOG	Fusion	Generation of singlet oxygen	Cytotoxic to human breast adenocarcinoma SK-BR-3 cells	Cytotoxic to breast ductal carcinoma BT-474 cancer cell subcutaneous tumors in mice	[39–41]
NLuc	SOPP3	Fusion	Generation of singlet oxygen		Cytotoxic to BT-474 cancer cell subcutaneous tumors and to SKOV3ip-Kat intraperitoneal tumors in mice	[42]
NLuc	miniSOG	Co-expression	Generation of singlet oxygen	Generation of Ca2+ oscillations in rat pancreatic acinar tumor cell AR4-2J		[44]
RLuc8	miniSOG	Fusion	Generation of singlet oxygen	Cytotoxic to various breast cancer cell lines		[45]

(continued)

Table 3.1 (continued)

Luciferase	Photoreceptor	Arrangement	Light triggers	Application—in vitro	Application—in vivo	Ref
RLuc8.6	KillerRed	Fusion	Generation of superoxide	Cytotoxic to various breast cancer cell lines and to primary triple-negative breast cancer cells from patients	Cytotoxic to triple-negative MDAMB-231 subcutaneous tumors in mice	[45]
FLuc	KillerRed	Fusion	Generation of superoxide	Targeting to F-actin-induced filopodia-like protrusions		[46]

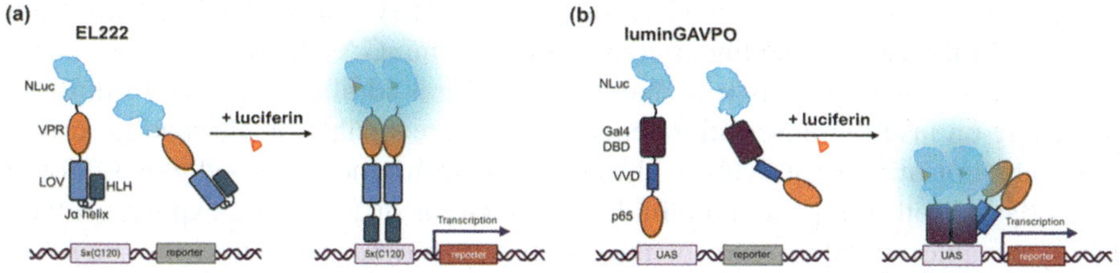

Fig. 3.1 Bioluminescence-driven transcription: single transactivator systems. (**a**) Natural light-sensing transcription factor EL222 (VPR, transcriptional activation domain; LOV, light-oxidation-voltage domain containing the chromophore; Jα helix; HLH, helix-loop-helix domain). In the dark, EL222 remains a monomer in an "off" conformation. Upon addition of luciferin and emission of light, the Jα helix unfolds, exposing DNA-binding residues of the HLH domain; the protein dimerizes and binds to the 5xC120 promoter, where the VPR initiates transcription of downstream genes. (**b**) Engineered light-sensing transcription factor luminGAVPO (Gal4DBD, DNA-binding domain; VVD, light-oxidation-voltage domain containing photoprotein; p65, transcriptional activation domain). Addition of luciferin and emission of light causes VVD to dimerize, enabling the DNA-binding domain of Gal4 to bind to its cognate UAS site and p65 to initiate transcription of the target gene

C120 promoter resulted in expression of RFP with the addition of coelenterazine (CTZ) and resulting bioluminescence.

While EL222 is a natural light-sensing transcription factor, the light sensitivity of GAVPO (Gal4-VVD-p65 activation domain, optimized by mutations within VVD) [12] was engineered, utilizing the yeast Gal4 transcription system and the *Neurospora crassa* photoprotein VIVID (VVD), the smallest LOV domain-containing protein. The GAL4 transcription factor comprises a DNA-binding domain, responsible for recognizing the UAS sequence upstream of a target gene, and the GAL4 activation domain, responsible for initiating transcription. The DNA-binding domain of Gal4 consists of a DNA-recognition element and a dimerization domain; dimerization enables binding to its cognate UAS site. By swapping the Gal4 dimerization domain with VVD, Gal4 dimerization and binding to UAS become light-dependent [12]. Li et al. [13] fused NanoLuc luciferase to the N-terminal of GAVPO to generate luminGAVPO (Fig. 3.1b). Upon induction by furimazine, the luminescence emitted from Nluc directly activates the proximal VVD domain, resulting in dimerization of luminGAVPO and binding to its cognate UAS promoter, thereby initiating transcription of the target gene. This blue light-inducible transgene expression system was first tested in various cell lines, where the addition of furimazine resulted in expression of a reporter *Gaussia* luciferase. in vivo, transfection of luminGAVPO and mCherry reporter plasmids into mouse livers and muscle enabled mCherry expression after furimazine administration.

Most importantly, the authors took advantage of the unique kinetics of bioluminescence-driven transcription. When furimazine is consumed, the activated luminGAVPO is gradually inactivated and dissociates from its cognate DNA-binding site, leading to inactivation of transcription. Thus, bioluminescence-induced transgene expression is pulsatile. Li et al. exploited this for regulating blood-glucose homeostasis in type 1 diabetic (T1D) mice [13]. They transplanted cell lines that expressed the insulin gene under control of the luminGAVPO transcription factor. Blood glucose homeostasis was maintained by oral administration of furimazine, switching on insulin transcription repetitively and reversibly by alternating furimazine induction at 12-h intervals, with sustainable restoration of blood glucose for up to 13 days. Both the pulse amplitude and duration of transgene expression were highly tunable via adjustment of the amount of furimazine, with lower doses of furimazine exhibiting shorter durations and smaller amplitudes of pulsatile activation, while higher doses of furimazine had longer durations and larger amplitudes.

Different from the single polypeptide chain design of EL222 and GAVPO, other systems for light-regulated gene expression rely on a two-hybrid strategy, employing split transcription factors that are composed of two separate polypeptide components. Applied to the Gal4/UAS system, the Gal4 transcription factor is split into two nonfunctional components: the DNA-binding domain (DBD) and the activation domain (AD). By fusing each separate component to light-sensing homo- or heterodimerizers, light will induce dimerization of the fusion proteins to reconstitute the transcription factor, enabling the binding of Gal4 to the UAS sequence and activating transcription. One example is fusing CRY2 and the Gal4 DNA-binding domain and CIB and the transcriptional activator sequence VP64 [14]. By further fusing or co-expressing a luciferase, this light-dependent Gal4 transcription system can be driven by bioluminescence (Fig. 3.2a) [10]. NLuc either co-transfected or fused to CRY2 in HEK293 cells expressing a UAS-FLuc construct led to transcription of the reporter FLuc in the presence of the luciferin furimazine. Interestingly, CRY/CIB showed consistently higher background levels with vehicle application compared to EL222 (see above), which had negligeable background transcription.

Alternative to the cryptochrome-based CRY2-CIB, Parag-Sharma et al. fused the Gal4 and transcription activator domains to the *Arabidopsis* photoreceptor Flavin

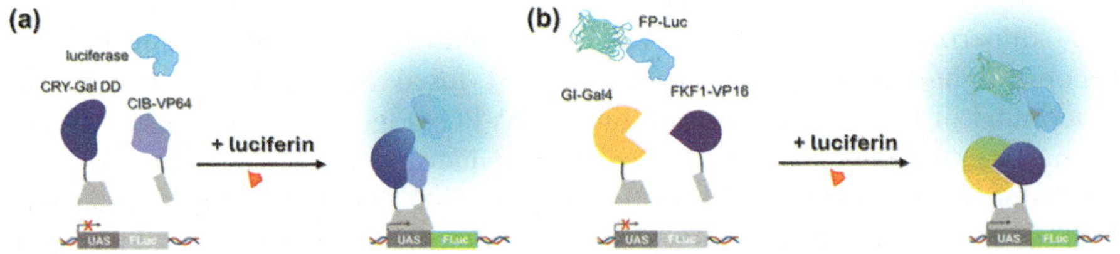

Fig. 3.2 Bioluminescence-driven transcription: two-hybrid systems. (**a**) Plant photoreceptor proteins CRY2/CIB are applied to a split Gal4 system, where the Gal4 DNA-binding domain (Gal DD) is fused to CRY2 and the activation domain VP64 is fused to the CIB protein. In the presence of blue light generated by luciferin application, CRY2 and CIB proteins dimerize and reconstitute the functional transcription factor. This reconstituted factor binds to the upstream activating sequence (UAS), activating the expression of the reporter gene downstream of it. (**b**) Plant photoreceptor proteins Flavin Kelch-repeat F-box 1 (FKF1) and its binding partner GIGANTEA (GI) are applied to the split Gal4 system, where the Gal4 DNA-binding domain (Gal4) is fused to GI and the activation domain VP16 is fused to FKF1. Blue light, generated by a fluorescent protein-luciferase fusion protein, results in GI/FKF1 dimerization and reconstitution of the functional transcription factor

Kelch-repeat F-box 1 (FKF1) and its binding partner GIGANTEA (GI) (Fig. 3.2b) [15]. Co-expression of a bright fluorescent protein–NanoLuc fusion protein as the light emitter with the Gal4-GI and FKF1-VP16 components in HEK293 cells resulted in bioluminescence-induced Gal4-GI/FKF1-VP16 interaction at the UAS promoter sequence and expression of the downstream located FLuc reporter in the presence of the NanoLuc luciferin furimazine. Bioluminescence-induced assembly of the transcription factor complex achieved ~40-fold gene induction, which is superior to that seen with many other Gal4 fusion transcription factors based on interaction with chemicals.

Yet other designs incorporate light-sensing molecule association with more complex assemblies. In one example, based on light-induced heterodimerization of cryptochromes, CRY2 was fused to a protease while the membrane-bound CIB was fused to a protease cleavage site followed by a transcription factor (Fig. 3.3a) [10]. Light emitted by the luciferase sbGLuc oxidizing CTZ induced protein translocation and released the transcription factor, leading to expression of reporter FLuc and dTomato in HEK293 cells. While bioluminescence robustly increased expression of FLuc, the presence of the transcription factor component alone resulted in considerable background signal, possibly due to spontaneous proteolysis. In another example, a dCas9-gRNA-based CRY2-CIB dimerization system based on LACE (light-activated CRISPR-Cas9 effector [16]) was designed for precise and targeted activation of gene expression (Fig. 3.3b) [15]. CIB was fused N- and C-terminally to the catalytically inactive dCas9 with four gRNAs targeting the gene of interest, while CRY2 was fused to the transactivation domain. The gRNAs would direct the binding of dCas9 to the gene of interest's promoter in both the light and dark. Transcription was induced only in the presence of blue light provided by a fluorescent protein–NLuc fusion in the presence of furimazine that colocalized the

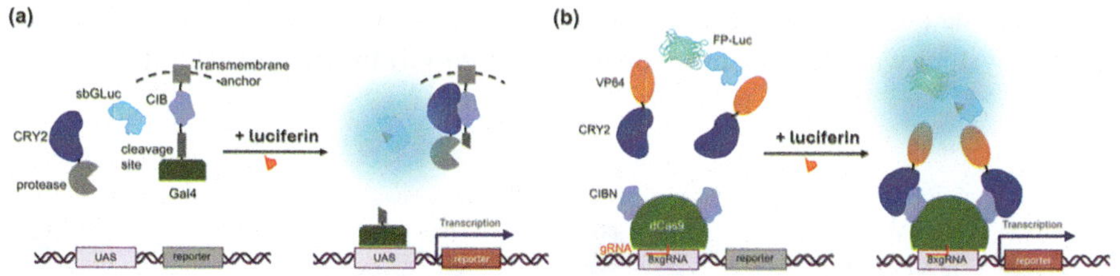

Fig. 3.3 Bioluminescence-driven transcription: multi-factor systems. (**a**) Bioluminescence-induced protease cleavage. In the dark, the transcription factor Gal4 is anchored to the cell membrane via a protease cleavage site, CIB, and a transmembrane anchor. In the presence of luciferin, light-induced heterodimerization of CRY2/CIB brings the protease attached to CRY2 close to its cleavage site, resulting in the release of the transcription factor and initiation of reporter gene expression. (**b**) Bioluminescence-induced light-activated CRISPR-Cas9 effector (LACE) system. The light-inducible heterodimerizing proteins CRY2 and CIBN are fused to a transactivation domain (VP64) and both the N-terminus and C-terminus of the catalytically inactive dCas9, respectively. The co-transfected gRNAs direct the binding of dCas9 to the 8xgRNA promoter in both the light and dark. With the application of luciferin and light emission, VP64 colocalizes with dCas9 via CRY2-CIBN interactions and induces transcription of the reporter

transactivator with dCas9 via CRY2-CIB interactions. The authors observed robust (~20-fold) induction of reporter gene expression, comparable to what others have observed for constitutively active dCas9-CRISPR transcription activation systems.

3.3 BL-OG Control of Recombination

Bioluminescence-driven optogenomic manipulation relies on photoactivatable recombinases that remove transcriptional STOP sequences to allow expression of a gene of interest. The recombinase is split into an N-terminal and C-terminal fragment, with each fragment fused to a homo- or heterodimerizing photoprotein, allowing light-induced reconstitution of a functional recombinase (Fig. 3.4). Parag-Sharma et al. exploited two versions of split light-dependent Cre recombinases that they activated with bioluminescence emitted by NanoLuc fused in-frame with a cyan fluorescent protein, mCerulean3, separated by a flexible five amino acid linker, DISGG (CeNLuc) [15]. Light output of this "LumiFluor" was estimated to be ~fourfold brighter than NanoLuc, sufficient to activate co-expressed photoreceptors. In the first application, the light-activated split Cre recombinase system [17] employed the CRY2-CIB dimerization partners to reconstitute a functional Cre recombinase molecule capable of bioluminescence-driven induction of *loxP* recombination and excision of the STOP cassette upstream of a FLuc reporter in HEK293 cells (Fig. 3.4a) [15]. The second application took advantage of an enhanced split Cre recombinase system based on the optimized, heterodimerizing VVD-Magnets light-responsive optogenetic system (Fig. 3.4b). Magnets are engineered VVD proteins that prevent the natural homodimerization of VVD and selectively induce heterodimerization of pMagnet and nMagnet [18]. While activation of Magnets by CeNLuc co-expressed in the cytoplasm only resulted in a modest induction of the reporter gene (~fourfold), the potency of CeNLuc-mediated activation of this optogenetic system was significantly enhanced (~tenfold) by colocalizing the biologic

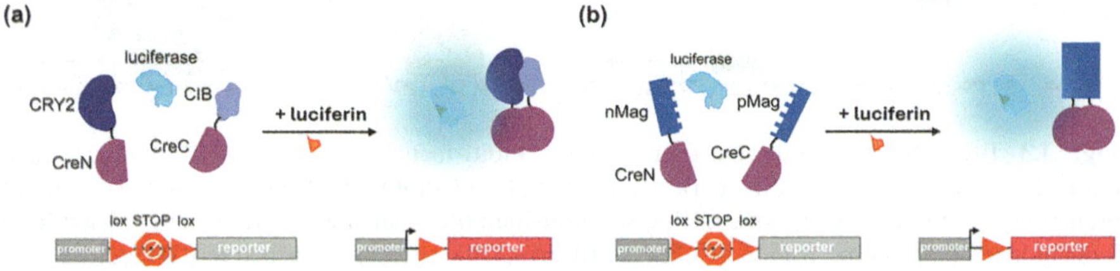

Fig. 3.4 Bioluminescence-driven recombination. (**a**) The N- and C-terminal parts of Cre recombinase are fused to CRY2 and CIB, respectively. In the presence of luciferin, light-induced heterodimerization of CRY2/CIB reconstitutes functional Cre, resulting in Cre-mediated removal of the lox-flanked STOP sequence and initiation of reporter gene expression. (**b**) The same principle as described in (a) applies, here with light-induced heterodimerization of the mutant VVD proteins nMagnet and pMagnet

light source through a nuclear localization signal with the light-responsive proteins within the same subcellular compartment.

A newer version of an optimized light-inducible VVD-based Cre recombinase system, iCreV [19], that uses improved homodimerizing wild-type VVDs instead of heterodimerizing VVD mutants (Magnets) was applied with co-expressed sbGLuc and a lox-stop-lox fluorescence reporter (tdTomato) in HEK293 cells [10]. CTZ application resulted in a robust increase of tdTomato expression over the background that was similar to that induced with LED application.

3.4 BL-OG Control of Protein Localization

There are several optogenetic systems available that control the physical location of proteins within cells, and for some of them, bioluminescence has been used to activate the light-sensing proteins, enabling this control. LOVTRAP and iLID are optogenetic protein localization tools with opposite functions: LOVTRAP induces protein dissociation, while iLID induces protein dimerization. Li et al. demonstrated bioluminescence activation of LOVTRAP (LOV2 trap and release of protein [20]), an optogenetic approach capable of repeatedly and reversibly controlling protein activity (Fig. 3.5a). The approach is based on a small protein, Zdark (Zdk), that binds selectively to the dark state of LOV2, a photo-sensor domain from *Avena sativa* phototropin 1. A mitochondria-anchored Zdk protein was co-expressed with a mCherry-LOV2-NLuc fusion protein; under dark conditions, mCherry fluorescence accumulated at mitochondria [13]. However, the mCherry fluorescence readily diffused to the cytosol within 5 s after the addition of furimazine, while mCherry fluorescence accumulated at mitochondria again once furimazine was consumed. Such translocation of mCherry could be triggered repetitively to the same magnitude with little or no loss in efficacy.

While LOVTRAP is based on light-induced dissociation, iLID achieves light-driven association of proteins. iLID, improved light-inducible dimer, is an engineered protein that embeds the bacterial SsrA peptide in the C-terminal helix of the blue-light-sensing LOV2 domain from *Avena sativa* [21]. In the dark, the SsrA peptide is sterically blocked from binding its natural binding partner, SspB. When activated with blue light, the C-terminal helix of the LOV2 domain undocks from the protein, allowing the SsrA peptide to bind with high affinity to the *E. coli* adaptor protein SspB. Here, iLID-SsrA was anchored in the outer mitochondrial membrane, while SspB-RFP was expressed in the cytoplasm (Fig. 3.5b) [15]. SsrA/SspB interaction and thus recruitment of the RFP from diffuse distribution in the cytoplasm to the mitochondria was achieved by bioluminescence from co-expressed NLuc and application of furimazine. NLuc expression had to be localized to the mitochondrial membrane to be effective, and then, it was as efficient as recruitment induced by conventional light sources. Modulating the luciferin dosing allowed the generation of dynamic kinetic temporal activation profiles from short-lived pulses to sustained

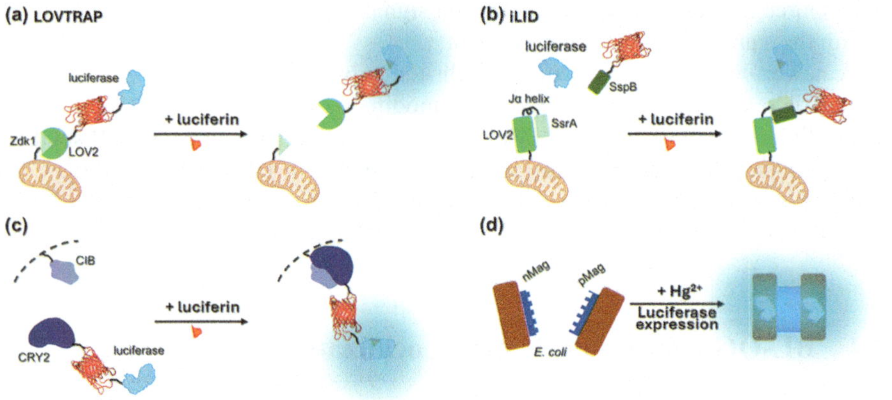

Fig. 3.5 Bioluminescence-driven protein localization (**a**) LOVTRAP: A protein of interest (mCherry) is fused to the LOV2 domain, and the binding partner Zdk1 is anchored to the mitochondria membrane. In the dark, Zdk1 binds to the LOV2 domain, "trapping" the protein of interest at the anchored location. When illuminated with blue light from the fused NanoLuc, the LOV2/Zdk interaction is broken, releasing the protein of interest. (**b**) iLID: LOV2 connected to SsrA is anchored to the mitochondrial membrane, and the protein of interest (red fluorescent protein) fused to SspB is separated in the cytoplasm. In the dark, SsrA is inaccessible. When illuminated with blue light from NanoLuc, the LOV2 domain undergoes a conformational change, revealing SsrA that recruits the SspB-fused protein of interest, forming a stable complex. (**c**) Protein localization using CRY2/CIB. CIB is anchored in the cell membrane, and CRY2 and NanoLuc are fused to the protein of interest (red fluorescent protein). Luciferin-induced bioluminescence triggers CRY2/CIB dimerization, recruiting the protein of interest from the cytoplasm to the cell membrane. (**d**) Bacterial aggregation. Bioluminescence triggers bacterial cell-cell adhesion by photoactivating heterodimerization of Magnets located on the bacterial surface. Bioluminescence is switched on by Hg^{2+}-dependent transcription of the self-sufficient bacterial bioluminescent system

and long-lasting activation. Low doses of furimazine could be repeatedly applied to generate pulsatile activation without significant loss in overall light output.

In an application using synthetic cells (SC), Adir et al. [11] achieved bioluminescence-driven iLID activation with the iLID component conjugated to the SC membrane and the other component (SspB-RFP) free in solution. Fusing a SC outer membrane-anchored *Gaussia* luciferase variant, sbGLuc, via a linker N-terminally to iLID favored efficient activation of the light-responsive domain through BRET and, at the same time, exposed the iLID to the extracellular environment with low steric interference. The addition of the luciferin coelenterazine caused bioluminescence-activated membrane recruitment of SspB-tagged RFP to sbGluc-iLID-labeled SCs with efficiencies exceeding those of laser illumination.

Li et al. (Li 2021) also used bioluminescence for CRY2/CIB light-driven dimerization to recruit CRY2-mCherry fused with NanoLuc to the plasma membrane through interaction with membrane-anchored CIB (Fig. 3.5c) [13].

Lastly, bioluminescence generated inside bacteria from the bacterial luciferase gene cassette (luxCDABE) was used to activate photoswitches located on the bacteria's surface [22]. The light-sensing components were Magnets, engineered from the fungal photoreceptor Vivid to develop pairs of nMag/pMag that turned homodimerization of VVD into heterodimerization pairs [18]. As the luciferin in bacteria is constitutively present, transcription of the bacterial luciferase gene cassette and thus bioluminescence was initiated by the presence of Hg^{2+} via a metalloregulatory protein. Cocultures of bacteria expressing the bacterial luciferase that either displayed nMag or pMag on the surface through fusion with outer membrane proteins aggregate into large clusters due to heterodimerization of the surface-displayed photoswitchable proteins activated by luminescence generated inside the bacteria (Fig. 3.5d).

3.5 BL-OG Control of Protein Allostery

Allosteric regulation of proteins is a mechanism where the functional site of a protein is modulated by conformational changes in a distant region (allosteric site) of the same protein. The light-induced undocking of the Jα helix of the conformation-switching LOV domain has been exploited as an allosteric photoswitch to regulate enzyme activity. Two examples of this engineered light-induced allostery are LOVTurbo [23] and LATeNT [24]. In both cases, the LOV2 domain was inserted at a predicted allosteric site, enabling reversible light control over enzymatic activity.

In LOVTurbo, the light-sensitive LOV domain was installed into TurboID, an artificial enzyme used for proximity-dependent biotinylation in living cells (Fig. 3.6a) [23]. TurboID uses biotin and ATP to generate a reactive biotinyl-AMP mixed anhydride that it releases from its active site to covalently tag nearby proteins on nucleophilic lysine sidechains. Allosteric engineering enabled optical control over TurboID with conventional light sources as well as bioluminescence. Exposure of HEK293 cells expressing a NLuc-LOVTurbo fusion protein to furimazine together with biotin for 10 min or 1 h resulted in biotinylation detectable on Western blots.

In LATeNT (light-activated TeNT, the light chain of tetanus neurotoxin), the light-sensitive LOV domain was inserted into a surface-exposed loop of the TeNT

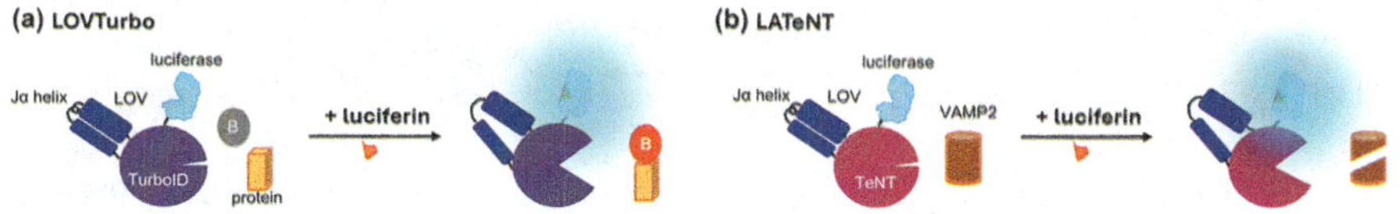

Fig. 3.6 Bioluminescence control of protein allostery. The allosteric site of a protein is clamped down by the LOV protein in the dark state, keeping the enzyme inactive. With light exposure, the Jα helix unwinds, and the resulting conformational change restores the enzyme to an active state. (**a**) LOVTurbo: upon bioluminescent light activation, the synthetic enzyme TurboID, a modified biotin ligase, labels nearby proteins with biotin. (**b**) LATeNT: bioluminescence activates the metalloprotease tetanus neurotoxin (TeNT), resulting in cleavage of VAMP2, inhibiting synaptic transmission as a downstream effect

protease to "clamp" the loop in the dark state (Fig. 3.6b) [24]. As the loop is allosterically coupled to TeNT's active site, clamping distorts the active site and renders the enzyme inactive. Blue light exposure causes the LOV domain to release its C-terminal Jα helix, "unclamping" the surface loop and restoring the active site to a native-like conformation and allowing TeNT to cleave the endogenous vesicle fusion protein VAMP2. Bioluminescence-driven activation of LATeNT was demonstrated in vitro and in vivo. Expression of a NLuc-LATeNT fusion protein in HEK293 cells together with a VAMP2 reporter showed that cells treated with fluorofurimazine (FFz [25]) for 60 min produced a similar degree of VAMP2 cleavage as cells exposed to blue light for 30 min. Moreover, tissue sections of mice expressing NLuc-LATeNT in the hippocampal CA3 region after receiving intraperitoneal injections of cephalofurimazine (CFz [26]) showed cleavage of endogenous VAMP2. Again, the increase in cleaved VAMP2 was comparable to that observed in mice subjected to light via an implanted optical fiber.

3.6 BL-OG Control of Second Messenger Signaling

Cyclic adenosine monophosphate (cAMP) is a ubiquitous intracellular second messenger involved in numerous biological processes, including regulation of cellular proliferation, differentiation, attachment, migration, and several other processes. Adenylyl cyclase is the enzyme that converts ATP into cAMP. The bacterium *Beggiatoa* encodes a photoactivated adenylyl cyclase (bPAC), where the cyclase is linked C-terminally to a flavin-binding BLUF (blue light receptor using FAD)-type light sensor domain [7]. Naim et al. fused bPAC N-terminally to NLuc for bioluminescence-driven bPAC activation (Fig. 3.7) [27]. They demonstrated that BL-OG activation of bPAC-NLuc can be fine-tuned through the use of different

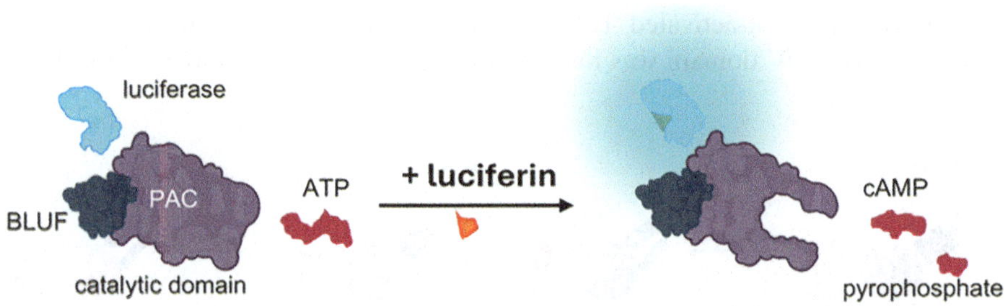

Fig. 3.7 Bioluminescence control of second messenger signaling. The photoactivated adenylyl cyclase (PAC) consists of the catalytic enzyme domain linked to a BLUF (blue light receptor using FAD) light sensor domain. Blue light from a nearby luciferase-luciferin reaction is absorbed by the BLUF domain, causing a conformational change in its flavin chromophore. This light-induced change in the BLUF domain is transmitted to the catalytic domain, where it triggers a larger conformational change, activating its enzymatic function. The activated catalytic domain then converts ATP into cAMP, which becomes available for downstream signaling

luciferins (hCTZ, Fz) and luciferin concentrations to mimic physiological levels and duration of cAMP production. Long-term bioluminescent stimulation of cAMP production induced cell proliferation in thyroid epithelial cells, similar to physiological TSH-stimulated cAMP signaling in this prototypical cAMP-responsive tissue. Among chemically activated tools bPAC–nLuc is unique in that it does not stimulate endogenous adenylyl cyclases.

Ohlendorf et al. (2024) used bPAC for converting bioluminescence signal to hemodynamic contrast that can be imaged noninvasively in deep tissue by MRI [28]. Upon administration of the luciferin coelenterazine (CTZ), bioluminescence emission from the *Gaussia* luciferase variant GLucM23 [29] activates bPAC that is expressed in vascular smooth muscle cells (VSMC) in the brain. Photoactivated bPAC increases cAMP levels in VSMCs and stimulates downstream signaling cascades that, in turn, cause local dilation of blood vessels and hemodynamic contrast detectable by MRI. Successful bioluminescent activation of bPAC was verified in CHO cells, HEK293 cells, and cultured VSMCs. To demonstrate BL-OG activation of bPAC and its downstream signal transduction driving VSMC relaxation and vasodilation in vivo, the authors photosensitized the brain vasculature in rats by AAV transduction with constructs targeting bPAC expression to VSMCs. Xenografts of HEK293 cells expressing GLucM23 were implanted into the photosensitized and control brain regions, and MRI scanning was performed before, during, and after CTZ substrate infusion. This strategy of BioLuminescence imaging Using Hemodynamics (BLUsH) induced large hemodynamic signals in the area of the xenografts implanted in photosensitized brain hemispheres, but not in control non-bPAC expressing regions. Further, the luminescent output correlated with luciferin concentration and could be elicited repeatedly by injecting luciferin over multiple distinct cycles, with the signal returning to baseline between injections. The authors also demonstrated the ability of BLUsH to visualize endogenous luciferase expression in animals that were virally transduced with a mixture of GLucM23 and bPAC. Through generating hemodynamic signals by activating a photosensitive enzyme, bioluminescence can be used to image luminescent implanted cells and virally induced endogenous luciferase expression, spatially resolved in deep tissue using MRI. While the reported experiments were carried out with intact luciferase, the designed approach might be amenable to use with bioluminescent sensors, opening possibilities for functional imaging.

3.7 BL-OG Control of Cell Death

There are different types of "programmed cell death," such as apoptosis, necroptosis, autophagy, and pyroptosis. Two studies engaged bioluminescence to control photoactivatable necroptosis and apoptosis, respectively.

He et al. took advantage of light-induced oligomerization of CRY2 by fusing it to a specific and crucial protein for necroptosis induction, mixed lineage kinase domain-like (MLKL) protein, generating MLKL-CRY2 or LiPOP (light-induced

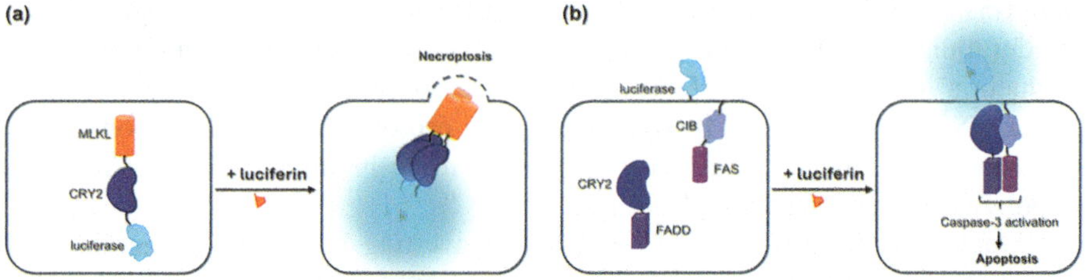

Fig. 3.8 Bioluminescence control of cell death. (**a**) MLKL, a crucial protein for necroptosis induction, is fused to CRY2 and a luciferase. Application of luciferin and bioluminescent light emission leads to CRY2 homodimerization and MLKL oligomerization, which induce plasma membrane permeabilization, necroptosis, and cell death. (**b**) Heterodimerization of CRY2 and CIB is used to dimerize two proteins, FAS and FADD, critical in activating the caspase-3 and apoptosis signaling pathway. CRY2-FADD is located in the cytoplasm, CIB-FAS is membrane-anchored, and the luciferase is attached to the outer cell membrane. Application of luciferin and bioluminescent light emission recruit CRY2-FADD in close proximity to CIB-FAS. Dimerization of FADD-FAS activates caspase-3, leading to apoptosis and cell death

non-apoptotic tool) (Fig. 3.8a) [30]. Cell death is triggered upon oligomerization of MLKL; oligomerized MLKL induces necroptosis by directly permeabilizing the plasma membrane. MLKL-CRY2 oligomerization and thus necroptosis were induced by bioluminescence via a C-terminally attached NanoLuc (LiPOP-NLuc). The authors demonstrated plasma membrane translocation of LiPOP-NLuc and necroptosis in three cancer cell lines in vitro (human HeLa cells, murine melanoma, B16 cells, and renal adenocarcinoma cells). Comparison of luciferin treatment and blue light stimulation suggested that the BL-OG approach is as effective as exogenous photostimulation. Importantly, they showed that bioluminescence effectively activates these CRY2-based optogenetic devices to kill tumor cells in vivo. Immunodeficient mice were subcutaneously inoculated with engineered human 786-O cells expressing LiPOP-NLuc. Established tumors received intratumoral injection of FFz multiple times, resulting in a significant reduction in tumor size compared to controls.

Light-induced CRY2 heterodimerization with CIB was used to dimerize fused FADD (Fas-associated via death domain) and FAS (death receptor) proteins, respectively, with FADD-FAS binding, activating the caspase-3 apoptotic signaling pathway (Fig. 3.8b) [31]. In this system, CRY2-FADD was located in the cytoplasm, CIB-FAS was anchored in the plasma membrane via a membrane localization signal, while NanoLuc was located on the outside of the targeted cell. Bioluminescence generated on the cell surface with the application of furimazine was able to trigger the apoptotic pathway and inhibit the growth of retinoblastoma Y79 cells in vitro and in vivo. In retinoblastoma-bearing mice receiving vitreous injections of the optogenetic components through a nanoparticle-based delivery system, together with furimazine, tumor growth was effectively inhibited. The bioluminescence-activated apoptotic pathway inhibited tumor growth with a higher therapeutic efficacy, i.e., reduction in ocular tumor size, compared to external blue light irradiation of the eyes. Most importantly, while external blue light irradiation induced retinal damage and corneal neovascularization, the BL-OG-driven approach preserved retinal structural integrity without corneal neovascularization.

3.8 BL-OG Control of Reactive Oxygen Species (ROS) Production

Early on, bioluminescence has attracted interest as an internal light source for theranostic applications [32]; specifically, photodynamic therapy (PDT) has been clinically approved as a noninvasive therapy technique for treating lung cancer, skin cancer, and esophageal cancer. PDT uses light irradiation of an appropriate wavelength to activate photosensitizers, inducing the production of reactive oxygen species (ROS), causing cell death of cancer cells in the presence of oxygen through mechanisms, including apoptosis, necrosis, necroptosis, and autophagy [33]. The photosensitizers can be chemicals, such as rose bengal or Chlorine e6, or proteins, such as miniSOG [34] or KillerRed [35] (Fig. 3.9). Light is commonly supplied by external light sources with their known limitations of penetration depth, safety, and efficiency. To overcome these limitations, bioluminescence has been explored as a light source in combination with chemical photosensitizers [36] and, more recently, with genetically encoded photosensitizers (see below). Interestingly, bioluminescence in ROS facilitation is divergent from its proposed original function to protect organisms as an antioxidant via luciferin-ROS interactions [37, 38].

The first bioluminescent optogenetics PDT strategy employed NanoLuc fused to the phototoxic flavoprotein miniSOG (Fig. 3.9a) [39]. The authors demonstrated that in human breast adenocarcinoma SK-BR-3 cells stably expressing a NanoLuc-GGGGS linker-miniSOG construct addition of the luciferin furimazine caused photoinduced cytotoxicity at a 48% cell death level. They suggested that generation of reactive oxygen species and cell death induction occurred through bioluminescent

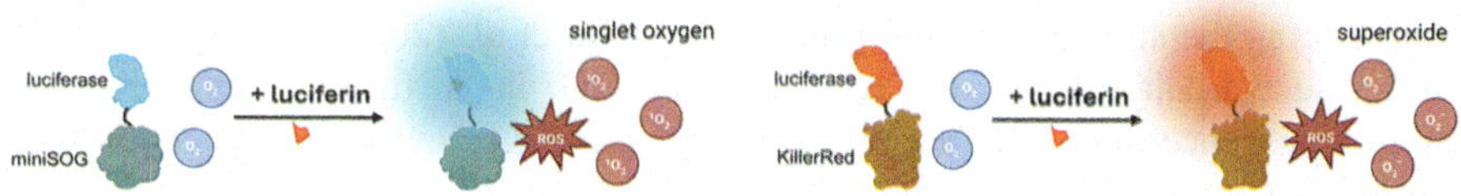

Fig. 3.9 Bioluminescence control of reactive oxygen species (ROS) production. ROS are produced from oxygen (O_2), ubiquitous in cells, where they function as signals in many biological processes, from support of cell survival to cell damage or death. (**a**) miniSOG is a genetically encoded photosensitizer derived from the phototropin 2 protein of *Arabidopsis thaliana*. When illuminated with blue light from fused luciferases such as NanoLuc or Renilla luciferase, miniSOG produces reactive oxygen species (ROS), primarily singlet oxygen ($^1O_2$). (**b**) KillerRed is a genetically engineered phototoxic protein that generates ROS, mainly superoxide radicals (O_2^-), upon green or orange light illumination from fused luciferases such as the red-shifted Renilla luciferase RLuc8.6 or firefly luciferase

resonance energy transfer to the genetically encoded photosensitizer. In further studies, the authors showed that the NanoLuc-furimazine bioluminescence system induces a photodynamic effect of miniSOG comparable to that of LED-induced miniSOG phototoxicity [40]. This was also demonstrated in vivo [41]. Here, NanoLuc-miniSOG was stably expressed in breast ductal carcinoma BT-474 cancer cells that were used to generate tumors in BALB/c nu/nu mice. Animals treated intravenously with furimazine trice per day for 10 days showed tumor-growth inhibition of 72%.

In a recent paper from this group [42], an improved version of miniSOG characterized by a high quantum yield of singlet oxygen generation, SOPP3 [43], was implemented as the photosensitizer. This improved NanoLuc-SOPP3 construct was applied in a mouse model of deep-seated intraperitoneal tumors, SKOV3ip-Kat intraperitoneal tumor xenografts. SKOV3ip-Kat ovarian carcinoma cells are characterized by the formation of multiple small tumor nodules in the abdominal cavity upon intraperitoneal implantation. Animals with deep-seated tumors that received bioluminescence-activated PDT via intraperitoneal administration of furimazine had, on average, fewer or smaller intraperitoneal nodules compared to animals in the conventional PDT group that were irradiated by blue light.

In another implementation, NanoLuc activates miniSOG with its singlet oxygen production, used to activate CCK1R [44]. The calcium-mobilizing G protein-coupled receptor cholecystokinin type 1 (CCK1R), widely expressed both in the central nervous system and in peripheral tissues, is permanently activated by singlet oxygen, resulting in calcium oscillations. Rat pancreatic acinar tumor cells, AR4-2J, were transformed with a miniSOG-IRES-NanoLuc plasmid, resulting in the expression of separate plasma membrane-localized miniSOG and cytoplasmic NanoLuc proteins. Addition of luciferin in perifused miniSOG-IRES-NanoLuc-AR4-2J cells induced strong bioluminescence that triggered long-lasting calcium oscillations through miniSOG generation of singlet oxygen, $^1O_2$, likely via a radiationless NanoLuc-to-miniSOG BRET process; $^1O_2$ then oxidatively activates the endogenously expressed CCK1R. The long-term goal of developing this tool is to study the photodynamic CCK1R physiology or pharmacology of CCK1R-expressing cells in vivo.

In the above examples, the photosensitizer miniSOG was combined with NanoLuc as the source for bioluminescence. Kim et al. (2020) employed Renilla luciferase variant RLuc8 to activate miniSOG in a fusion construct [45]. Bioluminescence-driven ROS generation had significant cytotoxic effects on various breast cancer cell lines in vitro.

The same group also explored the use of a different photosensitizer, KillerRed (Fig. 3.9b). Excitation of this protein by the red-shifted light emission of RLuc8.6 in a Rluc8.6-KillerRed fusion construct efficiently induced cytotoxic effects on various breast cancer cell lines and primary triple-negative breast cancer cells from patients in vitro [45]. In vivo, tumor growth was significantly inhibited when the fusion protein Rluc8.6-KillerRed was intratumorally injected into mice subcutaneously xenografted with triple-negative MDA-MB-231 breast tumor cells, followed by subcutaneous spread of the luciferin coelenterazine-h. Moreover, bioluminescence-driven ROS generation had superior effects in tumor growth suppression by comparison with laser light irradiation.

The photosensitizer protein KillerRed was also combined with firefly luciferase (FLuc) in a KillerRed-FLuc fusion construct, which generates ROS from KillerRed through bioluminescence resonance energy transfer from FLuc in response to luciferin treatment [46]. The motivation was the development of a tool for modulating target proteins in live cells with local ROS generation that offers an alternative to chromophore-assisted laser inactivation (CALI; [47]. The authors targeted the KillerRed-FLuc protein to F-actin, to investigate the effect of subcellular targeted exposure of ROS on actin polymerization. The consistent generation of very low-level ROS by KillerRed-FLuc in the vicinity of F-actin induced filopodia-like protrusions, cofilin-actin rods, as seen in patients with Alzheimer's disease. This subcellular bioluminescence-driven ROS generation allows the spatiotemporal control of molecular inactivation while avoiding the laser toxicity of CALI.

3.9 Conclusions

The use of bioluminescence to activate non-ion-moving photoreceptors has been extensively demonstrated, as detailed in the above sections. Most of the examples are proof-of-principle studies carried out in vitro, and some were demonstrations in vivo. There are six studies that applied BL-OG-driven photoreceptor activation in vivo in a therapeutic context: One is the control of glucose homeostasis in type 1 diabetic (T1D) mice [13]; two studies used BL-OG for the activation of cell death pathways in tumor treatment [30, 31]; and three studies leveraged ROS generation for the treatment of breast and ovarian tumors [41, 42, 45]. The application of BL-OG in controlling glucose homeostasis highlights the unique ability of bioluminescence-driven activation of photoreceptors for pulsatile regulation compared to chemical inducers, while the applications in cancer treatment underline the superior efficacy in reaching even deep-seated tumors compared to external light sources.

Bioluminescence for driving photoreceptors in vivo, either for proof-of-principle experiments or for therapeutic applications, has thus far been used in rodents. In principle, there are no limits preventing the approach from being applied to different species, from fish or flies to nonhuman primates and eventually in therapeutic applications in patients.

The approach of bioluminescent activation of photoreceptors has been demonstrated thus far for a limited number of luciferases and photosensory proteins, respectively, both in vitro and in vivo. It can be extended to a large variety of optogenetic systems that can be leveraged to control a multitude of biological processes in a spatiotemporally defined manner. Such expansion of the approach is further promoted by the continuous development of novel luciferases and luciferase-fluorescence protein pairs with much higher light emission than that of naturally occurring luciferases and with kinetic features tunable to different applications.

These advances are paralleled by the generation of novel luciferins, further adding to increased brightness and color palettes [48]. This tool platform offers applications to manipulate and investigate dynamic processes inside living cells, tissues, and organisms.

References

1. Losi A, Gardner KH, Moglich A. Blue-light receptors for optogenetics. Chem Rev. 2018;118(21):10659–709.
2. Vriz S, Ozawa T, editors. Optogenetics: light-driven actuators and light-emitting sensors in cell biology. Royal Society of Chemistry; 2018.
3. Zhou Y, Zhang K, editors. Opsin-free optogenetics. Routledge; 2023.
4. Förster T. Energiewanderung und Fluoreszenz. Naturwissenschaften. 1946;33(6):166–75.
5. Förster T. Zwischenmolekulare Energiewanderung Und Fluoreszenz. Ann Phys. 1948;2:55–75.
6. Nash AI, McNulty R, Shillito ME, Swartz TE, Bogomolni RA, Luecke H, et al. Structural basis of photosensitivity in a bacterial light-oxygen-voltage/helix-turn-helix (LOV-HTH) DNA-binding protein. Proc Natl Acad Sci U S A. 2011;108(23):9449–54.
7. Stierl M, Stumpf P, Udwari D, Gueta R, Hagedorn R, Losi A, et al. Light modulation of cellular cAMP by a small bacterial photoactivated adenylyl cyclase, bPAC, of the soil bacterium Beggiatoa. J Biol Chem. 2011;286(2):1181–8.
8. Motta-Mena LB, Reade A, Mallory MJ, Glantz S, Weiner OD, Lynch KW, et al. An optogenetic gene expression system with rapid activation and deactivation kinetics. Nat Chem Biol. 2014;10(3):196–202.
9. Hall MP, Unch J, Binkowski BF, Valley MP, Butler BL, Wood MG, et al. Engineered luciferase reporter from a deep sea shrimp utilizing a novel imidazopyrazinone substrate. ACS Chem Biol. 2012;7(11):1848–57.
10. Crespo EL, Bjorefeldt A, Prakash M, Hochgeschwender U. Bioluminescent optogenetics 2.0: harnessing bioluminescence to activate photosensory proteins in vitro and in vivo. J Vis Exp. 2021;174. https://doi.org/10.3791/62850.
11. Adir O, Albalak MR, Abel R, Weiss LE, Chen G, Gruber A, et al. Synthetic cells with self-activating optogenetic proteins communicate with natural cells. Nat Commun. 2022;13(1):2328.
12. Wang X, Chen X, Yang Y. Spatiotemporal control of gene expression by a light-switchable transgene system. Nat Methods. 2012;9(3):266–9.
13. Li T, Chen X, Qian Y, Shao J, Li X, Liu S, et al. A synthetic BRET-based optogenetic device for pulsatile transgene expression enabling glucose homeostasis in mice. Nat Commun. 2021;12(1):615.
14. Pathak GP, Spiltoir JI, Hoglund C, Polstein LR, Heine-Koskinen S, Gersbach CA, et al. Bidirectional approaches for optogenetic regulation of gene expression in mammalian cells using Arabidopsis cryptochrome 2. Nucleic Acids Res. 2017;45(20):e167.
15. Parag-Sharma K, O'Banion CP, Henry EC, Musicant AM, Cleveland JL, Lawrence DS, et al. Engineered BRET-based biologic light sources enable spatiotemporal control over diverse optogenetic systems. ACS Synth Biol. 2020;9(1):1–9.
16. Polstein LR, Gersbach CA. A light-inducible CRISPR-Cas9 system for control of endogenous gene activation. Nat Chem Biol. 2015;11(3):198–200.
17. Kennedy MJ, Hughes RM, Peteya LA, Schwartz JW, Ehlers MD, Tucker CL. Rapid blue-light-mediated induction of protein interactions in living cells. Nat Methods. 2010;7(12):973–5.
18. Kawano F, Suzuki H, Furuya A, Sato M. Engineered pairs of distinct photoswitches for optogenetic control of cellular proteins. Nat Commun. 2015;6:6256.

19. Yao S, Yuan P, Ouellette B, Zhou T, Mortrud M, Balaram P, et al. RecV recombinase system for in vivo targeted optogenomic modifications of single cells or cell populations. Nat Methods. 2020;17(4):422–9.
20. Wang H, Vilela M, Winkler A, Tarnawski M, Schlichting I, Yumerefendi H, et al. LOVTRAP: an optogenetic system for photoinduced protein dissociation. Nat Methods. 2016;13(9):755–8.
21. Guntas G, Hallett RA, Zimmerman SP, Williams T, Yumerefendi H, Bear JE, et al. Engineering an improved light-induced dimer (iLID) for controlling the localization and activity of signaling proteins. Proc Natl Acad Sci U S A. 2015;112(1):112–7.
22. Chen F, Warnock RL, Van der Meer JR, Wegner SV. Bioluminescence-triggered photoswitchable bacterial adhesions enable higher sensitivity and dual-readout bacterial biosensors for mercury. ACS Sens. 2020;5(7):2205–10.
23. Lee SY, Cheah JS, Zhao B, Xu C, Roh H, Kim CK, et al. Engineered allostery in light-regulated LOV-Turbo enables precise spatiotemporal control of proximity labeling in living cells. Nat Methods. 2023;20(6):908–17.
24. Roh H, Kim D, Kim B, Jeon Y, Kim Y, Jacko M, et al. Light-activated tetanus neurotoxin for conditional proteolysis and inducible synaptic inhibition in vivo. bioRxiv:2025.01.27.635161. https://doi.org/10.1101/2025.01.27.635161.
25. Su Y, Walker JR, Park Y, Smith TP, Liu LX, Hall MP, et al. Novel NanoLuc substrates enable bright two-population bioluminescence imaging in animals. Nat Methods. 2020;17(8):852–60.
26. Su Y, Walker JR, Hall MP, Klein MA, Wu X, Encell LP, et al. An optimized bioluminescent substrate for non-invasive imaging in the brain. Nat Chem Biol. 2023;19:731–9.
27. Naim N, White AD, Reece JM, Wankhede M, Zhang X, Vilardaga JP, et al. Luminescence-activated nucleotide cyclase regulates spatial and temporal cAMP synthesis. J Biol Chem. 2019;294(4):1095–103.
28. Ohlendorf R, Li N, Phi Van VD, Schwalm M, Ke Y, Dawson M, et al. Imaging bioluminescence by detecting localized haemodynamic contrast from photosensitized vasculature. Nat Biomed Eng. 2024;8(6):775–86.
29. Park SY, Song SH, Palmateer B, Pal A, Petersen ED, Shall GP, et al. Novel luciferase–opsin combinations for improved luminopsins. J Neurosci Res. 2020;98(3):410–21.
30. He L, Huang Z, Huang K, Chen R, Nguyen NT, Wang R, et al. Optogenetic control of non-apoptotic cell death. Adv Sci (Weinh). 2021;8(13):2100424.
31. Ding J, Lu J, Zhang Q, Xu Y, Song B, Wu Y, et al. Camouflage nanoparticles enable in situ bioluminescence-driven optogenetic therapy of retinoblastoma. ACS Nano. 2023;17(8):7750–64.
32. Kim H, Jung SO, Lee S, Lee Y. Bioluminescent systems for theranostic applications. Int J Mol Sci. 2024;25(14)
33. Algorri JF, Ochoa M, Roldán-Varona P, Rodríguez-Cobo L, López-Higuera JM. Light technology for efficient and effective photodynamic therapy: a critical review. Cancers (Basel). 2021;13(14)
34. Shu X, Lev-Ram V, Deerinck TJ, Qi Y, Ramko EB, Davidson MW, et al. A genetically encoded tag for correlated light and electron microscopy of intact cells, tissues, and organisms. PLoS Biol. 2011;9(4):e1001041.
35. Bulina ME, Chudakov DM, Britanova OV, Yanushevich YG, Staroverov DB, Chepurnykh TV, et al. A genetically encoded photosensitizer. Nat Biotechnol. 2006;24(1):95–9.
36. Theodossiou T, Hothersall JS, Woods EA, Okkenhaug K, Jacobson J, MacRobert AJ. Firefly luciferin-activated rose bengal: in vitro photodynamic therapy by intracellular chemiluminescence in transgenic NIH 3T3 cells. Cancer Res. 2003;63(8):1818–21.
37. Rees JF, de Wergifosse B, Noiset O, Dubuisson M, Janssens B, Thompson EM. The origins of marine bioluminescence: turning oxygen defence mechanisms into deep-sea communication tools. J Exp Biol. 1998;201(Pt 8):1211–21.
38. Widder EA. Bioluminescence in the ocean: origins of biological, chemical, and ecological diversity. Science. 2010;328(5979):704–8.

39. Shramova EI, Proshkina GM, Chumakov SP, Khodarovich YMDS. Flavoprotein miniSOG cytotoxisity can be induced by bioluminescence resonance energy transfer. Acta Naturae. 2016;8(4):118–23.
40. Proshkina GM, Shramova EI, Shilova ON, Ryabova AV, Deyev SM. Phototoxicity of flavoprotein miniSOG induced by bioluminescence resonance energy transfer in genetically encoded system NanoLuc-miniSOG is comparable with its LED-excited phototoxicity. J Photochem Photobiol B. 2018;188:107–15.
41. Shramova EI, Chumakov SP, Shipunova VO, Ryabova AV, Telegin GB, Kabashin AV, et al. Genetically encoded BRET-activated photodynamic therapy for the treatment of deep-seated tumors. Light Sci Appl. 2022;11(1):38.
42. Shramova EI, Frolova AY, Filimonova VP, Deyev SM, Proshkina GM. System for self-excited targeted photodynamic therapy based on the multimodal protein DARP-NanoLuc-SOPP3. Acta Naturae. 2023;15(4):100–10.
43. Westberg M, Bregnhøj M, Etzerodt M, Ogilby PR. No photon wasted: an efficient and selective singlet oxygen photosensitizing protein. J Phys Chem B. 2017;121(40):9366–71.
44. Li Y, Cui ZJ. NanoLuc bioluminescence-driven photodynamic activation of cholecystokinin 1 receptor with genetically-encoded protein photosensitizer MiniSOG. Int J Mol Sci. 2020;21(11)
45. Kim EH, Park S, Kim YK, Moon M, Park J, Lee KJ, et al. Self-luminescent photodynamic therapy using breast cancer targeted proteins. Sci Adv. 2020;6(37)
46. Ishimoto T, Mori H. A new bioluminescence-based tool for modulating target proteins in live cells. Sci Rep. 2019;9(1):18239.
47. Bulina ME, Lukyanov KA, Britanova OV, Onichtchouk D, Lukyanov S, Chudakov DM. Chromophore-assisted light inactivation (CALI) using the phototoxic fluorescent protein KillerRed. Nat Protoc. 2006;1(2):947–53.
48. Love AC, Prescher JA. Seeing (and using) the light: recent developments in bioluminescence technology. Cell Chem Biol. 2020;27(8):904–20.

Chapter 4
Bioluminescence for Cell-to-Cell Communication

Ute Hochgeschwender

Abstract Bioluminescence evolved in the natural world to enable communication among organisms via light in very dark environments, with one entity emitting and another sensing photons. It was unclear until recently whether bioluminescence is a feasible means of communication at the cellular level. Cell-to-cell communication via light transmission has been realized in several bioluminescent optogenetic applications. In the broadest sense, light-based communication between cells involves the generation of a light signal in one cell and its reception and translation into a response in another cell. Several different strategies have been employed to achieve such light-induced communication in synthetic cell systems. Bioluminescence-mediated communication between mammalian cells was used for the detection and transcriptional readout of trans-cellular interactions. Three strategies have been developed to control synaptic transmission between neurons. In two, luciferase-generated bioluminescence in a presynaptic neuron is used for activation of optogenetic elements in a postsynaptic neuron across the synaptic cleft, while another approach utilizes presynaptically released neurotransmitters to reconstitute a split luciferase for activating a fused opsin expressed in postsynaptic neurons. This overview chapter outlines the different approaches to light-induced communication. The three chapters following this overview chapter detail distinct strategies developed for bioluminescence-driven neuronal communication.

Keywords Cellular communication · Synaptic transmission · Bioluminescence · Photons · Synthetic cells · Mammalian cells

Communication between cells is essential for both unicellular populations and multicellular life. Cells need to exchange information regarding the density of their cellular community, their state, and the presence of pathogens or competing cells. Key processes requiring intercellular communication in multicellular organisms

U. Hochgeschwender (✉)
College of Medicine, Central Michigan University, Mt. Pleasant, MI, USA
e-mail: hochg1u@cmich.edu

U. Hochgeschwender (ed.), *BioLuminescent OptoGenetics*,
https://doi.org/10.1007/978-3-032-18121-3_4

include, for example, gene regulation during development, T cell activation, and synaptic transmission between neurons. Cells communicate through direct, physical contact or over a distance via soluble molecules. Chemical intercellular communication happens through molecular diffusion and requires the receiving cell to have transporters, receptors, or pores.

Using bioluminescence, researchers have created ways for nearby cells to communicate via light. In contrast to chemical diffusion, light is fast and functions without the need for transporters or pores in the receiving cell, as it passes through optically transparent cellular membrane bilayers [1]. Light signals are produced when the light reaction starts and persist as long as—and no longer than—the light is present. Light as a signal is highly specific because the response of a molecule to light is a unique feature and, unlike with chemical messengers, undesired cross-reactivity is very unlikely. Applied in vivo, this mode of intercellular communication has the potential to be an orthogonal way for controlling cellular interaction.

At the macroscopic scale, bioluminescence evolved in the natural world to enable communication between light-emitting and light-sensing organisms. For example, fireflies in the very dim light at night or marine species in the darkness of the deep sea communicate with others to attract mates or prey with the bioluminescent light they produce. While bioluminescence is used in nature for communication among organisms, it was unclear until recently whether it is a feasible means of communication at the cellular level.

In the broadest sense, light-based communication between cells involves the generation of a light signal in one cell and its reception and translation into a response in another cell. Three different strategies have been employed to achieve such light-induced communication in synthetic cell systems. In the first, luciferases from the sender cell emit light that induces direct cell-to-cell contact with a close-by receiver cell; the light-induced cellular contact then starts a cascade of molecular interactions between the two cells, resulting in lysis of the receiver cell [2]. In the second case, a split luciferase is used to restrict light emission to sites of cell contacts that reconstitute the split luciferase; the contact-induced light emission then activates a photoswitchable protein in the receiver cell [3]. In the third method, an intact luciferase in the sender cell emits light that activates a photoreceptor in the receiver cell [4]. All three cases are bioluminescent optogenetic applications for cell-to-cell communication.

Chakraborty et al. demonstrated that cell-to-cell signaling through bioluminescence is possible in a biomimetic artificial cell model derived from giant unilamellar vesicles (GUVs) [2] (Fig. 4.1). They created a synthetic predator–prey community, where the predator GUV encapsulates *Renilla* luciferase (RLuc) that generates an intracellular blue light signal in the presence of its luciferin, coelenterazine (CTZ). The outer membranes of the predator and prey GUVs contain the proteins iLID and Nano [5], respectively, that bind to each other under blue light and mediate adhesions between the GUVs. This light-triggered adhesion between predator and prey cells enables the contact-dependent transfer of calcium ions stored in the sender cell to the receiver cell, resulting in activation of phospholipase A2 (PLA2) and cleavage of the phospholipid membrane of the prey GUV, leading to its lysis.

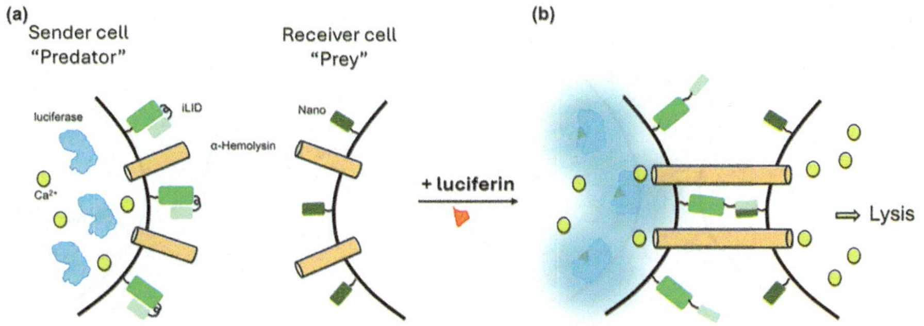

Fig. 4.1 Bioluminescence from sender cell induces cell adhesion with receiver cell to allow direct communication between sender and receiver. (**a**) The predator cell (sender) contains the luciferase, cell impermeable Ca^{2+} ions, iLID, and α-hemolysin (α-HL) pores. The prey cell (receiver) contains Nano, α-HL pores, and apo-PLA2 (not shown). In the absence of light, the α-HL pores are blocked by heptakis (2,3,6-tri-O-methyl)-beta-cyclodextrin, TRIMEB, preventing leakage of intracellular content. (**b**) In the presence of the luciferin, light emission from the sender cell induces the LOV-domain based dimerization of sender iLID (AsLOV2-ssrA) and receiver Nano (sspB), displacing the blocker of the α-HL pores at the adhesion site of the GUVs. This allows the transfer of Ca^{2+} ions from the predator to the prey cell, where it activates apo-PLA2. PLA2 cleaves the fatty acid tail of the phospholipid in the GUV's membrane and results in its lysis

The system described here combines cell-to-cell communication through bioluminescence with contact-dependent chemical communication. Luciferase activity in the sender cell generates the light needed to activate the light-responsive element in its membrane, causing the photoswitchable protein–protein interaction between iLID (sender cell) and Nano (receiver cell), resulting in sender-receiver cell adhesion. This light-induced adhesion initializes chemical signaling between the synthetic cells. While the signaling in these studies is controlled by light, it occurs through the transport of chemical calcium molecules. Thus, light is utilized as an intermediary means to trigger chemical communication.

In another GUV synthetic cell system [3], light is generated exclusively at the cell–cell contact interface and used as the primary signal for communication between sender and receiver cell (Fig. 4.2). In this case, proximity between two cell populations enables a SpyTag-SpyCatcher ligation, the irreversible conjugation of the SpyCatcher protein and the SpyTag peptide displayed on the outer membrane of the sender and receiver cell, respectively [6]. This dimerization strategy is combined with the NanoLuc Binary Technology (NanoBiT) [7], a structural protein complementation system composed of NanoLuc split into a large BiT (LgBiT; 18 kDa) subunit, fused to the sender cell's SpyCatcher, and a small BiT (SmBiT) complementary peptide, fused to the receiver cell's SpyTag. When the SpyTag-SpyCatcher proteins interact, the attached NanoLuc subunits come together to form an active enzyme. Sender/receiver cell–cell interaction enables split Nanoluc reconstitution and light emission in the presence of luciferin that, in turn, activates iLID-SspBNano dimerization in receiver cells, recruiting mCherry to the cell–cell contact interface.

In this application, light activation of the photoreceptor in the receiver cell is dependent on both the juxtaposition of the sender and receiver cells and the

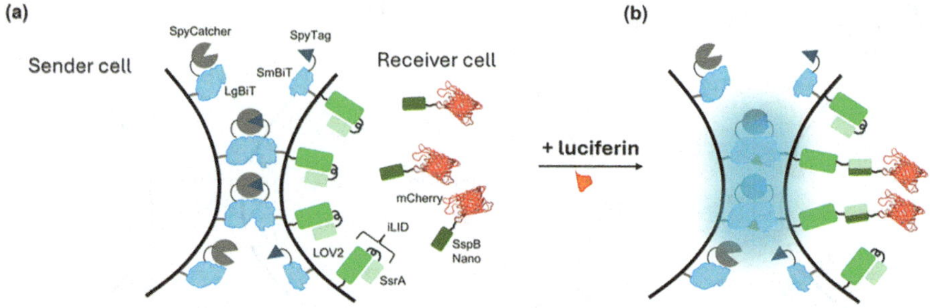

Fig. 4.2 Sender/receiver cell contact-mediated bioluminescence activates photoreceptors in receiver cell. (**a**) The sender cell displays on its outer membrane the anchored large part of NanoLuc (LgBiT) fused to the SpyCatcher protein. The receiver cell displays on its outer membrane the anchored small part of NanoLuc (SmBiT) fused to the SpyTag peptide as well as inner cell membrane-bound iLID and luminal sspBNano-mCherry. In the absence of light, sspBNano-mCherry is distributed in the cytoplasm of the receiver cell. (**b**) In the presence of luciferin those cells that are in close enough proximity to allow the covalent SpyTag-SpyCatcher bond to form along its membrane enable light emission from the reconstituted NanoLuc at the contact interface. The light emission in turn will activate the photoswitchable protein and lead to membrane recruitment of sspB-mCherry

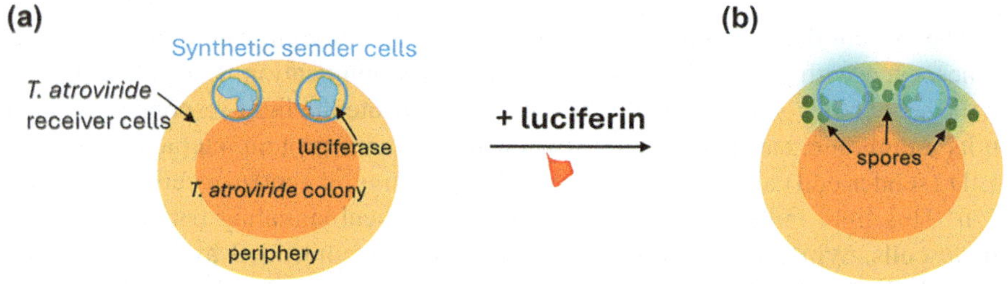

Fig. 4.3 Bioluminescence from sender cell activates photoreceptor in receiver cell. (**a**) The synthetic sender cells express high levels of blue-light-emitting luciferase. The fungal receiver cells in the periphery of a *T. atroviride* colony express two blue-light sensing regulator proteins (BLR-1, a LOV domain protein, and BLR-2; not shown); photo-activation in response to blue light triggers several downstream processes including conidiation (spore formation). In the absence of light, no conidiation is observed. (**b**) In the presence of the luciferin, light emission from the synthetic sender cells will activate the blue-light-sensing proteins in nearby fungal receiver cells and lead to conidiation

presence of luciferin. Light itself is the signal for activating light-sensing elements in the receiver cell.

A third application utilizes GUV synthetic sender cells and mycelial cells of the fungus *Trichoderma atroviride* as natural receiver cells [4] (Fig. 4.3). Synthetic sender cells generate light by expressing the *Gaussia* luciferase variant sbGLuc. In the presence of its luciferin (CTZ), bioluminescent light emitted by the sender activates photoconidiation, the natural light-dependent production of spores, in fungal recipient cells. The efficacy of fungal conidiation depends on the light emission from nearby cells and thus on the density of synthetic sender cells.

Here again, light itself is the signal emitted from the sender cell for activating light-sensing elements in the receiver cell. In the presence of luciferin, sender cells emit light that activates photoreceptors in only those receiver cells that are in close proximity.

Bioluminescence-mediated communication between mammalian cells was used for the detection and transcriptional readout of transcellular interactions [8] (Fig. 4.4). In this case, the activating light source was supplied by sender cells expressing NanoLuc (NLuc)-ICAM-Neurexin on the cell surface. These sender cells made use of the presynaptic adhesion protein Neurexin as a scaffold and the extracellular domain of ICAM-1 to extend NLuc further out from the cell surface. The receiver cell expressed SPARK2 (Specific Protein Association tool giving transcriptional Readout with rapid Kinetics), a molecular tool that links protein–protein interaction with transcription. A light-sensing G-protein coupled receptor (optoGPCR: oβ2AR, a chimera between bovine rhodopsin and β2AR) fused to a LOV domain containing a protease cleavage site (TEVcs) and a tethered transcription factor (Gal4) (optoGPCR-LOV-TEVcs-Gal4) is used to recruit its partnering protein (arrestin) fused to a luciferase (NLuc) and to a protease (TEVp) (NLuc-arrestin-TEVp). Light emission from the sender cell via the extracellular NLuc activates the GPCR and triggers recruitment of arrestin to oβ2AR, bringing NLuc and TEVp in

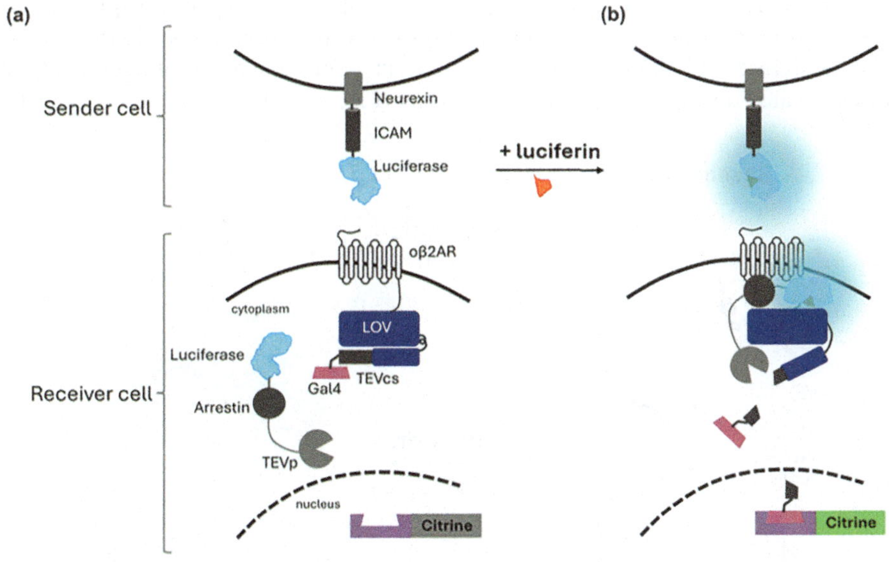

Fig. 4.4 Trans-cellular bioluminescence enables transcriptional readout of cell–cell contacts. (**a**) The sender cell expresses NanoLuc (NLuc)-ICAM-Neurexin. The receiver cell expresses a light-sensing G-protein coupled receptor (oβ2AR) fused to a LOV domain containing a protease cleavage site (TEVcs) and a tethered transcription factor (Gal4), as well as arrestin fused to NLuc and to a protease (TEVp), and the transcriptional reporter UAS-Citrine. (**b**) When sender and receiver cells are in close proximity, in the presence of luciferin, light emission from the extracellular NLuc of the sender cell triggers recruitment of arrestin to oβ2AR, allowing the intracellular NLuc to activate the LOV protein and expose the protease cleavage site for TEVp to release the transcription factor Gal4. Gal4–UAS interaction initiates Citrine expression

close proximity to the LOV protein. Bioluminescent light produced by the recruited intracellular NLuc exposes the protease cleavage site to the now close-by protease, allowing the release of the transcription factor.

In cocultures of HEK293 populations expressing either the sender cell components (NanoLuc-ICAM-Neurexin) or the receiver cell components (oβ2AR-LOV-TEVcs-Gal4, NLuc-arrestin-TEVp, UAS-Citrine reporter), application of luciferin to cultures enabled light emission from sender cells to activate the oβ2AR of nearby receiver cells, initiating the SPARK2 assembly and resulting in Citrine reporter expression in the receiver cells. Trans-cellular bioluminescent activation of this photoswitchable system enabled transcriptional readout of cell–cell contacts for further downstream imaging, selection, and characterization of the contacted cells.

Two applications of BL-OG, PhAST (photon-assisted synaptic transmission [9]) and Interluminescence (light between cells [10, 11]), have been developed to control quintessential cell-to-cell communication: synaptic transmission between neurons. In both, luciferase-generated bioluminescence in a presynaptic neuron is used for activation of optogenetic elements in a postsynaptic neuron across the synaptic cleft. The strategies implemented to achieve this bioluminescence-mediated neuronal communication are distinct, as are the experimental systems used to demonstrate their efficacy in vivo. Below is a brief description of the two strategies, with more detailed accounts provided in separate chapters (see Chaps. 5 and 6).

In PhAST, presynaptic neurons express a calcium (Ca^{2+})-dependent split luciferase. Activation of presynaptic neurons results in Ca^{2+} influx, reconstitution of the split luciferase, and—in the presence of luciferin—emission of bioluminescent light that activates opsins in postsynaptic neurons (Fig. 4.5). Postsynaptic optogenetic

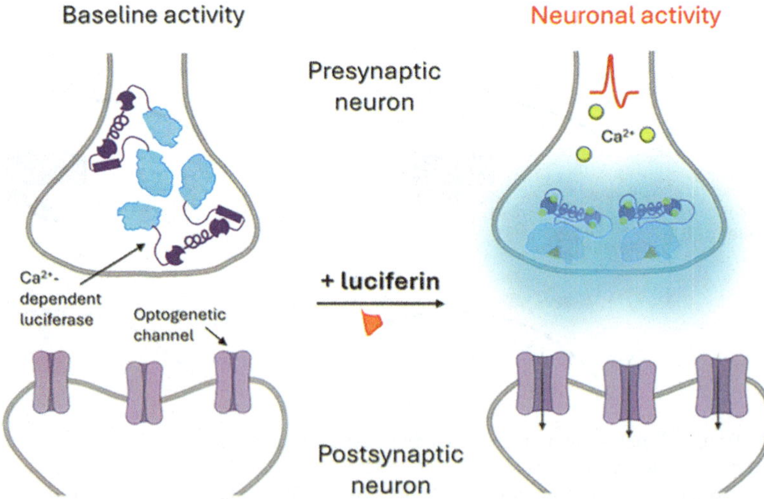

Fig. 4.5 PhAST (photon-assisted synaptic transmission). Presynaptic neurons express Ca^{2+}-dependent luciferase; postsynaptic neurons express optogenetic channels. Activity of presynaptic neurons causes Ca^{2+} influx, enabling reconstitution of a functional luciferase. In the presence of luciferin, bioluminescence emitted from presynaptic neurons activates postsynaptic channelrhodopsins, enhancing or dampening postsynaptic neuronal drive depending on the biophysical properties of the opsin

elements can be excitatory or inhibitory, depending on the intended synaptic modulation. This version of photon transmission for neural communication is always dependent on the activity of the presynaptic neuron. The system was established in and applied to different circuits in the nematode *C. elegans*. The concept was used to overcome the behavioral defect resulting from a disconnected functional neurotransmission between a sensory neuron and its corresponding postsynaptic interneurons, to suppress the endogenous pain response, and to rewire a circuit mediating attractive behavior into an avoidance response ([9] and Chap. 5).

In the Interluminescence application, intact luciferases expressed in presynaptic neurons are either targeted to synaptic vesicles or are tethered to the membranes of presynaptic neurons (Fig. 4.6). In the former version, Act-Int, the release of luciferases—and thus, in the presence of luciferin, the activation of postsynaptic optogenetic actuators—is dependent on the presynaptic neuron being activated. In this case, photon transmission across the synaptic cleft is analogous to natural synaptic transmission, where active presynaptic neurons release chemical neurotransmitters that bind to postsynaptic transmitter receptors. In the latter version, Persist-Int, the

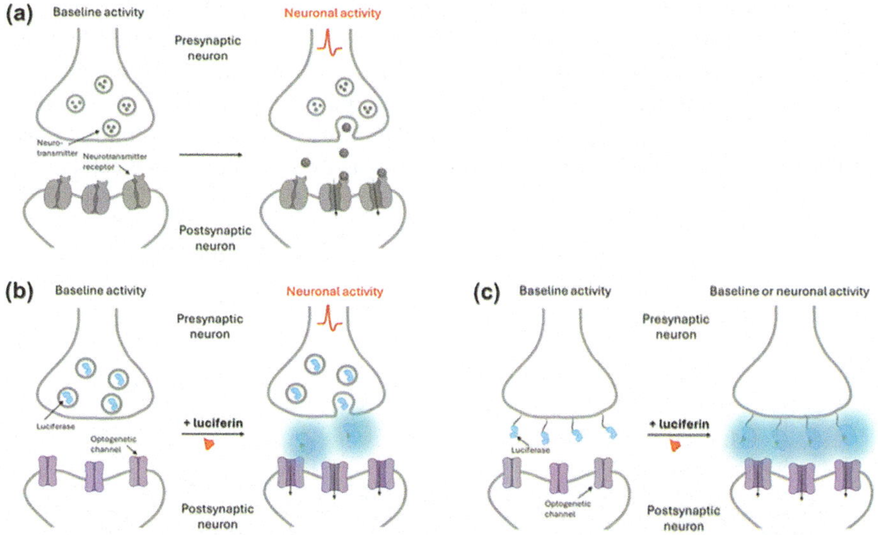

Fig. 4.6 Interluminescence (light between cells). (**a**) In classical chemical neurotransmission, presynaptic neurons contain synaptic vesicles holding neurotransmitters; postsynaptic neurons express the corresponding neurotransmitter receptors. Activation of presynaptic neurons leads to fusion of vesicles to the presynaptic membrane and release of neurotransmitters into the synaptic cleft, where they bind to their postsynaptic receptors. (**b**) Act-Int (activity-dependent Interluminescence): presynaptic neurons express luciferase targeted to presynaptic vesicles, postsynaptic neurons express optogenetic channels. Activation of presynaptic neurons leads to vesicle fusion and release of luciferase into the synaptic cleft. In the presence of luciferin, bioluminescence activates postsynaptic channelrhodopsins. (**c**) Persist-Int (activity independent, persistent Interluminescence): presynaptic neurons express luciferase anchored to the presynaptic neuron membrane facing the synaptic cleft. Application of luciferin and resulting bioluminescence will activate postsynaptic channelrhodopsins, regardless of the activity state of the presynaptic neuron

presynaptic luciferases are always present in the synaptic cleft, and administration of luciferin leads to postsynaptic photoreceptor activation, regardless of the activity state of the presynaptic neuron. Both strategies have been demonstrated to work in mice ([10, 11] and Chap. 6).

PhAST and Interluminescence use photons as universal transmitters to enable experimenter control of any genetically targeted synaptic partners. Another approach of bioluminescence-driven modulation of synaptic transmission is specific to synapses utilizing a particular neurotransmitter (Fig. 4.7). In this application, the sender cell (presynaptic neuron) releases its endogenous neurotransmitter. The receiver cell (postsynaptic neuron) expresses a neurotransmitter (NT)-dependent luciferase-opsin fusion protein (NT-LMO; Chap. 7). An optogenetic actuator, either excitatory or inhibitory, is fused to a luciferase that is split by a sensor for a specific NT. Reconstitution of a functional luciferase requires the presence of the NT; application of luciferin and resulting light emission will then activate the tethered opsin with changes in membrane potential and neural activity depending on the biophysical characteristics of the opsin. This allows for luciferin-gated modulation of postsynaptic neurons responding to a specific NT, amplifying or dampening transmission, or restoring pathologically disrupted transmission.

These applications of bioluminescence for cell-to-cell communication showcase the potential for using BL-OG to affect intercellular interactions. Applying bioluminescence for communication between cells requires high local light intensity attained through the proximity of the sender and receiver cells. Bioluminescence must be bright enough (produce enough photons) to cross distances and cell barriers

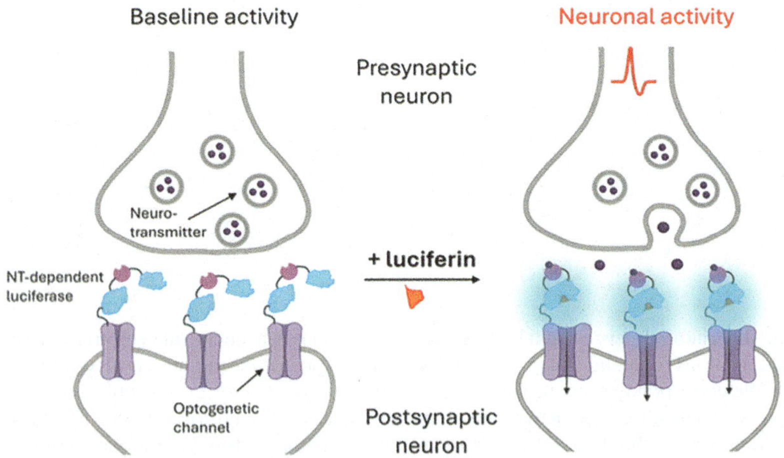

Fig. 4.7 Bioluminescence-modulated neurotransmission. Presynaptic neurons express their endogenous neurotransmitter (NT). Postsynaptic neurons are genetically targeted to express opsins fused to a luciferase split by an NT sensor. Activation of presynaptic neurons leads to the release of NTs and reconstitution of a functional luciferase. In the presence of luciferin, bioluminescence from the reconstituted luciferase will activate the channelrhodopsin, resulting in modulation of the NT effect on the postsynaptic neuron

to efficiently activate optogenetic elements. With the directed molecular evolution of brighter bioluminescence from luciferase-luciferin reactions and more sensitive photoreceptors, BL-OG activation across cells is now possible. Further improvements in the light-generating and light-responding elements will increase the flexibility of BL-OG for intercellular communication and contribute to realizing the unique opportunities of this synthetic communication system.

References

1. Glantz ST, Berlew EE, Jaber Z, Schuster BS, Gardner KH, Chow BY. Directly light-regulated binding of RGS-LOV photoreceptors to anionic membrane phospholipids. Proc Natl Acad Sci USA. 2018;115(33):E7720–7.
2. Chakraborty T, Wegner SV. Cell to cell signaling through light in artificial cell communities: glowing predator lures prey. ACS Nano. 2021;15(6):9434–44.
3. Moghimianavval H, Loi KJ, Hwang SW, Bashirzadeh Y, Liu AP. Light-based juxtacrine signaling between synthetic cells. Small Sci. 2025;5(1):2400401.
4. Adir O, Albalak MR, Abel R, Weiss LE, Chen G, Gruber A, et al. Synthetic cells with self-activating optogenetic proteins communicate with natural cells. Nat Commun. 2022;13(1):2328.
5. Guntas G, Hallett RA, Zimmerman SP, Williams T, Yumerefendi H, Bear JE, et al. Engineering an improved light-induced dimer (iLID) for controlling the localization and activity of signaling proteins. Proc Natl Acad Sci USA. 2015;112(1):112–7.
6. Zakeri B, Fierer JO, Celik E, Chittock EC, Schwarz-Linek U, Moy VT, et al. Peptide tag forming a rapid covalent bond to a protein, through engineering a bacterial adhesin. Proc Natl Acad Sci USA. 2012;109(12):E690–7.
7. Dixon AS, Schwinn MK, Hall MP, Zimmerman K, Otto P, Lubben TH, et al. NanoLuc complementation reporter optimized for accurate measurement of protein interactions in cells. ACS Chem Biol. 2016;11(2):400–8.
8. Kim CK, Cho KF, Kim MW, Ting AY. Luciferase-LOV BRET enables versatile and specific transcriptional readout of cellular protein-protein interactions. elife. 2019;8:e43826.
9. Porta-de-la-Riva M, Gonzalez AC, Sanfeliu-Cerdán N, Karimi S, Malaiwong N, Pidde A, et al. Neural engineering with photons as synaptic transmitters. Nat Methods. 2023;20(5):761–9.
10. Prakash M, Murphy J, St Laurent R, Friedman N, Crespo EL, Bjorefeldt A, et al. Selective control of synaptically-connected circuit elements by all-optical synapses. Commun Biol. 2022;5(1):33.
11. Slaviero AN, Prakash M, Bell E, Schumaker R, Crespo EL, Tree MO, et al. Control of synaptic communication through molecularly engineered bioluminescence light emission and sensing. bioRxiv. 2025;2025.10.27.684942.

Chapter 5
Reengineering the Connectome with Photon Assisted Synaptic Transmission (PhAST)

Michael Krieg

Abstract The emergence of comprehensive connectomes in organisms, like *Caenorhabditis elegans*, Drosophila, and the mouse, is transforming our understanding of neural circuit architecture. Manipulating these circuits noninvasively, however, remains a major challenge. Optogenetics is a promising neuromodulation technique but requires external light to activate neurons within the target tissue. To address this, we recently developed and deployed photon-assisted synaptic transmission, or short PhAST, a system of bioluminescence-driven optogenetics, where light is generated directly within presynaptic neuronal compartments through luciferase-luciferin reactions and activates postsynaptically expressed channelrhodopsins. Using the PhAST strategy in *C. elegans*, we demonstrated activity-dependent, transsynaptic activation of channelrhodopsins to restore disrupted signaling without external light. Photon emission from luciferases, broadly distributed along axons and gated by intracellular messengers such as calcium, enables flexible, synapse-independent information transfer, even under conditions of vesicular transmission failure. This modular system not only bypasses the physical limitations of traditional optogenetics but also sets the stage for programmable, stimulus-responsive networks that leverage intrinsic signaling dynamics for precision neuromodulation.

Keywords Bioluminescence · Connectome engineering · Optogenetics · Neuromodulation · Synaptic transmission

M. Krieg (✉)
ICFO – Institut de Ciències Fotòniques, Castelldefels, Spain

The Barcelona Institute of Science and Technology, Barcelona, Spain
e-mail: michael.krieg@icfo.eu

U. Hochgeschwender (ed.), *BioLuminescent OptoGenetics*,
https://doi.org/10.1007/978-3-032-18121-3_5

5.1 Introduction

Fueled by recent breakthroughs in large-scale connectome reconstructions from multiple species, unprecedented opportunities arise for understanding and manipulating brain functions [1–4]. Several strategies have been introduced in recent years that allow researchers to modulate neuronal activity through chemical, electric, magnetic, acoustic, and optical stimulation [5]. Whereas the first four methods leverage the endogenous sensitivity of the central nervous system to these stimuli, optical neuromodulation, commonly referred to as optogenetics, requires the neuronal expression of a dedicated light sensor.

This is commonly achieved using microbial channelrhodopsins, special ion channels which respond to blue, green, or orange light, and selectively conduct ions to depolarize or hyperpolarize the neuronal membrane potential [6]. Due to the poor light sensitivity of these channels, together with the high scattering properties of healthy brain tissue, optogenetics in the mammalian brain requires the implantation of a light diode [7]. However, this also confers exceptional precision, enabling control at the level of individual cell types—or even specific synapses [8]. The following table summarizes the pros and cons of four different neuromodulation techniques (Table 5.1).

5.2 Principle of PhAST

Considering that the risks of gene transfer are acceptable in severe health conditions, several strategies are currently discussed to improve light delivery to channelrhodopsins. Among those are red-shifted channelrhodopsins, such as ChRmine [9] and ChReef [10] and rhodopsins, with an extended open-state lifetime that confers an exceptional operational light sensitivity [11]. Although the mentioned options minimize invasiveness, depending on the location of the target neurons within the brain, they may require high intensities that could cause tissue damage due to heating and generate reactive oxygen species. To minimize energy

Table 5.1 Compact comparison of neural stimulation modalities

Modality	Advantages	Disadvantages
Chemical	Activates endogenous receptors	Slow; low spatial precision; off-target effects
Electrical	Clinically established; fast and reversible	Low spatial resolution; invasive; tissue damage
Magnetic	Noninvasive; deep penetration; clinically safe	Low precision; indirect; hard to focus
Acoustic	Noninvasive; deep targets w/focused ultrasound	Mechanism unclear; potential damage; calibration needed
Optical	Cell-type precision; high spatial/temporal resolution	Requires gene delivery and optrode implantation; phototoxic

deposition, an alternative approach is to generate light noninvasively within the target tissue, either through bioluminescence [12] or via luminescent nanoparticles with tunable and controlled emission [13–15].

As described extensively throughout this book, bioluminescence is a form of chemiluminescence mediated by biological systems, in which the enzymatic oxidation of a 'luciferin' by a luciferase leads to the formation of an electronically excited product that decays to the ground state via photon emission, typically in the 400–700 nm wavelength range [12]. Because this process requires additional substrates and cofactors, the light emission can be coupled to the presence of Ca^{2+}, ATP, and oxygen (see also Table 5.2). Furthermore, luciferases can be engineered to emit a photon only in the presence of high calcium concentrations that are typical for activated neurons [16, 17], for example, those that occur during synaptic transmission. These enzymes have been recognized as ideal calcium-dependent light switches that can be functionally coupled to channelrhodopsins [18] (Fig. 5.1) expressed in the same cell or on the postsynaptic compartment of the synapse between two neurons.

Table 5.2 Selection of bioluminescent sensors used for intracellular metabolite detection that are of interest for synaptic engineering purposes, including their target molecules, optical systems, key features, and reference

Sensor name	Target metabolite	Luciferase	Key features	Ref.
Nanolantern	Ca^{2+}	Renilla Luc BRET with FP	Low photon output	[21]
Enhanced Nanolantern	Ca^{2+}	Split NanoLuc with CaM/M13 and FP	High sensitivity; live-cell imaging.	[16, 22]
GLICO	Ca^{2+}	NanoLuc	2200% dynamic range; GCaMP6f	[23]
ReBLICO	Ca^{2+}	NanoLuc	Low affinity for ER imaging	[23]
LUCI-GECO1	Ca^{2+}	NanoLuc	GCaMP6s; ratiometric	[24]
BRIC	Ca^{2+}	teLuc		[25]
CaBLAM	Ca^{2+}	eKaz	15- to 20-fold change in cells	[26]
CaMBI3	Ca^{2+}	NanoLuc	Antares-based, 580 nm peak emission	[27]
BLING1.0	Glutamate	NanoLuc	Glt1 glutamate binding protein and PDGFRβ transmembrane anchor	[28]
LuCID	Ca^{2+}	Split NanoLuc + Gal4	Slow response; Coupled to gene expression	[29]
BTeam	ATP	NanoLuc BRET with YFP	High sensitivity; live- cell imaging	[30]
Nanolantern (cAMP1.6)	ATP	Renilla	Fused to fragment of EPAC1, 1.6 μM affinity	[21]
iNap-BL	NADPH	Lux operon	Tracks NADPH; high S/N ratio	[31]
LOTUS-Glc	Glucose	NanoLuc	MglB domain, ratio metric	[32]

For a more detailed compendium, refer to [33]

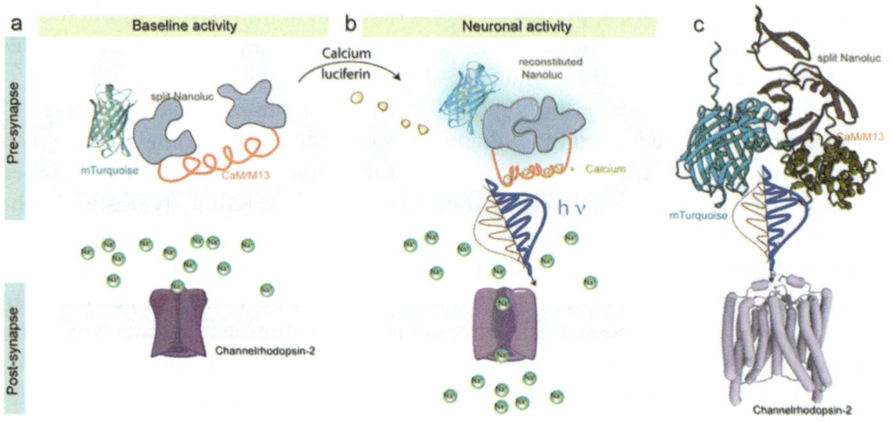

Fig. 5.1 Schematic of the PhAST principle based on split-enhanced Nanolantern and channelrhodopsin. (**a**) Under conditions of low baseline neuronal activity—and consequently low calcium levels in the presynaptic neuron—photon output is insufficient to trigger transsynaptic activation of channelrhodopsins. (**b**) High presynaptic neuronal activity, accompanied by a natural calcium influx, leads to the reconstitution of split luciferase enzymatic activity. The resulting increased quantum yield drives activation of the postsynaptic channelrhodopsin, leading to the postsynaptic depolarization and eventually signal transmission. Note, luciferases and channelrhodopsin do not require synaptic specialization but can also drive functional connections along axon-axon fascicles. (**c**) AlphaFold2 models of the presynaptic TeNL [19] and the postsynaptic channelrhodopsin (lilac)

PhAST, in contrast to interluminescence, which required the luciferase to diffuse across the synaptic cleft and get into a FRETting distance with the light receptor, activates the receptor through emission/reabsorption of the photons. Because the synaptic cleft (20–50 nm) is much lower than the wavelength of the emitted light (450 nm), photons are not scattered, and losses due to absorption are avoided. To demonstrate PhAST's versatility, the strategy was extended to multiple circuits in *C. elegans* to overcome transmission defects and to engineer ectopic, de novo connections between naturally unwired neurons. For example, transsynaptic activation of a photosensitive channelrhodopsin was demonstrated in *C. elegans*, leveraging its well-mapped connectome and the previously characterized nocifensive nose touch circuit [20]. This circuit consists of a pair of glutamatergic sensory neurons, called ASH, which synapse onto various interneurons, including AVA and AIB, which in turn instruct the motor circuit (Fig. 5.2a). In 75% of the cases, activation of ASH due to a mechanical stimulus causes a reversal response. In this system, photon emission was precisely controlled using a presynaptic, calcium-sensitive enhanced mTurquoise Nanolantern (TeNL) to restore functional connectivity in a glutamate-deficient mutant. This mutant lacked transmission between the ASH sensory neuron and its downstream interneurons, AVA and AIB [18] (Fig. 5.2).

Intriguingly, because luciferases are broadly distributed across the entire length of the neuron (unless they are specifically targeted to synapses using synaptogyrin tags [18]), PhAST does not require synaptic specializations for information transfer.

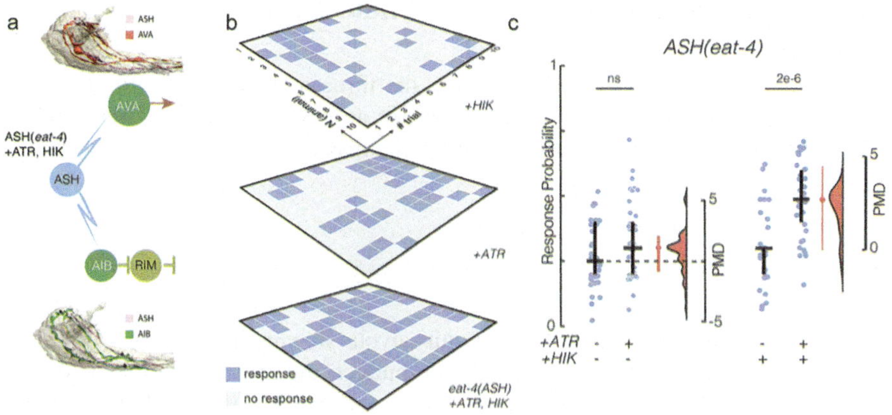

Fig. 5.2 Prosthetic, light-based neurotransmission replaces chemical neurotransmitter. Panel (**a**) is a flow chart illustrating neural pathways involving ASH lacking glutamate transmission (ASH:eat-4 mutant), AVA, AIB, and RIM neurons, with edges indicating synaptic connections colored by their neurotransmitter used. Upon reconstitution with all-trans retinal (ATR) and hikarazine (HIK), the functional, light-based connection is established (blue flashes). Blue flash, photons; red arrow, acetylcholine; yellow arrow, glutamate as neurotransmitter. (**b**) Representative outcome of a single experiment displayed as a raster diagram of the nose touch behavior experiment of the ASH-specific *eat-4* mutant in the presence of the necessary cofactors (Hik, ATR). Ten animals have been touched ten times successively, and each positive response is recorded as a blue square. (**c**) Summary of the average behavioral response of each animal for the different tested conditions presented a scatter plot with response probability on the y-axis and different conditions on the x-axis, showing data points, median values (±95% confidence interval), and statistical significance markers (ns and 2e-6). The floating axis indicates the paired median difference (PMD) between the two conditions to the left of it

This affords signaling even if vesicular release is deficient but may have the downside that signaling happens anywhere along the axon. A transient photon release anywhere in the axons can potentially activate neighboring channelrhodopsin-expressing neurons in an "en passant" synapse-like manner. This may also be harnessed for backpropagation, e.g., ascending information transfer from the postsynaptic (dendrite) to the presynaptic (axon) neurons.

Because light emission can also be coupled to other secondary messengers of cellular signaling cascades, such as ATP and cAMP (see Table 5.2), this principle is universal and affords the design of new-to-nature stimulus-responsive neuronal networks.

5.3 Requirements for Successful Photo-transmission

In the following sections, we will discuss the three main components of a light-based synapse, which consists of a light emitter, a light sensor, and luciferin, which is the actual chromophore required for photon emission.

5.3.1 Luciferases

Because luciferases in general have a duty cycle slower than that of fluorescent proteins and therefore have a lower photon budget, the most powerful luciferases are needed for PhAST. As mentioned above, the control of their light emission should be tight, meaning that the exposure to the cofactor is transient and the emission is of flash-type to avoid prolonged ChR2 activity without baseline activity.

Alternatively, if the cofactor is constitutively present, light emission may be coupled to a secondary messenger, such as ATP, calcium ions, or glucose concentration. Table 5.2 already exemplifies the current version of genetically encoded sensors of various metabolites that are poised to be deployed in synthetic circuits, together with their color-matched channelrhodopsin. To link pairs or cohorts of neurons based on presynaptic activity, reconstituting light emission in a split luciferase with a calcium-sensitive domain provided a useful platform for functional PhAST signaling [18]. The calcium dependence of the luciferase ensures that light emission is synchronized with neuronal activity, thereby conserving substrate and avoiding issues associated with constitutively active luciferases, such as substrate depletion [34] and enzyme inactivation from rapid catalysis [35]. A high dynamic range and high absolute photon numbers are desired parameters. If combined with slow-closing channelrhodopsins, on/off kinetics may be of secondary choice. For PhAST, enhanced calcium-sensitive Nanolanterns with a Kd = 250 nM of the calcium sensor have been successfully used, but the newer version promises to enter the field [26]. Brighter luciferases could enable the use of faster channels, thereby preserving the rapid dynamics of synaptic signaling, reflecting the inherent change in neuronal circuit dynamics.

5.3.2 Luciferins and Cofactors

All luciferases require a chemical substrate, a so-called luciferin, that is enzymatically oxidized. Chemical structures of luciferins differ depending on the luciferase used. In the experiments reported below, we utilize luciferins optimized for NanoLuc, such as fluorofurimazine (FFz) and Hikarazine (HIK). The luciferase-luciferin reaction requires oxygen, and sometimes a cellular metabolite as a secondary cofactor. This metabolite could be ATP, cAMP, NADH, calcium, or other more exotic molecules [36] that are already present in the cell. In heterologous systems, the luciferin, in contrast, needs to be added externally. This has advantages and disadvantages depending on the desired outcome. The advantage is that the background activity of the transgenic circuit expressing the luciferase and the channelrhodopsin can be determined in the absence of the light emission without the cofactor. The addition of luciferin, hence, initiates light emission, which provides a direct evaluation of the photon emission to bridge synaptic defects. The disadvantage of this approach is that the bioavailability of many luciferins, especially those used by marine luciferases, may be a limiting factor [12]. Therefore, special media formulations that improve solubility and tissue transport and high concentrations may be required.

5.3.3 Channelrhodopsins

For PhAST, the postsynaptic light sensor is typically an ion channel that coordinates a light-sensitive group, most commonly the vitamin A derivative all-trans retinal (ATR). *C. elegans*, different from mammals, does not produce ATR itself, and thus ATR has to be added as a cofactor. When channelrhodopsins (ChRs) absorb light, the light-sensitive group changes shape from an all-trans form to a 13-cis form, which is coupled to conformational changes in the protein that render it more conductive for a specific ionic species. This shape change starts a series of steps that eventually return the molecule to its original state—this full process is called the photochemical reaction cycle. Compared to typical ion channels, ChRs conduct much smaller amounts of current. Their single-channel conductance is only tens to hundreds of femtoSiemens (fS), while most regular ion channels have conductances in the range of 5 to 500 picoSiemens (pS, see also Table 5.3). Because of this, ChRs behave more like transporters than typical fast ion channels and are lousy neuronal actuators. Combined with the poor quantum yield of the luciferases, this is a deadly combination. Luckily, the low conductance can be offset by longer open state lifetimes, which extends time the channel resides in the ion-conducting conformation, effectively increasing the operational light sensitivity. This does not mean that the ion channel is more sensitive, but the charge transported per photon absorption is higher. Thus, even in photon-starved conditions and a low probability of activation, significant current can be produced [11]. Several naturally occurring and synthetic light-gated ion channels are known to conduct a specific ionic species, for example, H^+, Na^+, Ca^{2+}, K^+, or Cl^-. The former three depolarize neuronal membrane potential and can be used to activate neuronal signaling, whereas the latter two commonly hyperpolarize and therefore suppress neuronal activity. In contrast to presynaptic channelrhodopsins, where a high calcium conductance more effectively triggers synaptic release, postsynaptic channelrhodopsins should favor fast sodium channels to induce rapid spiking and prevent calcium-mediated adaptation mechanisms.

Table 5.3 Different channelrhodopsins and their properties

Variant	Peak abs. (nm)	Conductance	Primary ion selectivity	τ_{open}	Ref.
ChR2 (WT)	~470	40–70 fS	Nonselective cation (Na^+, H^+)	~13.5 ms	[37]
ChR2-H134R	~470	Increased vs. WT	Na^+	~17.9 ms	[38]
ChR2-XXL (D156C)	~480	750 fS	Nonselective cation	≈70 s	[39]
ChR2-HRDC (H134R+D156C)	~480	?	Nonselective cation	Tens of seconds	[18, 40]
SSFO (C128S, D156A, T159C)	488	?	Na^+	30 min	[11]
Kalium ChR (HcKCR1)	540–550	700 pS	K^+-selective	≈25 ms	[41]
HcKCR1-C110A	540–550	Not specified	K^+-selective	≈25 ms	[42]
ChRmine	~530	110 fS	Na^+-selective	>300 ms	[10, 43]
ChReef	~530	110 fS	Na^+-selective		[10]
GtACR1	515	1.2 pS	Anion (Cl^-)	Fast (~5–10 ms)	[44]
GtACR2	470	0.7 pS	Anion (Cl^-)	Fast (<10 ms)	[44]

ChR2 channelrhodopsin, *ChReef* channelrhodopsin that excites efficiently, *GtACR Guillardia* theta anion channelrhodopsins, *HcKCR Hyphochytrium catenoides* potassium channelrhodopsin

5.3.4 Source-Sink Distance

Contrary to volumetric neurotransmission, which by definition is long-lasting and long-range, PhAST requires close contacts between the photon-emitting cell and the photosensitized receiver cells. Even though we have not formally established the maximum distance between two neurons at which PhAST can still successfully initiate signaling, we performed simulations that model photon propagation through biological tissues. These results suggest a steep drop in signal intensity due to absorption and scattering, and most notably due to geometric dilution (because as the light travels outward, it spreads over the surface of a growing sphere centered at the source). In *C. elegans*, which does not have hemoglobin, the intensity drops by a factor of 10^4 over a few hundred μm [18], while this factor is likely much larger in the mammalian brain [45].

Our subsequent experiments further suggest that synaptic specializations are not required, as two neighboring cells expressing PhAST can influence each other. This is an asset when the activity state shall be transferred onto an anatomically unconnected neuron but may cause unwanted cross talk if multiple neurons express PhAST components.

5.4 Excitatory Synaptic Transmission

Repairing 'broken' circuits PhAST has been successfully implemented in *C. elegans* to restore synaptic transmission deficits both between multiple neurons and at the neuromuscular junction. In both cases, the system was used to restore effective signaling: first, by compensating for a glutamatergic transmission defect between defined neurons and, second, by overcoming synaptic silencing induced by tetanus toxin expression in neurons presynaptic to egg-laying muscles. In both cases, presynaptic emission of the mTurquoise-coupled Nanoluciferase (TeNL) light emission from the presynaptic neuron was used to trigger postsynaptic channelrhodopsin activity. In the first case, neuronal activity in the mechanosensory ASH neuron was triggered by a mechanical stimulus delivered to the nose of the animal, either by an obstacle, which induced a behavioral avoidance response, or in a microfluidic device, in which the activity of the postsynaptic neuron can be followed by calcium imaging as an indicator of neurotransmission. In successful cases, the calcium increase on the postsynaptic neuron was correlated with a behavioral change of the animals, directly demonstrating that photons can act as synaptic neurotransmitters in vivo.

Engineering a gain-of-function Intriguingly, this ectopic photon-based pathway can not only rescue a transmission defect but also be used to engineer a gain-of-function within the same circuit and sensitize naturally insensitive animals to external mechanical stimuli. For example, in the nose touch circuit, no synapses between ASH and AVA have been found in the male connectome [1], and males are less sensitive to chemical [46] and mechanical stimuli (Fig. 5.3a and Ref. [18]). Indeed, AVA in males did not show elevated Ca^{2+} levels in response to mechanical stimulation via the Trap'N'Slap device (Fig. 5.3b), while AIB responded, even though to a lesser degree than hermaphrodites. Behavioral tests confirmed this sexual dimorphism: While males still avoided the stimulus when touched on the nose, their responses were significantly reduced compared to hermaphrodites (Fig. 5.3c).

To enhance this male sensitivity to nose touch, we expressed PhAST in ASH, AVA, and AIB and performed both mechanical (Fig. 5.3c) and optical avoidance assays [18]. With both cofactors (the luciferin Hikarazine, HIK, and the ChR chromophore all-trans retinal, ATR) present, males responded nearly indistinguishably from hermaphrodites, demonstrating that ectopic connections can sensitize neurons and create new behaviors (Fig. 5.3c).

PhAST at the excitatory neuromuscular junction In the second application, PhAST reinstated excitatory transmission at the neuromuscular junction after synaptic vesicle release was selectively inhibited in motoneurons by tetanus toxin (TeTx) expression. We used the egg-laying circuit of *C. elegans* as a model circuit for NMJ function [47], in which VC motorneurons and the hermaphrodite specific neurons HSN form synaptic contacts with the vulval muscles, vm2 (Fig. 5.4a and Ref. [48]). Under normal conditions, the VC and HSN motor neurons cooperatively induce calcium transients in the vulval muscles (vm), (Fig. 5.4a ii, and Ref. [18]). The expression of tetanus toxin (TeTx) in HSN, VC, or both neurons significantly

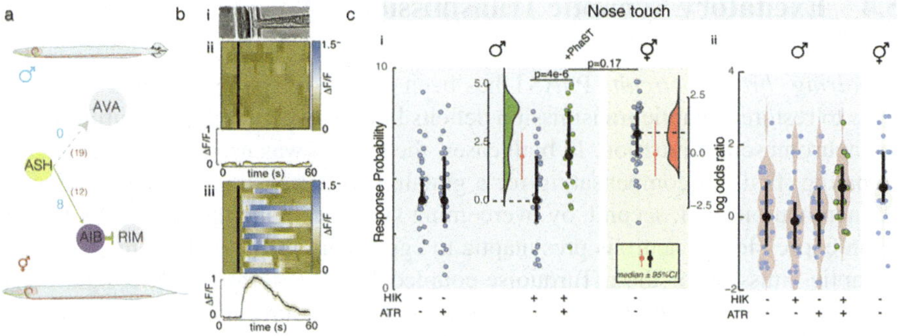

Fig. 5.3 PhAST confers gain-of-function to nociception in a dimorphic circuit. (**a**) Sexually dimorphic connectivity of the ASH nociceptive avoidance circuit, with the number of connections between ASH and AVA or AIB observed in males (blue numbers) or the ones from hermaphrodites (brown font). (**b**) (*i*) Image of an animal inside the Trap'N'Slap device. Stacked kymographs of multiple neuronal calcium recordings from AVA after direct mechanical stimulation of ASH in a microfluidic device. (ii) Response for male animals, (iii) shows the response in hermaphrodites. Blue indicates high neuronal activity; brown indicates baseline activity. Graph below the kymograph indicates the average calcium response of all recordings in the kymograph. Reproduced with permission from [18]. (**c**) Outcome of the nose touch experiment. (*i*) Response probability of male animals expressing the PhAST system in the presence and absence of the ATR and Hikarazine (HIK) cofactors, compared to untreated hermaphrodites of the same genotype ($N = 30$ animals). Moss green data points = functional PhAST after cofactor delivery. Floating axis represents the paired mean difference between control males and PhAST males (green histogram to the left), and between PhAST males (moss green) and control hermaphrodites (red histogram to the right). (*ii*) Violin plots of the log-odds ratio of detecting a positive response in the indicated animals compared to untreated control males (leftmost violin). Moss green data points = functional PhAST after cofactor delivery. Rightmost column indicates the log-odds ratio for detecting a positive response of hermaphrodite control animals compared to untreated males. Filled circle indicates the median; vertical bar indicates 95% CI of the median

reduced spontaneous vm calcium activity (Fig. 5.4b). Notably, this reduction could not be rescued by PhAST when both neurons lacked endogenous transmission, regardless of whether PhAST was expressed in HSN, VC, or both (Fig. 5.4d and Ref. [18]).

However, when VC transmission remained intact, PhAST-mediated restoration of HSN activity successfully rescued vm calcium transients (Fig. 5.4c and Ref. [18]). These complex and seemingly counterintuitive results align with a recently proposed disinhibitory circuit, where VC activity suppresses the inhibitory influence of the neuroendocrine uv1 cells on HSN. In this model, loss of VC neurotransmission leads to sustained uv1 activity, which tonically suppresses HSN excitability and functionally silences its contribution to vulval muscle (vm) activation, even if HSN retains the intrinsic ability of producing depolarizing output.

Our findings, supported by a kinetic model that incorporates this disinhibitory interaction (Fig. 5.4a-d *iv* and Ref. [18]), underscore both the therapeutic potential of synaptic engineering and its utility to uncover functional connectivity in poorly characterized neuronal circuits.

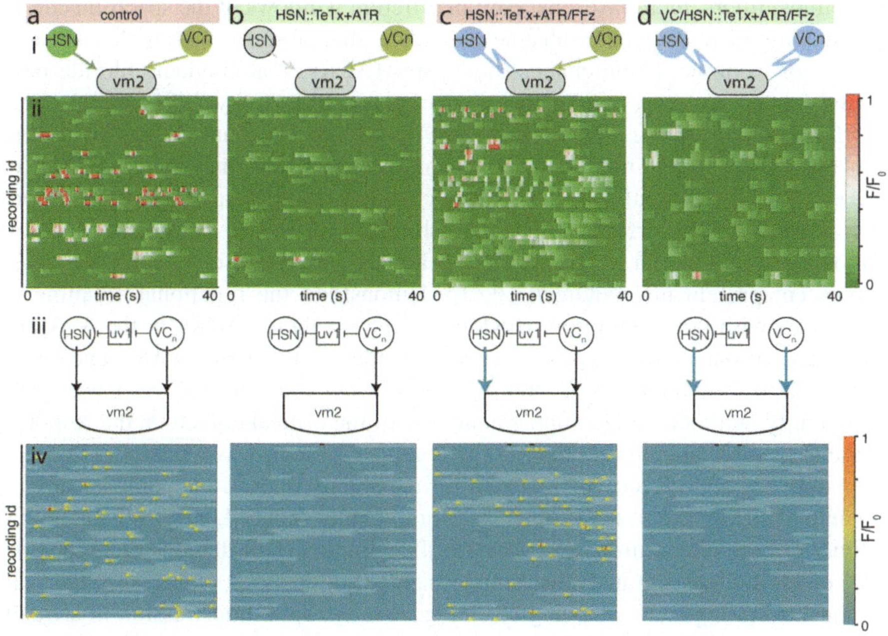

Fig. 5.4 PhAST at the neuromuscular junction. PhAST can be used to dissect the wiring pattern of an unknown neuronal circuit by stepwise expression and activation of photon emission in the individual parts. For all panels: (*i*) Schematics of the egg-laying circuit. Nodes are color-coded by the neurotransmitter used; edges indicate functional connection using chemicals (arrows) or photons (flashes). Green, serotonin; yellow, acetylcholine; blue, photons. (*ii*) Kymographs showing spontaneous calcium activity in vm2 from PhAST animals supplemented with ATR and the luciferin fluorofurimazine, FFz. Color scale indicates the normalized neuronal activity. (*iii*) Schematic of the simulated egg-laying circuit incorporating a possible disinhibitory edge connecting VC and HSN through the neuroendocrine cells uv1. HSN and VC are connected by an AND gate to the vm2 calcium activity. Only if both neurons are "on" and provide input, vm2 responds with an increase in activity. Black edges are normal, wild-type connections; missing edges represent mutant conditions (or tetanus toxin (TeTx) expression), and blue edges represent PhAST connections. (*iv*) Stacked kymographs of the outcome of the simulation of the vm activity, considering the circuit pattern shown above. Yellow indicates high; blue indicates low activity. Data taken and replotted from [18]. **a** vm2 activty in unperturbed control animals. (**b**) vm2 activity in animals expressing TeTx in HSN to interfere with presynaptic signaling. (**c**) Outcome of functional PhAST in HSN on vm2 activity in animals expressing TeTx in HSN to interfere with presynaptic signaling. (**d**) Outcome of functional PhAST in HSN and VC in animals expressing TeTx in both neurons. Note the absence of vm2 activity cannot be rescued in this condition

5.5 Inhibitory Synaptic Transmission

The construction of neuronal networks is based on a fine balance between the excitatory and inhibitory circuits. The bipartite nature of PhAST, composed of color-matched photon emitters and photosensors, affords a modular design that combines luciferases and ion channels with differing functionalities. Similar to different neurotransmitters, luciferases that emit in different windows of the spectrum can be paired with their "cognate" channelrhodopsin that absorbs within this spectral range. For example, anion channelrhodopsins (ACRs) that conduct chloride have recently been used to efficiently silence neuronal activity [49]. ACR1 from *Guillardia theta*, among the most powerful light-gated ion channels, has an absorption maximum at 520 nm, significantly more red-shifted than the maximum of 470 nm for the commonly used excitatory channels (Table 5.3, Ref. [44]). For optimal activation, this required a green-shifted luciferase, e.g green-enhanced Nanolantern (GeNL), a fusion of the calcium-sensitive luciferase with the green fluorescent protein mNeonGreen 16. To demonstrate the functional coupling of luciferases with other inhibitory rhodopsins, we expressed ACR1 in the AVA and AIB interneurons of wild-type *C. elegans*, together with GeNL in ASH and tested their efficacy in suppressing the nose-touch evoked reversal response [18]. Importantly, without dietary supplementation of the critical cofactors, the response of the transgenic animals expressing the complete system was wildtype. However, upon delivery of ATR and the luciferin furimazine to reconstitute photosensitivity, the animal's response to nose touch was significantly reduced, as was the neuronal activity of both interneurons secondary to the stimulus. This demonstrated the efficacy of bioluminescent activated anion channelrhodopsin in suppressing the pain response resulting from silencing of endogenous synaptic transmission in a freely behaving animal [18].

5.6 Synthetic Neuronal Connections Between Unrelated Circuits

We also sought to engineer a de novo synthetic connection by rewiring the diacetyl (DA)-sensing AWA neuron with the avoidance circuit mediated by AVA and AIB. DA, a volatile attractant, activates AWA and normally suppresses reversal behavior through AIA interneurons [50]. To repolarize this attraction into an aversive response, we aimed to functionally link AWA to AVA using PhAST. While AWA and AVA lack direct synaptic connections, their axons run in parallel within the nerve ring (Fig. 5.5a).

To that end, we repurposed the AIB/AVA::ChR2-HRDC line and expressed TeNL specifically in AWA (Fig. 5.5a). To assess the behavioral relevance of these neuronal responses, we examined the ability of transgenic animals to navigate diacetyl (DA) gradients in a chemotaxis assay. In this assay, the behavioral test arena was divided into four quadrants, two of which contained DA and the other two a mock solution [51]. Importantly, PhAST transgenic animals displayed chemotaxis toward DA that was indistinguishable from wild-type (N2) animals (Fig. 5.5b, [18]), regardless of whether they received ATR or HIK alone, while odr-7 mutants that cannot sense DA failed to properly chemotax in these arenas (Fig. 5.5c). Notably, when both cofactors, HIK and ATR, were administered, the transgenic PhAST

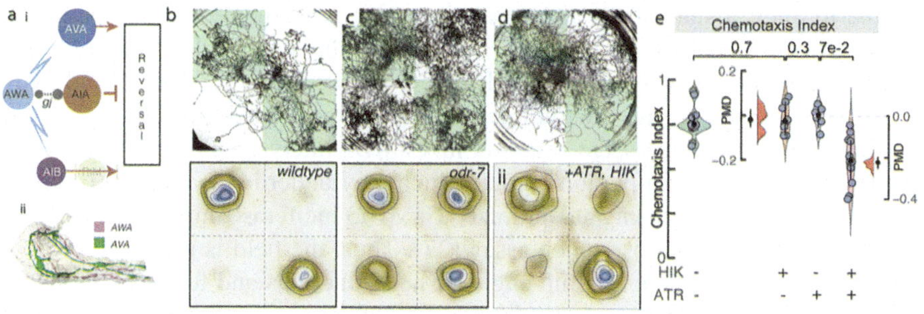

Fig. 5.5 Olfactory confusion experiment through competition between attractive and avoidance circuit. (**a**) (*i*) Sketch of the attractive AWA olfactory circuit. gj: gap junctions. Blue flashes indicate intended PhAST connections. (*ii*) 3D rendering showing the location of AWA and AVA as reconstructions from EM datasets [52]. (**b–d**) Top row: Outcome of a representative olfactory confusion experiment. Maximum intensity projections of a 30 min video recording showing the crawling trajectories of (**b**) wild type, (**c**) *odr-7*, and (**d**) PhAST animals on diacetyl-containing plates (applied on the green quadrant). Wild-type animals crawl efficiently toward the DA source, whereas *odr-7* cannot distinguish the odorant. PhAST animals are confused and do not chemotax efficiently toward the odorant. Bottom row: 2D probability density function (PDF) of the animal distribution on agar plates for wild-type**,** *odr-7* mutant and PhAST transgenics, where DA was administered on the top-left and bottom-right quadrants. (**e**) Distribution of the CI for independent chemotaxis experiments, indicated as individual dots. p-value displayed above the brackets from two-sided KS test. The floating axes indicate the bootstrapped distribution of the paired median difference (PMD) between Hikarazine and ATR single-treated animals (left) and ATR single and double treated animals (right)

animals exhibited chemotaxis behavior similar to that of odr-7 mutants, as evidenced by progressively random movement resulting in less animals successfully navigating toward the DA sources (see statistics in Fig. 5.5d) and a globally reduced chemotaxis index (Fig. 5.5e). Together, this indicates that PhAST can be used to successfully connect neighboring but otherwise unconnected neurons functionally together.

5.7 Color Tuning: PhASTg and PhASTr

To fully reengineer the connectome, it would be desirable to repurpose different luciferases that emit photons with spectrally separable wavelengths and that are color-matched to specific channelrhodopsins, e.g., in the green- or red-shifted spectrum (PhASTg, PhASTr, respectively). In our previous work, we applied the cyan luciferase TeNL with an mTurquoise emitter, a green-emitting luciferase with an mNeonGreen emitter (see also Sect. 5.5 above), and an orange-emitting luciferase with the long Stokes shift cyOFP (CaMBI) [17]. Whereas cyan and green luciferases were able to noticeably trigger their cognate channelrhodopsin, CaMBI was unable to elicit a statistically significant response with the red-shifted pump-like channelrhodopsin ChRmine [18]. The lack of signaling may be due to the relatively low operational light sensitivity of ChRmine compared to ChR2-HRDC and ACR1 (Table 5.3). However, the low bioluminescent energy transfer efficiency between the luciferase and fluorescent proteins in CaMBI likely results in an additional caveat in the insufficient photon output, which may underlie the absence of detectable signaling. Indeed, using an optimized CaMBI3 luciferase improved the PhAST results in the classical nose touch assay [27]. Thus, the field is poised to combine different luciferases to bottom-up engineer new functions and feedback loops that rely on nuanced excitatory and inhibitory input into existing structural connectomes. Unfortunately, a caveat of this approach is the broad emission spectrum of luciferases and the wide absorption range of rhodopsins, which increases the risk of spectral cross talk between different color channels. Thus, multiplexing different spectral pairs in the same circuit, especially when the wiring is compact and the targets are close together, may require the development of narrow-band emitters and sensors [53].

5.8 Summary

PhAST (photon-assisted synaptic transmission) is a synaptic engineering approach designed to replace chemical neurotransmission to create light-based connections, enabling precise information exchange between neurons within neural circuits. Unlike traditional optogenetic methods that target individual neurons, PhAST enables cell-specific directional communication by coupling presynaptic light emission to postsynaptic light sensing.

The core of the PhAST system involves engineering two separate neuronal populations: one expressing a luciferase enzyme (such as NanoLuc) and the other expressing a light-sensitive ion channel or pump, such as channelrhodopsin or ChRmine. Upon application of a luciferin substrate and the necessary cofactors, the luciferase in the presynaptic neuron emits bioluminescent photons. These photons, in turn, activate the opsin in the postsynaptic neuron, triggering a physiological response. This photon-mediated activation mimics synaptic transmission but is entirely synthetic and programmable.

Crucially, PhAST does not depend on the presence of natural synaptic contacts between paired neurons, allowing researchers to rewire circuits and create novel pathways that do not exist in the native connectome. By spatially restricting luciferase and opsin expression, PhAST enables highly specific and temporally precise modulation of activity between defined cell types, even in freely behaving animals.

This strategy has been successfully applied in *C. elegans* to several problems, including sensitization and desensitization of endogenous circuits and the conversion of an attractive cue into an aversive response in the absence of direct synaptic specializations. This flexibility makes PhAST a powerful tool for dissecting the logic of neural circuits, testing hypotheses about connectivity, and exploring how reconfigured communication between neurons alters behavior. Looking ahead, PhAST offers a promising platform for building programmable neural networks in vitro and in vivo, enabling the design of custom information processing pathways for both basic neuroscience and therapeutic applications.

References

1. Cook SJ, Jarrell TA, Brittin CA, Wang Y, Bloniarz AE, Yakovlev MA, et al. Whole-animal connectomes of both Caenorhabditis elegans sexes. Nature. 2019;571:63–71.
2. Ripoll-Sánchez L, Watteyne J, Sun HS, Fernandez R, Taylor SR, Weinreb A, et al. The neuropeptidergic connectome of *C. elegans*. Neuron. 2023;11(111):3570–89.e5.
3. Dorkenwald S, Matsliah A, Sterling AR, Schlegel P, Chieh Yu S, McKellar CE, et al. Neuronal wiring diagram of an adult brain. Nature. 2024;10(634):124–38.
4. Bae JA, Baptiste M, Baptiste MR, Bishop CA, Bodor AL, Brittain D, et al. Functional connectomics spanning multiple areas of mouse visual cortex. Nature. 2025;4(640):435–47.
5. Chen R, Canales A, Anikeeva P. Neural recording and modulation technologies. Nat Rev Mater. 2017;1:2.
6. Emiliani V, Entcheva E, Hedrich R, Hegemann P, Konrad KR, Lüscher C, et al. Optogenetics for light control of biological systems. Nat Rev Methods Primers. 2022;2:55.
7. Aravanis AM, Wang LP, Zhang FA, Meltzer L, Mogri MZ, Schneider MB, et al. An optical neural interface: in vivo control of rodent motor cortex with integrated fiberoptic and optogenetic technology. J Neural Eng. 2007;4:S143–56.
8. Rabinowitch I, Colón-Ramos DA, Krieg M. Understanding neural circuit function through synaptic engineering. Nat Rev Neurosci. 2024;2(25):131–9.
9. Chen R, Gore F, Nguyen QA, Ramakrishnan C, Patel S, Kim SH, et al. Deep brain optogenetics without intracranial surgery. Nat Biotechnology. 2021;39:161–4.
10. Alekseev, A., Hunniford, V., Zerche, M. et al. Efficient and sustained optogenetic control of sensory and cardiac systems. Nat. Biomed. Eng (2025). https://doi.org/10.1038/s41551-025-01461-1
11. Gong X, Mendoza-Halliday D, Ting JT, Kaiser T, Sun X, Bastos AM, et al. An ultra-sensitive step- function opsin for minimally invasive optogenetic stimulation in mice and macaques. Neuron. 2020;107:38–51.e8.
12. Porta-de-la Riva M, Morales-Curiel LF, Gonzalez A, Krieg M. Bioluminescence as a functional tool for visualizing and controlling neuronal activity in vivo. Neurophotonics. 2024;11:1–26.
13. Wang W, Wu X, Tang KWK, Pyatnitskiy I, Taniguchi R, Lin P, et al. Ultrasound-triggered in situ photon emission for noninvasive optogenetics. J Am Chem Soc. 2023;145:1097–107.
14. Wu X, Zhu X, Chong P, Liu J, Andre LN, Ong KS, et al. Sono-optogenetics facilitated by a circulationdelivered rechargeable light source for minimally invasive optogenetics. Proc Natl Acad Sci U S A. 2019;116:26332–42.
15. Matsubara T, Yanagida T, Kawaguchi N, Nakano T, Yoshimoto J, Sezaki M, et al. Remote control of neural function by X-ray-induced scintillation. Nat Commun. 2021:12.
16. Suzuki K, Kimura T, Shinoda H, Bai G, Daniels MJ, Arai Y, et al. Five colour variants of bright luminescent protein for real-time multicolour bioimaging. Nat Commun. 2016;7:1–10.
17. Oh Y, Park Y, Cho JH, Wu H, Paulk NK, Liu LX, et al. An orange calcium-modulated bioluminescent indicator for non-invasive activity imaging. Nat Chem Biol. 2019;15:433–6.
18. Porta-de-la Riva M, Gonzalez AC, Sanfeliu-Cerdán N, Karimi S, Malaiwong N, Pidde A, et al. Neural engineering with photons as synaptic transmitters. Nat Methods. 2023;20:761–9.
19. Mirdita M, Schütze K, Moriwaki Y, Heo L, Ovchinnikov S, Steinegger M. ColabFold: making protein folding accessible to all. Nat Methods. 2022;6(19):679–82.

20. Piggott BJ, Liu J, Feng Z, Wescott SA, Xu XZS. The neural circuits and synaptic mechanisms underlying motor initiation in *C. elegans*. Cell. 2011;147:922–33.
21. Takai A, Nakano M, Saito K, Haruno R, Watanabe TM, Ohyanagi T, et al. Expanded palette of nano-lanterns for real-time multicolor luminescence imaging. Proc Natl Acad Sci U S A. 2015;112:4352–6.
22. Morales-Curiel LF, Gonzalez AC, Castro-Olvera G, Lin LCL, El-Quessny M, Porta-de-la Riva M, et al. Volumetric imaging of fast cellular dynamics with deep learning enhanced bioluminescence microscopy. Commun Biol. 2022;5:1330.
23. Farhana I, Hossain MN, Suzuki K, Matsuda T, Nagai T. Genetically encoded fluorescence/bioluminescence bimodal indicators for Ca2+ imaging. ACS Sens. 2019;4:1825–34.
24. Qian Y, Rancic V, Wu J, Ballanyi K, Campbell RE. A bioluminescent Ca 2+ indicator based on a topological variant of GCaMP6s. ChemBioChem. 2019;20:516–20.
25. Tian X, Zhang Y, Li X, Xiong Y, Wu T, Ai HW. A luciferase prosubstrate and a red bioluminescent calcium indicator for imaging neuronal activity in mice. Nat Commun. 2022:13.
26. Lambert GG, Crespo EL, Murphy J, Boassa D, Luong S, Celinskis D, et al. CaBLAM: A high-contrast bioluminescent Ca2+ indicator derived from an engineered Oplophorus gracilirostris luciferase. Nature Methods. 2026;23:205–15.
27. Zhao Y, Lee S, Campbell RE, Lin MZ. Directed evolution of a genetically encoded bioluminescent Ca2+ Sensor. Eng Proc. 2023;35:20.
28. Petersen ED, Lapan AP, Fillion AJ, Crespo EL, Lambert GG, Zanca AT, et al. Bioluminescent genetically encoded glutamate indicator for molecular imaging of neuronal activity. ACS Synth Biol. 2023;12:2301–9.
29. Erdenee E, Ting AY. A dual-purpose real-time indicator and transcriptional integrator for calcium detection in living cells. ACS Synth Biol. 2022;3(11):1086–95.
30. Yoshida T, Kakizuka A, Imamura H. BTeam, a novel BRET-based biosensor for the accurate quantification of ATP concentration within living cells. Sci Rep. 2016;12:6.
31. Gregor C, Pape JK, Gwosch KC, Gilat T, Sahl SJ, Hell SW. Autonomous bioluminescence imaging of single mammalian cells with the bacterial bioluminescence system. Proc Natl Acad Sci U S A. 2019;116:26491–6.
32. Tanaka R, Sugiura K, Osabe K, Hattori M, Nagai T. Genetically encoded bioluminescent glucose indicator for biological research. Biochem Biophys Res Commun. 2025;1:742.
33. Yeh HW, Ai HW. Development and applications of bioluminescent and chemiluminescent reporters and biosensors. Annu Rev Anal Chem. 2019;12:129–50.
34. Syed AJ, Anderson JC. Applications of bioluminescence in biotechnology and beyond. Chem Soc Rev. 2021;50:5668–705.
35. Dijkema FM, Nordentoft MK, Didriksen AK, Corneliussen AS, Willemoës M, Winther JR. Flash properties of Gaussia luciferase are the result of covalent inhibition after a limited number of cycles. Prot Sci. 2021;30:638–49.
36. Yeh HW, Xiong Y, Wu T, Chen M, Ji A, Li X, et al. ATP-independent bioluminescent reporter variants to improve in vivo imaging. ACS Chem Biol. 2019;14:959–65.
37. Nagel G, Szellas T, Huhn W, Kateriya S, Adeishvili N, Berthold P, et al. Channelrhodopsin-2, a directly light-gated cation-selective membrane channel. Proc Natl Acad Sci. 2003;100:13940–5.
38. Nagel G, Brauner M, Liewald JF, Adeishvili N, Bamberg E, Gottschalk A. Light activation of Channelrhodopsin-2 in excitable cells of caenorhabditis elegans triggers rapid behavioral responses. Curr Biol. 2005;15:2279–84.
39. Dawydow A, Gueta R, Ljaschenko D, Ullrich S, Hermann M, Ehmann N, et al. Channelrhodopsin-2- XXL, a powerful optogenetic tool for low-light applications. Proc Natl Acad Sci. 2014;111:13972–7.
40. Bergs A, Schultheis C, Fischer E, Tsunoda SP, Erbguth K, Husson SJ, et al. Rhodopsin optogenetic toolbox v2.0 for light-sensitive excitation and inhibition in Caenorhabditis elegans. PLoS One. 2018;13:1–24.

41. Govorunova EG, Gou Y, Sineshchekov OA, Li H, Lu X, Wang Y, et al. Kalium channelrhodopsins are natural light-gated potassium channels that mediate optogenetic inhibition. Nat NeuroScience. 2022;7(25):967–74.
42. Morizumi T, Kim K, Li H, Nag P, Dogon T, Sineshchekov OA, et al. Structural insights into lightgating of potassium-selective channelrhodopsin. Nat Commun. 2025;12(16):1283.
43. Sineshchekov OA, Govorunova EG, Li H, Wang Y, Melkonian M, Wong GKS, et al. Conductance mechanisms of rapidly desensitizing cation channelrhodopsins from cryptophyte algae. mBio. 2020;11.
44. Govorunova EG, Sineshchekov OA, Janz R, Liu X, Spudich JL. Natural light-gated anion channels: a family of microbial rhodopsins for advanced optogenetics. Science (New York, NY). 2015;349:647–50.
45. Durduran T, Zhou C, Edlow BL, Yu G, Choe R, Kim MN, et al. Transcranial optical monitoring of cerebrovascular hemodynamics in acute stroke patients. Opt Exp. 2009;17:3884.
46. Pechuk V, Goldman G, Salzberg Y, Chaubey AH, Bola RA, Hoffman JR, et al. Reprogramming the topology of the nociceptive circuit in *C. elegans* reshapes sexual behavior. Curr Biol. 2022;32:4372–85.e7.
47. Schafer WR. Egg-laying. WormBook: the online review of C elegans biology. 2005:1–7.
48. Collins KM, Bode A, Fernandez RW, Tanis JE, Brewer JC, Creamer MS, et al. Activity of the *C. elegans* egg-laying behavior circuit is controlled by competing activation and feedback inhibition. eLife. 2016;5:1–24.
49. Mahn M, Gibor L, Patil P, Malina KCK, Oring S, Printz Y, et al. High-efficiency optogenetic silencing with soma-targeted anion-conducting channelrhodopsins. Nat Commun. 2018;12:9.
50. Larsch J, Flavell SW, Liu Q, Gordus A, Albrecht DR, Bargmann CI. A circuit for gradient climbing in *C. elegans* chemotaxis. Cell Rep. 2015;12:1748–60.
51. Margie O, Palmer C, Chin-Sang I. *C. elegans* chemotaxis assay. J Vis Exp. 2013:1–6.
52. Witvliet D, Mulcahy B, Mitchell JK, Meirovitch Y, Berger DR, Wu Y, et al. Connectomes across development reveal principles of brain maturation. Nature. 2021;596:257–61.
53. Carrasco-López C, Lui NM, Schramm S, Naumov P. The elusive relationship between structure and colour emission in beetle luciferases. Nat Rev Chem. 2021;5:4–20.

Chapter 6
Interluminescence: Control of Synaptic Communication by Biological Light

Ute Hochgeschwender

Abstract Synapse-selective control of neural communication requires two independent components for pre- and postsynaptic partners and whose interaction can be controlled at the time selected by the experimenter. Interluminescence meets these criteria by placing a light-emitting luciferase in the presynaptic and a light-sensing opsin in the postsynaptic neuron, with engagement of this "optical synapse" dependent on the application of luciferin. Two distinct versions of Interluminescence have been developed. For activity-dependent Act-Int, luciferases targeted to synaptic vesicles are released upon firing of presynaptic neurons, while for activity-independent Persist-Int, the luciferase is tethered to the membrane of presynaptic neurons. Both strategies can activate or silence postsynaptic neurons with similar efficacy, and both have been demonstrated to work in mice. Interluminescence has strong potential as a modular platform technology that permits the use of luciferases and opsins ranging in brightness and light sensitivity for synapse-specific optical modulation of neural communication.

Keywords Bioluminescent optogenetics · Cell communication · Synaptic transmission · Luciferase · Channelrhodopsin

6.1 Interluminescence Principles

In luminopsins, an opsin is activated by light emitted from a tethered luciferase. With the availability of opsins with increasing light sensitivity and luciferases with increasing light emission, it has become feasible to activate photoreceptors with a physically separate light emitter and light sensor. Expression of the light emitter in one cell and the light sensor in another cell opens BL-OG to be used as a tool for cell-to-cell communication via light, or Interluminescence (light between cells).

U. Hochgeschwender (✉)
College of Medicine, Central Michigan University, Mt. Pleasant, MI, USA
e-mail: hochg1u@cmich.edu

U. Hochgeschwender (ed.), *BioLuminescent OptoGenetics*,
https://doi.org/10.1007/978-3-032-18121-3_6

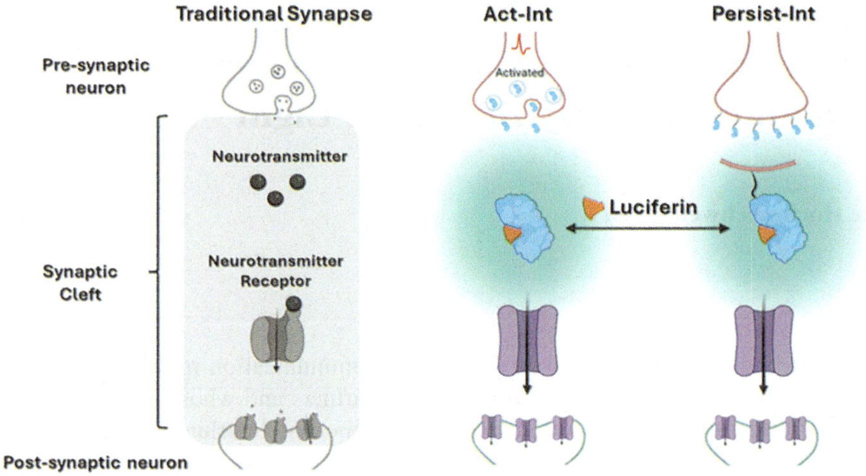

Fig. 6.1 Principle of Interluminescence. Neurotransmitter-mediated synaptic transmission is compared to activity-dependent (Act-Int) and activity-independent (Persist-Int) forms of Interluminescence. By comparison to neurotransmitter-mediated synaptic transmission, Interluminescence depends on light emitted from a luciferin-luciferase chemical reaction to travel across the synaptic cleft and activate an opsin expressed in the postsynaptic cell. In Act-Int, the luciferase is in synaptic vesicles and released (along with neurotransmitter) during synaptic vesicle exocytosis. In Persist-Int, the luciferase is in the synaptic cleft, anchored to the extracellular face of the presynaptic cell. Persist-Int is independent of presynaptic vesicle release and is activated whenever the luciferin is present, whereas Act-Int depends on presynaptic vesicle release and the presence of the luciferin. Compared to neurotransmission, which is mediated by neurotransmitter diffusion across the synaptic cleft and activation of postsynaptic neurotransmitter receptors, Persist-Int and Act-Int are independent of neurotransmitters, but both require luciferin, and the resulting postsynaptic response depends on the nature of the opsin expressed. (Figure from Ref. [2])

The Interluminescence strategy has been developed using neuron-to-neuron communication as the test case [1, 2]. Two distinct realizations of this principle for synaptic transmission have been developed (Fig. 6.1): Act-Int, activity-dependent Interluminescence, where the luciferase is targeted to synaptic vesicles and released into the synaptic cleft upon activation of the presynaptic neuron, and Persist-Int, activity-independent persistent Interluminescence, where the luciferase is tethered to the presynaptic neuron membrane and is continuously available in the synaptic cleft. For both approaches, activation of these versions of an "optical synapse" is dependent on the application of the luciferin and thus is controlled by the experimenter.

6.2 Act-Int

To allow for the activity-dependent release of the luciferase from presynaptic neurons, the vesicle-targeting sequence from the human pro-opiomelanocortin prohormone (hPOMC1-26) was fused to the bright *Gaussia* luciferase variant sbGLuc [1]. Postsynaptic excitatory and inhibitory opsins were ChR2(C128S) and hGtACR2, two optogenetic actuators with high light sensitivity. Multielectrode array recordings, from cultured primary cortical neurons expressing the luciferase and striatal neurons expressing the opsin, demonstrated luciferin-dependent increase or decrease of neural firing. In addition to showing that Act-Int enables excitation and inhibition of postsynaptic neurons, these experiments established core features of Act-Int, such as depending on the activity of presynaptic neurons, requiring intact synaptic connections, occurring independent of classical neurotransmission, requiring active presynaptic vesicle release, and requiring functional opsins in postsynaptic neurons [1]. Most importantly, multichannel electrophysiology recordings in mice revealed the efficacy of Act-Int to drive changes in gamma oscillations in vivo [1] (Fig. 6.2).

The original Act-Int strategy releases an untethered luciferase into the synaptic cleft. There are alternate strategies for activity-dependent luciferase exposure into the synaptic cleft. By attaching the luciferase to the luminal end of synaptobrevin/VAMP2 [3], the luciferase is exposed to the cleft with activity-driven vesicle fusion into the presynaptic membrane and available for light activation of the postsynaptic opsin in an activity-dependent manner but without releasing it (Fig. 6.3a). Similarly, tagging luciferases onto vesicular amino acid transporters, for example, attaching the luciferase to the luminal C-terminus of the vesicular GABA transporter (VGAT) [4], is a strategy that requires activity of the presynaptic neuron to expose the luciferase to the synaptic cleft. We found no consistent difference in the efficacy of

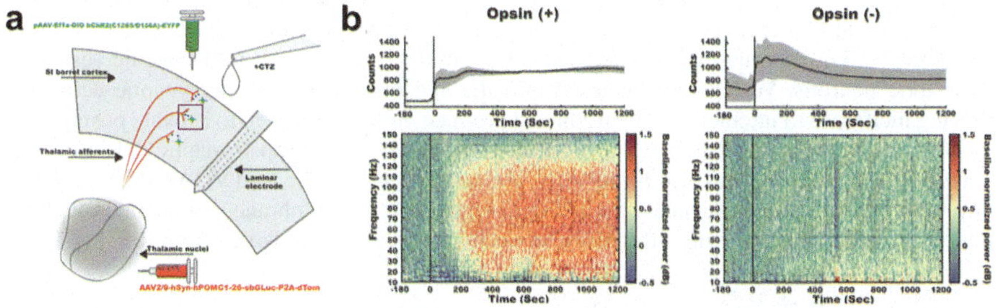

Fig. 6.2 Act-Int in vivo. (**a**) Schematic of the in vivo Interluminescence configuration targeting somatosensory thalamic nuclei and SI Barrel cortex. (**b**) Upper panels: The average bioluminescence signal with ±1 SEM in semi opaque bands for the opsin (+) and opsin (−) group, both expressing the presynaptic luciferase and with (+) or without (−) the postsynaptic opsin. Both groups show robust bioluminescent light emission. Lower panels: Time-frequency spectrograms averaged across all laminar electrode contacts for the opsin (+) (left panel) and opsin (−) (right panel) cohorts, demonstrating Act-Int driving changes in gamma oscillations when postsynaptic neurons express the optogenetic actuator. Time zero refers to the onset of the bioluminescence signal. (Figure adapted from Ref. [1])

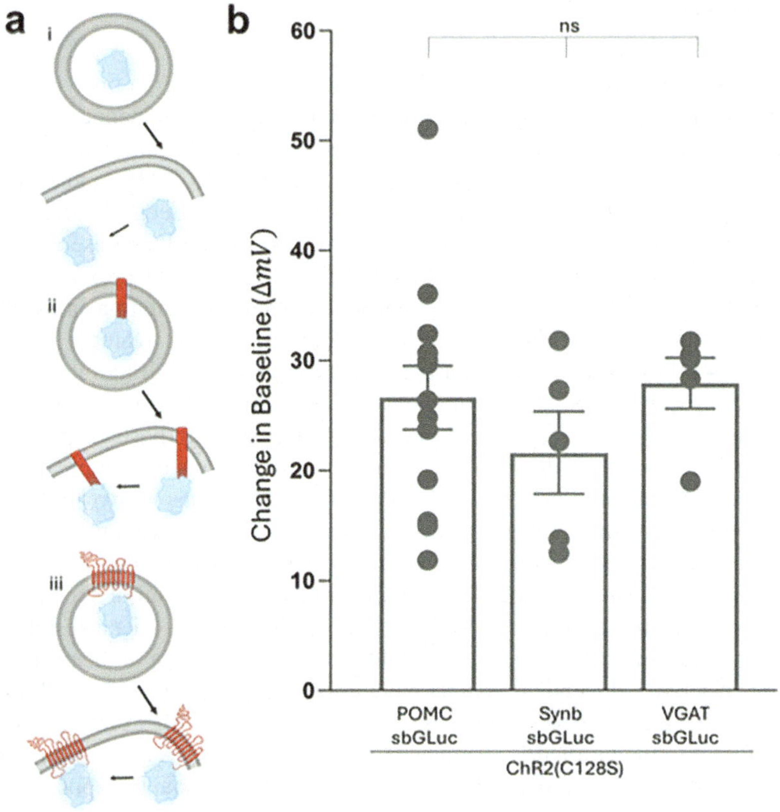

Fig. 6.3 Different mechanisms for activity-dependent Interluminescence. (**a**) Different versions of Act-Int were designed using vesicle targeting sequences: (*i*) the human proopiomelanocortin (hPOMC) signal sequence, (*ii*) a truncated synaptobrevin2 TMD (Synb), (*iii*) the vesicular inhibitory amino acid transporter (VGAT). All sequences shuttle the luciferase into synaptic vesicles in presynaptic neurons. Vesicles are released into the synaptic cleft upon presynaptic activation, allowing luciferase to interact with luciferin and produce bioluminescence to activate postsynaptic opsin-expressing neurons. (**b**) Change in baseline membrane voltage in postsynaptic ChR2(C128S)-expressing neurons was measured for hPOMC sbGLuc, Synb-sbGLuc, and VGAT-sbGLuc Act-Int constructs. All targeting sequences resulted in large changes in membrane voltage upon luciferin application, without significant differences between groups. (Figure adapted from Ref. [2])

presynaptic POMC sbGLuc, Synb-sbGLuc, and VGAT-sbGLuc in Act-Int experiments to couple presynaptic depolarization to a postsynaptic opsin-mediated response [2] (Fig. 6.3b). Utilizing neuropeptide tags such as hPOMC1-26 will release luciferases mainly via large dense core vesicles (LDCVs) [1, 5], while vesicular amino acid transporters such as VGAT preferentially target small clear vesicles (SCVs) [6], and synaptobrevin is present in both vesicle types [7, 8]. This opens the possibility of targeting different secretory organelles, taking advantage of the different kinetic properties of activity-dependent presynaptic release of luciferase.

We also compared the efficiency of presynaptically released intact luciferases with split luciferases, where a "small bit" (smb) is released from presynaptic vesicles and the "large bit" (lgb) is tethered to the postsynaptic opsin [2] (Fig. 6.4). Using the hPOMC vesicle targeting sequence, we compared the classical POMC sbGLuc paired with postsynaptic neurons expressing ChR2(C128S) and POMC $SSLuc_{smb}$ (amino acids 159–169 of SSLuc [9]) paired with postsynaptic neurons expressing a lgb (amino acids 1–158 of SSLuc)-ChRmine fusion (Fig. 6.4a). Presynaptic neurons co-expressed the excitatory opsin ChrimsonR to induce presynaptic activation with red light. Postsynaptic responses evoked by Act-Int bioluminescence with intact and smb luciferase were comparable [2] (Fig. 6.4b). Luciferin addition without activation of presynaptic neurons did not alter the membrane voltage of the postsynaptic cell. Utilizing this split luciferase design offers the advantages of an increased number of small bits that fit into a vesicle compared to

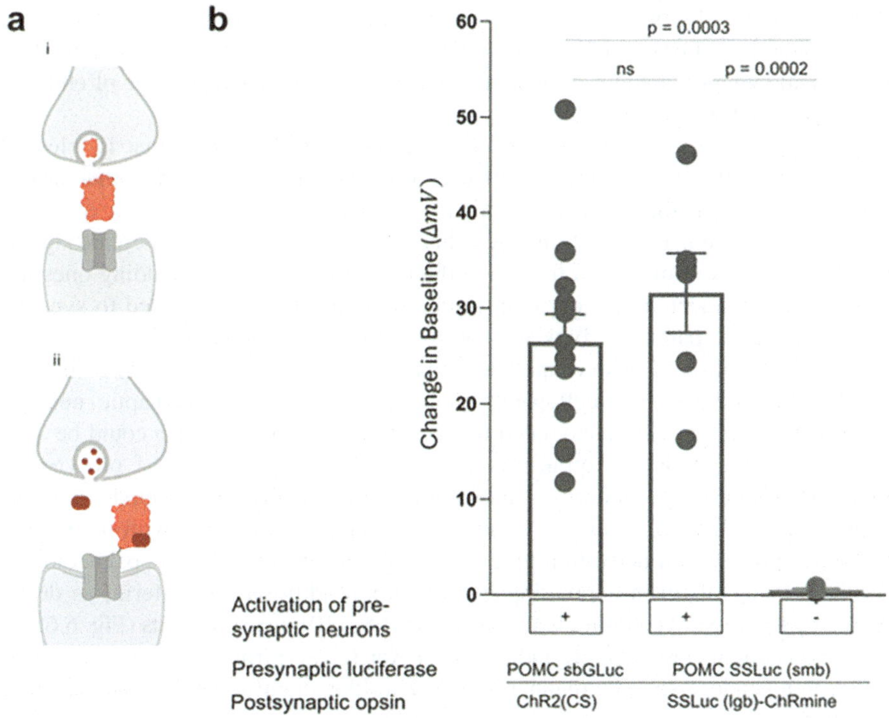

Fig. 6.4 Intact versus split luciferase for presynaptic release. (**a**) Act-Int with intact luciferase (*i*) versus split luciferase (*ii*). (**b**) Whole-cell recordings from cocultures of pre- and postsynaptic neurons in the presence of luciferin and synaptic blockers and with or without red light stimulation of presynaptic neurons. Postsynaptic responses with intact sbGLuc and SSLuc_smb were comparable. No change in baseline membrane voltage (ΔmV) was measured in cocultures when presynaptic neurons were not active (no red-light stimulation), indicating that the small bit of the split luciferase, SSLucsmb, requires vesicle fusion and release from activated neurons. (Figure adapted from Ref. [2])

intact luciferases and the increased kinetics of on/offset of bioluminescence due to the rapid removal of the small bit from the large bit at the postsynaptic site.

Experimental applications of Act-Int may include boosting or counteracting endogenous synaptic activities. The user of this neural modulation technology might want to know how much presynaptic activity is required to engage the "optical synapse." Experiments conducting whole-cell recordings in primary cultured neurons addressed this question [2]. Postsynaptic striatal neurons expressing endogenous neurotransmitter receptors and excitatory blue-light-sensitive opsins were synaptically connected to presynaptic cortical neurons expressing a vesicle-targeted luciferase and the red light-sensitive opsin ChrimsonR. Postsynaptic action potentials were recorded after red light stimulation of presynaptic neuron activity for different lengths of time, with increasing time resulting in higher numbers of postsynaptic action potentials (Fig. 6.5). The initial recording reflects synaptic transmission by endogenous neurotransmitters. This was followed by red light stimulation in the presence of synaptic blockers and luciferin, isolating the postsynaptic firing response elicited by bioluminescence-mediated activation of opsins. For each neuron, the number of postsynaptic action potentials was highly similar, indicating that the optical synapse engages at presynaptic activity levels characteristic of endogenous synaptic transmission [2].

During Act-Int-mediated synaptic transmission, the luciferase that is released into the synaptic cleft eventually diffuses out, potentially reaching other opsin-expressing neurons that are not synaptically connected to the luciferase-expressing presynaptic neuron. The question is whether Act-Int resembles neuropeptide signaling that operates through a volumetric effect on other neurons, including ones not synaptically connected, or neurotransmitter signaling that is restricted to synaptically connected partners. Two sets of experiments demonstrated that Act-Int-mediated activation of postsynaptic neurons is restricted to synaptically connected neurons [2]. First, whole-cell patch clamp recordings of postsynaptic neurons showed Interluminescence-mediated firing only in neurons that also could be activated by endogenous neurotransmitters released upon activation of presynaptic luciferase-expressing neurons, indicating synaptically connected neurons (Fig. 6.6a–e). Second, transfer of culture medium "conditioned" with luciferases released upon neuron activation to a culture of neurons expressing opsins did not result in firing of opsin-expressing neurons upon addition of luciferin, as determined by fluorescent calcium indicator jGCaMP8s [10] measurements (Fig. 6.6f–i). Both experiments are strongly indicative of Act-Int performing like a neurotransmitter in a synapse-specific manner, rather than like a neuropeptide via volumetric effects.

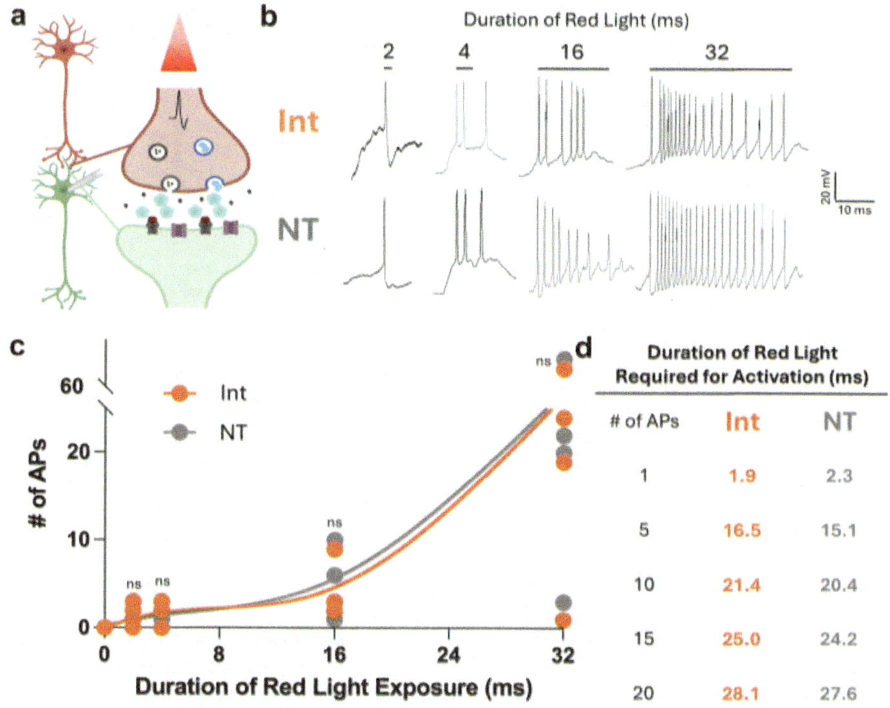

Duration of Red Light Required for Activation (ms)		
# of APs	Int	NT
1	1.9	2.3
5	16.5	15.1
10	21.4	20.4
15	25.0	24.2
20	28.1	27.6

Fig. 6.5 Similar presynaptic activity required for Act-Int and neurotransmitter-mediated synaptic transmission. (**a**) Act-Int schematic: Presynaptic luciferase-expressing neurons were activated by red light, releasing vesicle bound luciferase and neurotransmitter (NT) molecules. To isolate Act-Int effects without NT effects on postsynaptic neurons, synaptic blockers (SB) together with the luciferin CTZ were added. (**b**) Experimental design: Presynaptic neurons were activated by exposure to red light of varying durations (2–32 ms) and postsynaptic action potentials (APs) elicited by Act-Int (Int) or by endogenous NT were recorded. (**c**) Postsynaptic neurons were recorded after a single exposure to red light of a given length ($N = 3$ to 5 recordings for each light exposure time, each data point is one neuron, NT and Int data for each respective time point are from the same neuron). Restricted cubic smoothing spline with 2000 points was used to interpolate data. No significant differences in the resulting number of action potentials between Int and NT recordings were found at each time point. (**d**) Using interpolation data points of a smoothing spline, the durations of red light activation required to elicit 1, 5, 10, 15, and 20 Int and NT action potentials were identified. (Figure from Ref. [2])

6.3 Persist-Int

In addition to the activity-dependent modulation of synaptic transmission, there are applications for a strategy that provides activity-independent, sustained modulation, for example, across a behavioral session or for inverting the direction of synaptic transmission. For this “Persist-Int” strategy, the luciferase is continuously present in the synaptic cleft by tethering it to the presynaptic membrane; light emission and activation of postsynaptic opsins occur whenever luciferin is present. To target the

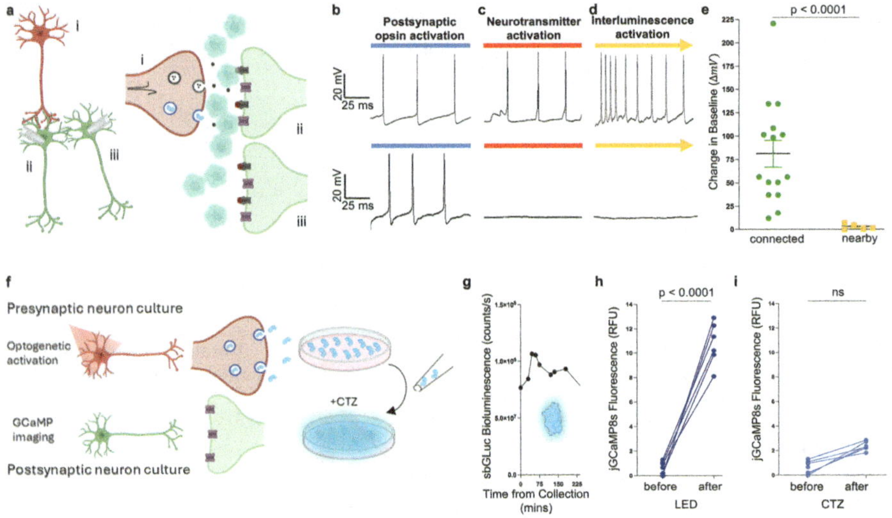

Fig. 6.6 Synapse specificity of Act-Int. (**a**) Experimental design: Postsynaptic opsin-expressing neurons were identified by their green fluorescence and patched. Neurons were either synaptically connected to luciferase and ChrimsonR-expressing neurons or not. (**b**) Recording from opsin-expressing neurons under blue light (480 nm) exposure revealed their level of opsin expression. (**c**) Recording from opsin-expressing neurons under red light (580 nm) exposure showed the postsynaptic neuron's neurotransmitter response to synaptically connected (upper trace) versus non-connected (lower trace) presynaptic neurons. (**d**) Recording from opsin-expressing neurons under red light exposure in the presence of synaptic blockers and CTZ allows isolation of the postsynaptic neuron's response to bioluminescence-mediated opsin activation. A response is elicited only in the synaptically connected (upper trace) but not in the nearby non-connected (lower trace) neuron. (**e**) Summary of recordings conducted in the presence of synaptic blockers and exposure to red light (1 s) showing the change in baseline membrane potential when CTZ was added to the recording bath. Changes in membrane potential are only seen in synaptically connected neurons in the presence of CTZ. (**f**) Schematic of conditioned media transfer experiment. Cultures of neurons expressing presynaptic construct hPOMC sbGLuc p2a ChrimsonR dTom were stimulated with continuous red light and media containing vesicle released sbGLuc was collected and added to separate cultures of neurons expressing postsynaptic construct ChR2(C128S) as well as jGCaMP8s, luciferin was added and change in jGCaMP8s fluorescence as a proxy for neuronal activation was measured. (**g**) Presynaptic media containing sbGLuc was removed from cultures and bioluminescence was measured at different time points from collection, demonstrating stability of the luciferase. (**h**) Control cultures of postsynaptic opsin-expressing neurons were stimulated with blue LED (470 nm) to test jGCaMP8s fluorescence in response to standard optogenetic activation. There was a significant increase in fluorescence compared to baseline. (**i**) jGCaMP8s fluorescence of wells with postsynaptic opsin-expressing neurons given presynaptic sbGLuc containing media and CTZ was measured and compared to baseline. There was no significant increase in fluorescence. (Figure from Ref. [2])

luciferase to the extracellular face of the presynaptic membrane, we used a fusion protein of mNeonGreen and the NanoLuc variant eKL9h (GeNL_eKL9h, a bright molecularly evolved FRET construct [9]) connected to ICAM-Nrxn3b [11] via the 15-amino acid linker used in luminopsins [12]. Nrxn3b (a presynaptic adhesion protein neurexin variant) tethered GeNL_eKL9h to the extracellular presynaptic membrane, and the extracellular domain of ICAM-1 was included to extend the light emitter further into the synaptic cleft [11, 13]. In multielectrode array recordings from cultured primary cortical neurons expressing the luciferase and striatal neurons expressing excitatory or inhibitory opsins, addition of luciferin caused the predicted change in neural activity [2]. In vivo efficacy of the Persist-Int approach was demonstrated in neural activity reporter mice. The TRAP2 (c-fos-CreER):: lox-stop-lox-EYFP double transgenic mouse model expresses EYFP when c-fos expression increases [14, 15]. We employed viral delivery, using stereotaxic surgery, to express the luciferase construct (AAV2/9-hSyn-GeNL_SS-linker-CD4-Nrxn3b) in the cortex (Fig. 6.7a, b) and the excitatory opsin construct (AAV2/9-hSyn-ChR2(C128S)), together with AAV2/9-CAG-tdTomato in the dorsal striatum of double transgenic mice (Fig. 6.7a, c). To maximize Interluminescence for in vivo experiments, we used GeNL_SS, a molecularly evolved bright FRET construct consisting of mNeonGreen fused to NanoLuc variant SSLuc [9]. Administration of the luciferin fluorofurimazine (FFz), but not of vehicle, resulted in robust expression of the EYFP reporter in opsin-expressing striatal neurons, indicating Persist-Int-driven activation [2] (Fig. 6.7c, d).

While for Act-Int the luciferase can cross the synaptic cleft and interact with the opsin close enough to reach FRET-distance, for Persist-Int, the distance between luciferase and opsin is determined by the extent to which the luciferase reaches from the presynaptic neuron membrane into the synaptic cleft. Given this, two parameters need to be taken into consideration when designing Persist-Int constructs: the brightness of the luciferase light emission and the length of the sequence reaching into the synapse (Fig. 6.8). NanoLuc-ICAM-Nrxn3b, GeNL_eKL9h-ICAM-Nrxn3b, and GeNL_SS-ICAM-Nrxn3b contain luciferases of increasing brightness (SSLuc > eKL9h > NanoLuc; SSLuc and eKL9h are brighter variants of NanoLuc [9]) (Fig. 6.8a). The largest ChR2(C128S) postsynaptic responses induced by light were recorded from GeNL_SS-ICAM-Nrxn3b expressing cocultures, and luciferase brightness was positively correlated with the size of membrane depolarization induced by bioluminescent stimulation [2] (Fig. 6.8b). We also varied the type and length of sequences between luciferase and the membrane targeting protein. The constructs used were GeNL_SS-ICAM-Nrxn3b, GeNL_SS-CD4-Nrxn3b, and SSLuc-B7, and overall construct lengths ranged from 1292 to 272 amino acids (Fig. 6.8c). Whole-cell postsynaptic responses in ChR2(C128S)-expressing cells were largest for the longest ICAM sequence. Sequence length was positively correlated with the size of the membrane depolarization induced by bioluminescent light stimulation (Fig. 6.8d).

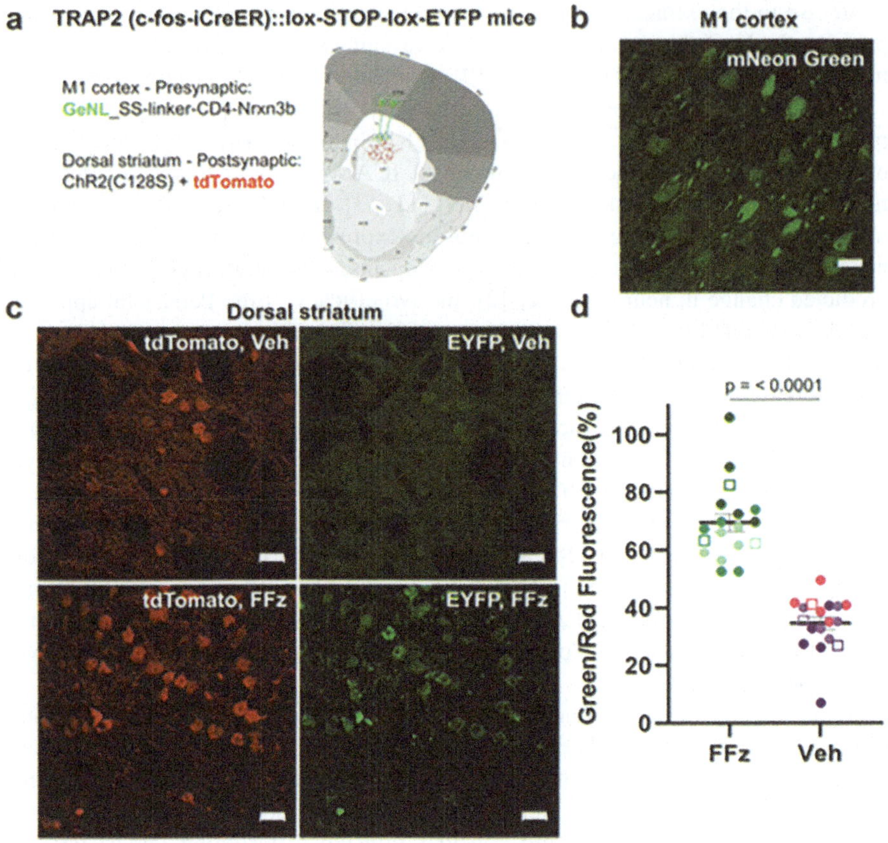

Fig. 6.7 Persist-Int *in vivo* (**a**) Experimental design. TRAP2::EYFP mice were transduced into presynaptic neurons in the M1 motor cortex with AAV2/9-hSyn carrying the sequences for Persist-Int, GeNL_SS-linker-CD4-Nrxn3b, and into postsynaptic neurons of the dorsolateral striatum with a mixture of AAV2/9-hSyn of ChR2(C128S) and the fluorescent reporter tdTomato. Two weeks later, mice were injected intraperitoneally with tamoxifen and luciferin (FFz) or vehicle. Brains were collected and sections were imaged under a fluorescence microscope. (**b**) Fluorescent image of the M1 cortex showing mNeonGreen positive presynaptic neurons. (**c**) Application of luciferin (FFz) results in a significant increase in postsynaptic neuron firing as reflected in the increased EYFP signal in opsin-expressing neurons in the dorsolateral striatum receiving synaptic input from Persist-Int expressing neurons (lower panel) compared to mice receiving vehicle (upper panel). (**d**) EYFP signal was quantified as the percent green fluorescence over red fluorescence from the tdTomato reporter. $N = 3$ mice per group, five sections per mouse. Measurements from individual sections are displayed by dots in 3 different color shades, averages from five sections used for statistical analysis are displayed as open squares. (Figure from Ref. [2])

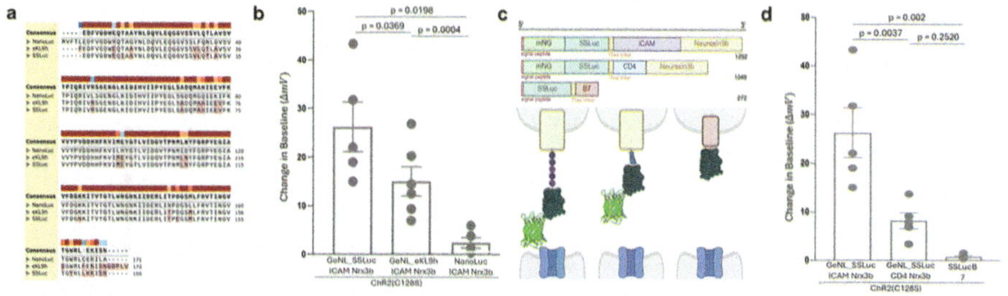

Fig. 6.8 Optimizing Persist-Int constructs. (**a**) Comparison of sequences of NanoLuc, eKL9h, and SSLuc. Mutations were introduced into NanoLuc to yield brighter variants eKL9h and SSLuc. Luciferase brightness is as follows: NanoLuc < eKL9h < SSLuc. (**b**) Change in baseline membrane voltage in postsynaptic ChR2(C128S)-expressing neurons was compared for different presynaptic Persist-Int constructs with the format of Luciferase-ICAM-Nrx3b. Increasing luciferase brightness was positively correlated with an increase in the change in membrane voltage. (**c**) Different sequences were used to tether SSLuc to the extracellular surface of the presynaptic membrane and facing into the synaptic cleft. Sequences differed in length and composition (top) and potentially reached different distances across the synaptic cleft (bottom schematic). (**d**) The change in baseline membrane voltage in postsynaptic neurons expressing ChR2(C128S) induced by luciferin-mediated light emission from Persist-Int presynaptic constructs. The change in the postsynaptic response in GeNL-SSLuc-ICAM-Nrx3b expressing cocultures was significantly higher compared to GeNL-SSLuc-CD4-Nrx3b and SSLuc-B7 expressing cocultures, while postsynaptic responses in GeNL_SS-CD4 and SSLuc-B7 co-cultures were not significantly different. (Figure adapted from Ref. [2])

6.4 Effects on Postsynaptic Neurons

To make Interluminescence useful to a broad community, users need to know the impact of Act-Int and Persist-Int on individual neurons. To this end, recordings were conducted from individual postsynaptic neurons in vitro, comparing neurotransmitter-induced and Interluminescence-induced action potentials (APs) [2] (Fig. 6.9). We used whole-cell recording to monitor postsynaptic membrane potentials in cocultures of presynaptic cortical (luciferase-expressing) and postsynaptic striatal (opsin-expressing) neurons. Cortical neurons were nucleofected with the presynaptic constructs for Act-Int (pcDNA-CAG-POMC sbGLuc-P2A-Chrimson-dTom) or Persist-Int (pcDNA-CAG-GeNL_eKL9h-linker-ICAM-Nrxn3b-p2a-dTom), and striatal neurons were nucleofected with the construct for the excitatory opsin ChR2(C128S) (pcDNA CAG ChR2(C128S) EYFP), unless indicated otherwise. We compared features of postsynaptic APs mediated by Act-Int and Persist-Int (Int-APs) to control postsynaptic APs generated by neurotransmitters in the same recordings (NT-APs) (Fig. 6.9a–c). We did not find statistically significant differences in the properties of postsynaptic APs mediated by Act-Int and Persist-Int, and most parameters were similar between Int-APs and NT-APs, including AP latency, threshold, and peak amplitude [2]. To further compare Act-Int and Persist-Int we quantified the change in base membrane potential induced by each method by measuring the difference between resting membrane potential (i.e., membrane

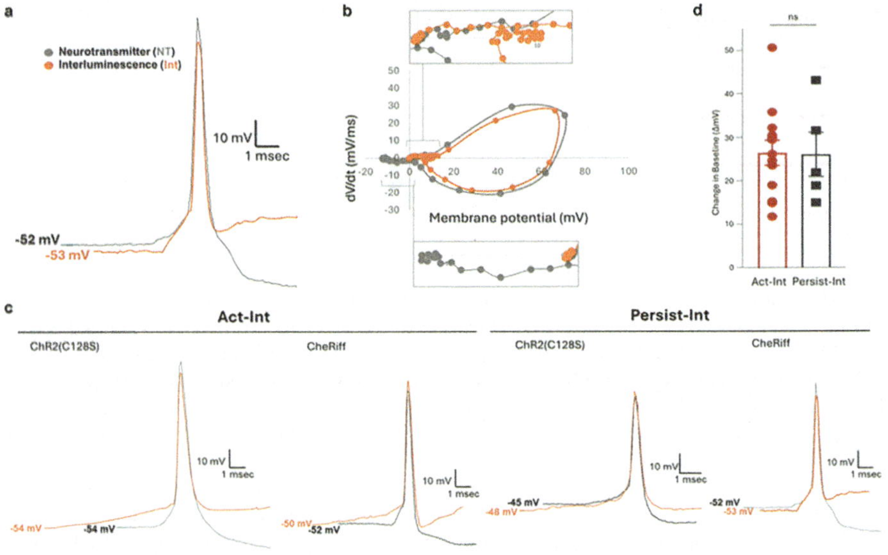

Fig. 6.9 Interluminescence- and neurotransmitter-induced postsynaptic action potentials. (**a**) Representative APs for Int (orange) and NT (gray) and (**b**), their dV/dt plots. (**c**) Representative APs elicited through Act-Int and Persist-Int (orange) in comparison to NT-induced (gray) in postsynaptic neurons expressing different excitatory opsins. (**d**) For recordings of ChR2(C128S)-expressing postsynaptic neurons, the change in baseline membrane voltage (ΔmV) for Act-Int (POMC sbGLuc) and Persist-Int (GeNL_SS-ICAM) is not significantly different. (Figure adapted from Ref. [2])

potential before Int onset) and shifted baseline membrane potential at the saturation level during Int, (ΔmV) [16, 17]. We recorded the change in baseline (in mV) after Int of ChR2(C128S)-expressing postsynaptic neurons paired with either POMC sbGLuc (Act-Int) or GeNL_SS-ICAM-Nrxn3b (Persist-Int). No significant difference in postsynaptic activation was observed between the two Int paradigms (Fig. 6.9d).

6.5 Summary

Interluminescence is a versatile and modular platform providing synapse-specific transmission strategies that can be used to achieve transsynaptic communication between partners of choice that is either activity-dependent or activity-independent. The modularity of this method allows expansion into combining the growing arsenal of potent luciferases with the ever-increasing toolset of optogenetic elements, affording experiment-specific tuning of approaches. Each version of Int offers unique features for applications. Act-Int is based on the presynaptic neuron's activity and thus is embedded in the brain's endogenous circuit dynamics, synthetically altering the dynamics and flow of internally generated neuronal activity, while Persist-Int can override any status of the presynaptic component, either activity or inactivity. Both modalities allow a new regime of experiments for changing or rerouting information flow.

While tested in neurons, the use of genetically encoded light sources and light sensors for communication between cells is a powerful tool for modulating information exchange in potentially many biological systems. Light emitter and light sensor can be targeted to cells that are normally partnered, or new interactions can be set up. Communication between cells can enhance or counteract endogenous information flows, or the direction of information can be reversed. The technology platform provides a new means of interrogating communication in biological systems.

References

1. Prakash M, Murphy J, St Laurent R, Friedman N, Crespo EL, Bjorefeldt A, et al. Selective control of synaptically-connected circuit elements by all-optical synapses. Commun Biol. 2022;5(1)
2. Slaviero AN, Prakash M, Bell E, Schumaker R, Crespo EL, Tree MO, et al. Control of synaptic communication through molecularly engineered bioluminescence light emission and sensing. bioRxiv. 2025:2025.10.27.684942.
3. Rizo J, Xu J. The synaptic vesicle release machinery. Annu Rev Biophys. 2015;44(1):339–67.
4. Martens H, Weston MC, Boulland JL, Grønborg M, Grosche J, Kacza J, et al. Unique luminal localization of VGAT-C terminus allows for selective labeling of active cortical GABAergic synapses. J Neurosci. 2008;28(49):13125–31.
5. Cawley NX, Li Z, Loh YP. 60 YEARS OF POMC: biosynthesis, trafficking, and secretion of pro-opiomelanocortin-derived peptides. J Mol Endocrinol. 2016;56(4):T77–97.
6. Chaudhry FA, Reimer RJ, Bellocchio EE, Danbolt NC, Osen KK, Edwards RH, et al. The vesicular GABA transporter, VGAT, localizes to synaptic vesicles in sets of glycinergic as well as GABAergic neurons. J Neurosci. 1998;18(23):9733–50.
7. Baumert M, Maycox PR, Navone F, De Camilli P, Jahn R. Synaptobrevin: an integral membrane protein of 18,000 daltons present in small synaptic vesicles of rat brain. EMBO J. 1989;8(2):379–84.
8. Winkler H. Membrane composition of adrenergic large and small dense cored vesicles and of synaptic vesicles: consequences for their biogenesis. Neurochem Res. 1997;22(8):921–32.
9. Lambert GG, Crespo EL, Murphy J, Boassa D, Luong S, Celinskis D, et al. CaBLAM! A high-contrast bioluminescent Ca(2+) indicator derived from an engineered Oplophorus gracilirostris luciferase. Nat Methods. 2026;23(1):205–15.
10. Zhang Y, Rózsa M, Liang Y, Bushey D, Wei Z, Zheng J, et al. Fast and sensitive GCaMP calcium indicators for imaging neural populations. Nature. 2023;615(7954):884–91.
11. Kim CK, Cho KF, Kim MW, Ting AY. Luciferase-LOV BRET enables versatile and specific transcriptional readout of cellular protein-protein interactions. Elife. 2019;8:e43826.
12. Berglund K, Birkner E, Augustine GJ, Hochgeschwender U. Light-emitting channelrhodopsins for combined optogenetic and chemical-genetic control of neurons. PLoS One. 2013;8(3):e59759.
13. Talay M, Richman EB, Snell NJ, Hartmann GG, Fisher JD, Sorkaç A, et al. Transsynaptic mapping of second-order taste neurons in flies by trans-Tango. Neuron. 2017;96(4):783–795.e4.

14. DeNardo LA, Liu CD, Allen WE, Adams EL, Friedmann D, Fu L, et al. Temporal evolution of cortical ensembles promoting remote memory retrieval. Nat Neurosci. 2019;22(3):460–9.
15. Srinivas S, Watanabe T, Lin CS, William CM, Tanabe Y, Jessell TM, et al. Cre reporter strains produced by targeted insertion of EYFP and ECFP into the ROSA26 locus. BMC Dev Biol. 2001;1:4.
16. Marciante AB, Farmer GE, Cunningham JT. G(q) DREADD activation of CaMKIIa MnPO neurons stimulates nitric oxide activity. J Neurophysiol. 2020;124(2):591–609.
17. Lee JI, Werginz P, Kameneva T, Im M, Fried SI. Membrane depolarization mediates both the inhibition of neural activity and cell-type-differences in response to high-frequency stimulation. Commun Biol. 2024;7(1):734.

Chapter 7
Neurotransmitter Sensor-Driven Neuromodulation

Kaylee Taylor, Jenna Covell, Zina Pinderi, Hunter Galvin, Brevin St. Onge, and Eric D. Petersen

Abstract Neurotransmitters (NTs) play an important role in modulating global brain processes, such as arousal, attention, or emotion, and thus behavior. As a result, they are central to the pathogenesis and treatment of human neurological and psychiatric disorders, including Parkinson's disease, Alzheimer's disease, schizophrenia, depression, and addiction. In an effort to understand these pathways, sensors have been developed for NTs, including glutamate (Glu), γ-aminobutyric acid (GABA), acetylcholine (ACh), dopamine (DA), and serotonin. Thus far, most published genetically encoded NT sensors have been based on fluorescent proteins. Our lab has been working to develop NT sensors based on bioluminescence to take advantage of several qualities, including not requiring an excitation light source to enable minimally invasive imaging of a population of cells, lack of phototoxicity, and the ability for these light-emitting sensors to drive optogenetic actuators. We previously reported the development of a bioluminescent glutamate sensor to expand current capabilities into mesoscale recording of glutamate dynamics with a sensor that increases bioluminescent light emission in response to increased glutamate levels. We recently expanded on this initial glutamate sensor to accomplish reporting of ACh, GABA, serotonin, and glucose. The ultimate goal is to not only use bioluminescent NT sensors to report neuronal activity but also to achieve NT-dependent modulation of neural circuits. By pairing the bioluminescent NT sensors that we have developed with light-sensitive ion channels or pumps, we create NT-dependent luminopsins (NT-LMOs). The goal is to apply these NT-LMOs for selective synaptic modulation to either amplify or inhibit the postsynaptic response of a specific NT.

Keywords Synapse · Synaptic transmission · Sensor · Luciferase · Bioluminescence

---

K. Taylor · J. Covell · Z. Pinderi · H. Galvin · B. St. Onge · E. D. Petersen (✉)
College of Medicine, Central Michigan University, Mt. Pleasant, MI, USA
e-mail: peter1ed@cmich.edu

U. Hochgeschwender (ed.), *BioLuminescent OptoGenetics*,
https://doi.org/10.1007/978-3-032-18121-3_7

7.1 Introduction

Neural transmission of information relies on voltage, calcium, and neurotransmitter dynamics. All of which can be used to report changes in activity with genetically encoded sensors. While genetically encoded indicators of voltage [1–3] and especially of calcium have been under continuous iterative development for decades, neurotransmitter (NT) sensors have only more recently become available [4]. Reflecting their important role in modulating global brain processes such as arousal, attention, or emotion, and thus behavior as well as their involvement in the pathogenesis or treatment of neurological and psychiatric disorders, including Parkinson's disease, schizophrenia, depression, migraines, and addiction tools are needed to study NT dynamics. Recently, sensors have been developed for neurotransmitters, including glutamate [5, 6], GABA [7, 8], acetylcholine [9, 10], dopamine [11, 12], and serotonin [13]. Thus far, the majority of published NT sensors have been based on fluorescent proteins.

Our lab has been working to create a series of genetically encoded NT sensors based on bioluminescence. With bioluminescence, light is produced by an enzymatic reaction of luciferases with their substrates (luciferins). This feature negates the need for an excitation light source, thereby eliminating major limitations of fluorescent imaging, such as phototoxicity, autofluorescence, and excitation light-scattering to overcome depth and volume barriers (Fig. 7.1a). In addition, bioluminescence does not require the use of a fiber for an excitation light source when accessing deep brain regions. This makes it much less invasive and allows population-wide recording because the volume of imaging is not limited to the area

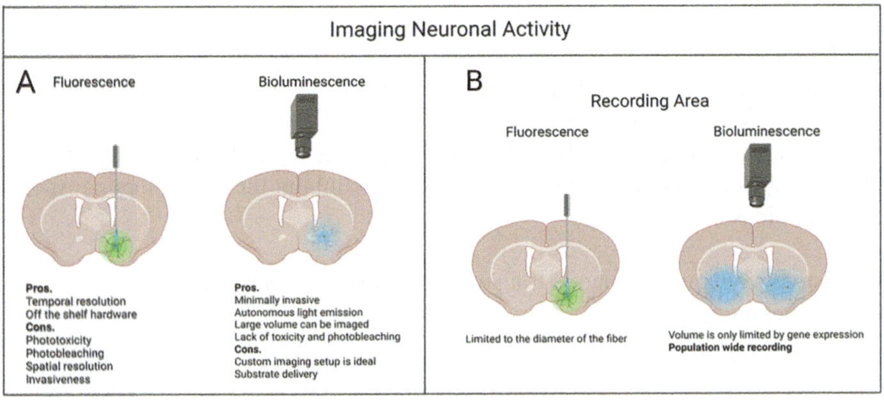

Fig. 7.1 Imaging neuronal activity with fluorescence vs bioluminescence. (**a**) Bioluminescence has many advantages over fluorescence, including lower invasiveness, autonomous light emission, a large recording volume, and a lack of phototoxicity and photobleaching. (**b**) The recording area that is possible with fluorescence is limited to the diameter of the fiber or field of view, whereas with bioluminescence, a whole population of cells can be recorded from, as gene expression of the cells is the only limiting factor

around the tip of the fiber and is only limited by the volume of gene expression (Fig. 7.1b). Critically, the use of bioluminescence can enable noninvasive imaging of subcortical structures such as the hippocampus [14, 15], basal ganglia [16–18], and the hypothalamus [19]. Bioluminescent sensors can enable noninvasive imaging at the population level in deep regions of the brain at 10 hertz or greater [20]. Given the attractive features of using bioluminescence to report activity, multiple groups have been working on creating various bioluminescent sensors. Current bioluminescent sensors used for recording neuronal activity primarily include calcium and voltage [21–26]. Despite the attractive advantages of using bioluminescence for recording NT dynamics, bioluminescent probes have received less attention than their fluorescent counterparts. However, we believe that bioluminescence imaging is poised to become an enabling technology due to advances in high-throughput screening, novel ultra-bright luciferases, innovations in chemical modifications of luciferins, and the development of low-cost one-photon mini-scopes.

7.2 Engineering Bioluminescent Neurotransmitter Sensors

We recently reported the development of a bioluminescent NT sensor to expand current capabilities into mesoscale recording of glutamate dynamics, with an increase in bioluminescent light emission in response to increased glutamate levels [27]. We recently expanded on our initial glutamate sensor to accomplish reporting of acetylcholine, GABA, serotonin, and glucose to accomplish meso- to macroscale reporting of activity and neuromodulation based on local levels of each specific neurotransmitter (Fig. 7.2). The sensors have a split-luciferase design, where a small portion of the luciferase is separated from the rest, and they are placed on either side of an NT binding domain using linkers. This allows the sensor to produce light in the presence of luciferin only when the NT binds and causes a conformational change that allows the luciferase to come together. The sensors are designed in a modular manner, so that each part can be removed and swapped with another component to improve the NT response or change what the sensor responds to. For example, the binding domain can be changed so that the sensor detects serotonin instead of ACh. To improve NT response, the length and rigidity of the linkers, changing the affinity of the small part of the luciferase to its other half, and the transmembrane domain can all be interchanged to find the best combination for the highest response.

7.3 Example Imaging Applications

In the adult brain, GABA acts as a major inhibitory neurotransmitter and works together with Glu to control neuronal activity. In the developing brain, however, the roles of GABA and Glu are reversed due to high intracellular chloride levels. This

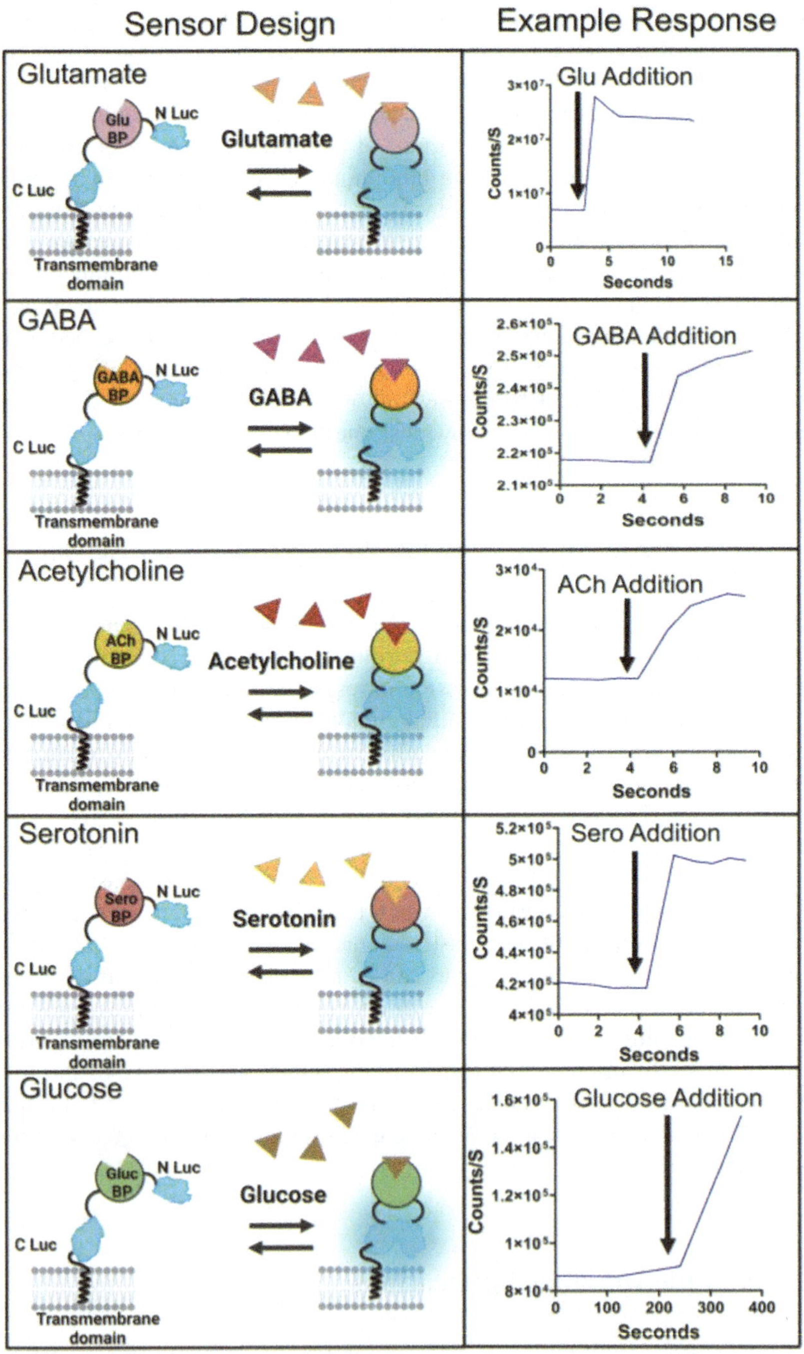

Fig. 7.2 We have developed sensors for glutamate, GABA, acetylcholine, serotonin, and glucose. In the presence of luciferin, these sensors are dim in the unbound state. When their respective neurotransmitter bind to its specific binding domain, it brings together the two halves of the luciferase, resulting in increased light emission. For each sensor, this is shown in the brightness over time graphs to the right (Figure adapted from Ref. [27])

makes GABA very important for proper development, and early abnormalities in GABA are seen in periventricular leukomalacia and schizophrenia [28]. In adults, GABA abnormalities play a role in many neurodegenerative disorders like Parkinson's disease, and mature brain injury such as transient ischemia, stroke, and post-traumatic epilepsy [28, 29]. The ability to image large populations of GABAergic neurons with our GABA sensor can help provide insight into how they interact with other neurons in the brain to give us a better understanding of disease pathology. ACh can act as a neuromodulator, changing the state of neurons based on environmental conditions. This makes it very important in the regulation of many behaviors and learning and memory in discrete pathways in the brain, and cholinergic dysregulation is implicated in many neurodegenerative diseases like Alzheimer's disease and mood disorders, such as stress and depression [30]. Because of how widespread it is, being able to noninvasively image ACh activity in the brain would give us crucial insights into its effects on the brain and how that plays into disease. Serotonin detection specifically could prove useful in migraine pathophysiology, a neurological disease process that is common but not well understood. Current research allows for the peripheral detection of serotonin and its metabolites, which cannot be used as a direct comparison to deep-brain processes [31, 32]. An in vivo analysis of serotonin in the migraine brain could allow novel information about the pathogenesis of the disease to be revealed, and further treatment options. Serotonin is also involved in other neuropsychiatric disorders, such as autism spectrum disorder, obsessive-compulsive disorder, attention-deficit hyperactivity disorder, and carcinoid syndrome, so this sensor could be used to further study the wide variety of serotonin-related diseases [33, 34].

7.4 NT Sensors for Optogenetic Control

While recording neural activity is essential for broadening our understanding of the neurobiology underlying behavior, control of neural activity enables targeted interrogation of mechanistic concepts and testing of specific hypotheses. Our ultimate goal is to not only use our bioluminescent NT sensors to report activity but also to achieve NT-dependent modulation of neural circuitry. Currently, there is no other technology that enables NT-specific manipulation of neuronal activity in genetically defined populations that also relies on external experimenter control. We have developed such an approach by using NT-dependent light emission to activate optogenetic elements where any of our sensors can be tethered to either an excitatory or inhibitory opsin (Fig. 7.3). This builds on our and others' prior work developing and implementing BioLuminescent OptoGenetics (BL-OG), where luciferases are used to activate opsins either when tethered to the opsin in luminopsins (LMOs) [35–50] or to enable synaptic communication via interluminescence with the luciferase expressed pre- and the opsin postsynaptically [51]. These approaches have been used to study brain circuitry [48] and show promise when applied to

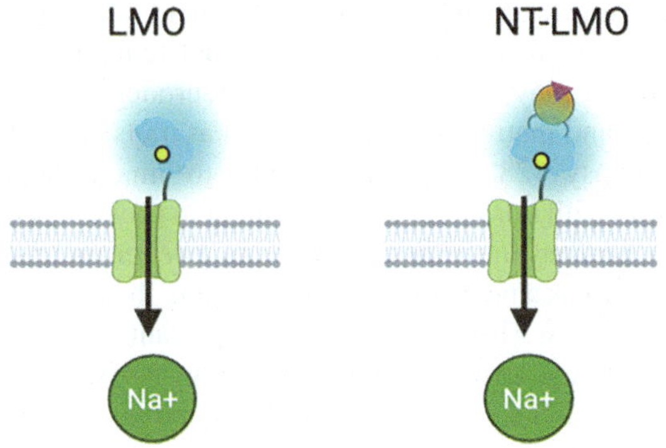

Fig. 7.3 Design of NT-LMOs. A traditional Luminopsin, LMO, pictured on the left, consists of an optogenetic channel, in this case excitatory, fused to an intact luciferase. Our NT-LMOs consist of a bioluminescent neurotransmitter sensor fused to the optogenetic channel so that the channel is now only activated in the presence of both the luciferin and the NT the sensor responds to

neurodegenerative diseases, such as spinal cord injury [46] and Parkinson's disease models [38, 40].

These NT-LMOs allow the modulation of activity that can be controlled by the delivery of luciferin because the channel can only be activated when both NT and luciferin are present to allow the sensor to produce light in response to the NT of interest and thus activate the optogenetic channel. This system functions as a logic gate, requiring both the presence of the luciferin AND the endogenous release of the NT to hyperactivate or hypoactivate the circuit of interest. Experimenter-driven temporal control is achieved by determining the time frame during which the luciferin is provided and thus the window for neuromodulation to occur, while spatial control over subcircuits is provided by restricting the expression of NT-LMOs to pathways of interest. This provides a unique opportunity to interrogate as well as tune NT-specific neuromodulation. For example, for our excitatory ACh-LMO, we tested using patch clamp in HEK 293 cells (Fig. 7.4a). This data from an excitatory ACh-LMO shows inward current conduction with luciferin alone with a much larger amplitude for luciferin and ACh with a 3x greater response to ACh+luciferin compared to luciferin alone (Fig. 7.4b–e).

7.5 Neuromodulation Applications

The glutamatergic, GABAergic, serotonergic, and cholinergic systems are involved in a large variety of brain functions, from learning and memory, sleep and anxiety regulation to motor output and control. Neurotransmitter dysfunction is a hallmark of neurodegenerative diseases, like Alzheimer's, Parkinson's, and Huntington's disease. We have been working to pair each NT sensor above with both excitatory and inhibitory channels for a comprehensive toolset of NT-LMOs.

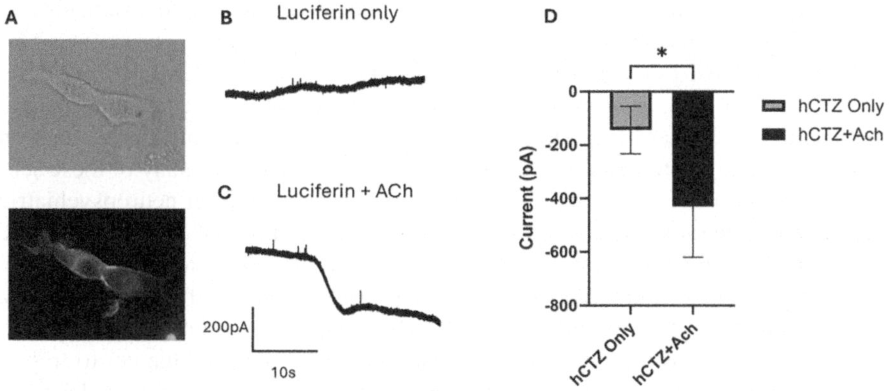

Fig. 7.4 Patch clamp data of ACh sensor paired with an excitatory channel (**a**) Brightfield (top) and fluorescent (bottom) image showing the cell is expressing our EYFP-tagged ACh-NT-LMO. (**b**) Current response to luciferin (hCTZ) alone (18.3 pA). (**c**) Current response to luciferin and ACh (253.3 pA). (**d**) Quantification of patch recordings showing that luciferin plus ACh induces about 3x larger inward current compared to luciferin alone ($p < 0.05$)

NT actuators enable several crucial lines of investigation. First, they allow functional testing of synaptic information flow. Changing the valence of connections in networks and making typically excitatory connections into suppressive ones or vice versa allow researchers to selectively suppress or amplify specific biological events, a unique approach to understanding the role of those processes. An example for application is the interrogation, in mice, of synaptic inputs to defined populations of cells across brain regions. By expressing the NT actuator in defined postsynaptic cell populations, each can be sampled for its functional significance on information flow by interrogating the effects on animal behavior. Such functional testing of NT-specific synaptic information flow will ultimately contribute toward functional connectivity maps. Second, co-transmission of neuropeptides and neurotransmitters represents a fundamental characteristic of neurons. However, their differential effects on postsynaptic neurons remain poorly understood. The use of NT actuators will allow us to disambiguate co-transmission of NTs and neuropeptides at specific synapses, experiments essential for understanding the processes and modulation underlying higher brain functions. Finally, NT actuators will facilitate a better understanding of pathway-specific neurochemical dysregulation in pathological states. Most pharmaceutical treatments for treating neuropsychiatric disorders either mimic or block the action of NTs and work to correct partially characterized or hypothesized dysfunction in the complex regulation of brain chemistry. The lack of knowledge about the functional effects of many of these pharmaceuticals on neuronal activity presents a significant gap in neuropsychiatric research. Further, NT actuators allow us to study functional under- or over-connectivity, which might underlie neuropsychiatric disorders, aiding in furthering our understanding of the pathophysiology of these disorders. Lastly, applied in animal models of neurological and psychiatric diseases, NT actuators can be used to rewire networks to overcome failures in communication, probing underlying disease mechanisms and potentially be used as therapeutics.

While we have pioneered the concept by developing an ACh-LMO, GABA-LMO, and Glu-LMO, we expect this approach to be broadly adaptable to neurotransmitters of interest, having broad applications for studying brain function while also having potential clinical applications. We have tested many of these sensors and opsin pairings that will have applications in a variety of neuropsychiatric and neurodegenerative diseases. For example, the excitatory ACh-LMO could be applied in Alzheimer's disease to amplify cholinergic transmission to compensate for the degenerated cholinergic circuitry and Myasthenia gravis if applied in neuromuscular junctions. Our inhibitory GABA-iLMO, where, in the presence of GABA, the GABA sensor would produce light that then causes an ion channel to hyperinhibit the postsynaptic neuron, could be used to treat a disease where low levels of GABAergic activity play a role, like Parkinson's disease, by compensating for the lost GABAergic drive. In another example, by pairing a Glu sensor with an inhibitory channel, the excessive Glu that would cause a seizure could be applied to transform pathologically high levels of glutamate to an inhibitory action, suppressing a seizure (Fig. 7.5).

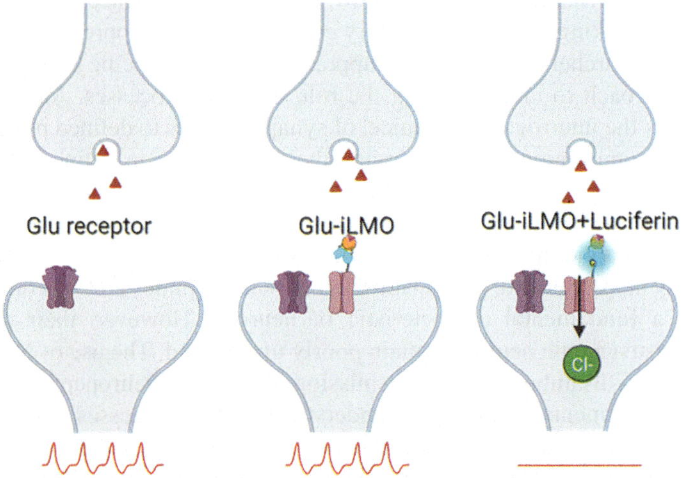

Fig. 7.5 An example of how the glutamate (Glu) inhibitory luminopsin (Glu-iLMO) can be applied to suppress transmission. Left panel: normal activity of the native glutamate receptor. Middle panel: expression of the Glu-iLMO does not alter the function of the synapse without luciferin present. Right panel: in the presence of luciferin, the Glu sensor produces light, which opens the light-sensitive inhibitory channels, dampening the postsynaptic response and thereby counteracting the normal valence of this synapse

7.6 Summary

The ability to noninvasively image neurotransmitter levels from distinct populations of neurons in the brain is important for the study of many diseases, including Alzheimer's, Parkinson's, epilepsy, and depression. Bioluminescent NT sensors enable this because they do not need an excitation light source, eliminating the need for implanted hardware within the brain. We designed sensors using a split-luciferase model, where two pieces of luciferase are tethered on either side of an NT binding domain. This allows the sensor to produce light when NT binding causes a conformational change that brings the luciferase halves together to reconstitute enzyme function. Taking advantage of the modular design of this sensor, we have used this same general design to create sensors for Glu, GABA, ACh, serotonin, and glucose. In addition to using these sensors for imaging, they can be paired with optogenetic channels to create NT-dependent LMOs that can modulate neuronal activity NT- and synapse-specifically to either excite or inhibit postsynaptic neural activity. These NT-LMOs have many applications, depending on which NT sensor and opsin pairing is used. We envision this to be an enabling approach to study neural circuitry and also have the potential to be applied to correct underlying NT pathologies present in a wide variety of neurodegenerative diseases.

Acknowledgments This work was supported by grants from the Michigan Economic Development Corporation (ADVANCE Grant), Diabetes Research Connection, and the National Institutes of Health (NIH R35GM156968).

References

1. Chamberland S, Yang HH, Pan MM, Evans SW, Guan S, Chavarha M, et al. Fast two-photon imaging of subcellular voltage dynamics in neuronal tissue with genetically encoded indicators. Elife. 2017;6:e25690.
2. Yang HH, St-Pierre F, Sun X, Ding X, Lin MZ, Clandinin TR. Subcellular imaging of voltage and calcium signals reveals neural processing in vivo. Cell. 2016;166(1):245–57.
3. Yang HH, St-Pierre F. Genetically encoded voltage indicators: opportunities and challenges. J Neurosci. 2016;36(39):9977–89.
4. Sabatini BL, Tian L. Imaging neurotransmitter and neuromodulator dynamics in vivo with genetically encoded indicators. Neuron. 2020;108(1):17–32.
5. Takikawa K, Asanuma D, Namiki S, Sakamoto H, Ariyoshi T, Kimpara N, et al. High-throughput development of a hybrid-type fluorescent glutamate sensor for analysis of synaptic transmission. Angew Chem Int Ed Engl. 2014;53(49):13439–43.
6. Marvin JS, Borghuis BG, Tian L, Cichon J, Harnett MT, Akerboom J, et al. An optimized fluorescent probe for visualizing glutamate neurotransmission. Nat Methods. 2013;10(2):162–70.
7. Masharina A, Reymond L, Maurel D, Umezawa K, Johnsson K. A fluorescent sensor for GABA and synthetic GABA(B) receptor ligands. J Am Chem Soc. 2012;134(46):19026–34.
8. Marvin JS, Shimoda Y, Magloire V, Leite M, Kawashima T, Jensen TP, et al. A genetically encoded fluorescent sensor for in vivo imaging of GABA. Nat Methods. 2019;16(8):763–70.
9. Schena A, Johnsson K. Sensing acetylcholine and anticholinesterase compounds. Angew Chem Int Ed Engl. 2014;53(5):1302–5.

10. Jing M, Zhang P, Wang G, Feng J, Mesik L, Zeng J, et al. A genetically encoded fluorescent acetylcholine indicator for in vitro and in vivo studies. Nat Biotechnol. 2018;36(8):726–37.
11. Sun F, Zeng J, Jing M, Zhou J, Feng J, Owen SF, et al. A genetically encoded fluorescent sensor enables rapid and specific detection of dopamine in flies, fish, and mice. Cell. 2018;174(2):481–96 e19.
12. Patriarchi T, Cho JR, Merten K, Howe MW, Marley A, Xiong WH, et al. Ultrafast neuronal imaging of dopamine dynamics with designed genetically encoded sensors. Science. 2018;360(6396)
13. Unger EK, Keller JP, Altermatt M, Liang R, Matsui A, Dong C, et al. Directed evolution of a selective and sensitive serotonin sensor via machine learning. Cell. 2020;183(7):1986–2002.e26.
14. Dunn-Meynell AA, Dowling P, Marchese M, Rodriguez E, Blumberg B, Choi YB, et al. In vivo bioluminescence imaging used to monitor disease activity and therapeutic response in a mouse model of tauopathy. Front Aging Neurosci. 2019;11:252.
15. Fricke IB, Schelhaas S, Zinnhardt B, Viel T, Hermann S, Couillard-Despres S, et al. In vivo bioluminescence imaging of neurogenesis – the role of the blood brain barrier in an experimental model of Parkinson's disease. Eur J Neurosci. 2017;45(7):975–86.
16. Kim DE, Tsuji K, Kim YR, Mueller FJ, Eom HS, Snyder EY, et al. Neural stem cell transplant survival in brains of mice: assessing the effect of immunity and ischemia by using real-time bioluminescent imaging. Radiology. 2006;241(3):822–30.
17. Tung JK, Gutekunst CA, Gross RE. Inhibitory luminopsins: genetically-encoded bioluminescent opsins for versatile, scalable, and hardware-independent optogenetic inhibition. Sci Rep. 2015;5:14366.
18. Aswendt M, Vogel S, Schafer C, Jathoul A, Pule M, Hoehn M. Quantitative in vivo dual-color bioluminescence imaging in the mouse brain. Neurophotonics. 2019;6(2):025006.
19. Cordonier EL, Liu T, Saito K, Chen SS, Xu Y, Fukuda M. Luciferase reporter mice for in vivo monitoring and ex vivo assessment of hypothalamic signaling of Socs3 expression. J Endocr Soc. 2019;3(7):1246–60.
20. Silvagnoli A, Taylor A, Slaviero A, Petersen E. A method for optimizing imaging parameters to record neuronal and cellular activity at depth with bioluminescence. Neurophotonics. 2024;11(2):024206.
21. Oh Y, Park Y, Cho JH, Wu H, Paulk NK, Liu LX, et al. An orange calcium-modulated bioluminescent indicator for non-invasive activity imaging. Nat Chem Biol. 2019;15(5):433–6.
22. Farhana I, Hossain MN, Suzuki K, Matsuda T, Nagai T. Genetically encoded fluorescence/bioluminescence bimodal indicators for Ca2+ imaging. ACS Sens. 2019;4(7):1825–34.
23. Inagaki S, Agetsuma M, Ohara S, Iijima T, Yokota H, Wazawa T, et al. Imaging local brain activity of multiple freely moving mice sharing the same environment. Sci Rep. 2019;9(1):7460.
24. Inagaki S, Tsutsui H, Suzuki K, Agetsuma M, Arai Y, Jinno Y, et al. Genetically encoded bioluminescent voltage indicator for multi-purpose use in wide range of bioimaging. Sci Rep. 2017;7:42398.
25. Suzuki K, Kimura T, Shinoda H, Bai G, Daniels MJ, Arai Y, et al. Five colour variants of bright luminescent protein for real-time multicolour bioimaging. Nat Commun. 2016;7:13718.
26. Qian Y, Rancic V, Wu J, Ballanyi K, Campbell RE. A bioluminescent Ca(2+) indicator based on a topological variant of GCaMP6s. Chembiochem. 2019;20(4):516–20.
27. Petersen ED, Lapan AP, Castellanos Franco EA, Fillion AJ, Crespo EL, Lambert GG, et al. Bioluminescent genetically encoded glutamate indicators for molecular imaging of neuronal activity. ACS Synth Biol. 2023;12(8):2301–9.
28. Wu C, Sun D. GABA receptors in brain development, function, and injury. Metab Brain Dis. 2015;30(2):367–79.
29. Błaszczyk JW. Parkinson's disease and neurodegeneration: GABA-collapse hypothesis. Front Neurosci. 2016;10:269.
30. Picciotto MR, Higley MJ, Mineur YS. Acetylcholine as a neuromodulator: cholinergic signaling shapes nervous system function and behavior. Neuron. 2012;76(1):116–29.

31. Ferrari MD, Saxena PR. On serotonin and migraine: a clinical and pharmacological review. Cephalalgia. 1993;13(3):151–65.
32. Ribeiro CA, Cotrim MD, Morgadinho MT, Ramos MI, Santos ES, de Macedo TR. Migraine, serum serotonin and platelet 5-HT2 receptors. Cephalalgia. 1990;10(5):213–9.
33. Pourhamzeh M, Moravej FG, Arabi M, Shahriari E, Mehrabi S, Ward R, et al. The roles of serotonin in neuropsychiatric disorders. Cell Mol Neurobiol. 2022;42(6):1671–92.
34. Gade AK, Olariu E, Douthit NT. Carcinoid syndrome: a review. Cureus. 2020;12(3):e7186.
35. Berglund K, Birkner E, Augustine GJ, Hochgeschwender U. Light-emitting channelrhodopsins for combined optogenetic and chemical-genetic control of neurons. PloS One. 2013;8(3):e59759.
36. Berglund K, Clissold K, Li HE, Wen L, Park SY, Gleixner J, et al. Luminopsins integrate opto- and chemogenetics by using physical and biological light sources for opsin activation. Proc Natl Acad Sci U S A. 2016;113(3):E358–E67.
37. Birkner E, Berglund K, Klein ME, Augustine GJ, Hochgeschwender U. Non-invasive activation of optogenetic actuators. Proc SPIE Int Soc Opt Eng. 2014;8928:89282F.
38. Park SY, Song S-H, Palmateer B, Pal A, Petersen ED, Shall GP, et al. Novel luciferase–opsin combinations for improved luminopsins. J Neurosci Res. 2020;98(3):410–2139.
39. Tung JK, Shiu FH, Ding K, Gross RE. Chemically activated luminopsins allow optogenetic inhibition of distributed nodes in an epileptic network for non-invasive and multi-site suppression of seizure activity. Neurobiol Dis. 2018;109:1–10.
40. Zenchak JR, Palmateer B, Dorka N, Brown TM, Wagner L-M, Medendorp WE, et al. Bioluminescence-driven optogenetic activation of transplanted neural precursor cells improves motor deficits in a Parkinson's disease mouse model. J Neurosci Res. 2020;98:458–68.
41. Gutekunst C-A, Tung JK, Gross RE. Inhibitory luminopsins: genetically-encoded bioluminescent opsins for versatile, scalable, and hardware-independent optogenetic inhibition. Sci Rep. 2015;5:14366.
42. Tung JK, Berglund K, Gutekunst C-A, Hochgeschwender U, Gross RE. Bioluminescence imaging in live cells and animals. Neurophotonics. 2016;3(2):025001.
43. Berglund K, Tung JK, Higashikubo B, Gross RE, Moore CI, Hochgeschwender U. Combined optogenetic and chemogenetic control of neurons. Methods Mol Biol (Clifton, NJ). 2016;1408:207–25.
44. Yu SP, Tung JK, Wei ZZ, Chen D, Berglund K, Zhong W, et al. Optochemogenetics stimulation of transplanted iPS-NPCs enhances neuronal repair and functional recovery after ischemic stroke. J Neurosci. 2019:2010-18.
45. Prakash M, Medendorp WE, Hochgeschwender U. Defining parameters of specificity for bioluminescent optogenetic activation of neurons using in vitro multi electrode arrays (MEA). J Neurosci Res. 2020;98(3):437–47.
46. Petersen ED, Sharkey ED, Pal A, Shafau LO, Zenchak-Petersen J, Peña AJ, et al. Restoring function after severe spinal cord injury through bioluminescent-optogenetics. Front Neurol. 2022;12:792643.
47. Gomez-Ramirez M, More AI, Friedman NG, Hochgeschwender U, Moore CI. The bioluminescent-optogenetic in vivo response to coelenterazine is proportional, sensitive and specific in neocortex. J Neurosci Res. 2020;98(3):471–80.
48. Medendorp WE, Bjorefeldt A, Crespo EL, Prakash M, Pal A, Waddell ML, et al. Selective postnatal excitation of neocortical pyramidal neurons results in distinctive behavioral and circuit deficits in adulthood. iScience. 2021;24(3):102157.
49. Berglund K, Fernandez AM, Gutekunst CN, Hochgeschwender U, Gross RE. Step-function luminopsins for bimodal prolonged neuromodulation. J Neurosci Res. 2020;98(3):422–36.
50. Ikefuama EC, Kendziorski GE, Anderson K, Shafau L, Prakash M, Hochgeschwender U, et al. Improved locomotor recovery in a rat model of spinal cord injury by bioluminescent-optogenetic (BL-OG) stimulation with an enhanced luminopsin. Int J Mol Sci. 2022;23(21):12994.
51. Prakash M, Murphy J, St Laurent R, Friedman N, Crespo EL, Bjorefeldt A, et al. Selective control of synaptically-connected circuit elements by all-optical synapses. Commun Biol. 2022;5(1):33.

Chapter 8
Bioluminescence for Integration of Cellular Processes

Ute Hochgeschwender

Abstract A unique and powerful application of BioLuminescent OptoGenetics is its use for the integration of cellular processes. Here, locally produced bioluminescence couples sensing of intracellular chemical fluctuations with modulation of the cell's response. The most versatile approach to integration is the use of split luciferases, where two parts of the luciferase, separated by a sensing module, are reconstituted into a functional enzyme upon binding of the intracellular chemical to the sensing moiety. Resulting bioluminescent light emission activates photoreceptors, causing discrete cellular responses. Intracellular events that have been detected include calcium increase, protein-protein interactions, G-protein-coupled receptor activation, pH change, and protein phosphorylation. Cellular outcomes that have been integrated with these events are transcription, proximity labeling of proteins, protein cleavage, and activation of opsins. This paradigm allows investigators to selectively detect, and then suppress or amplify, specific biological events for studies aimed at understanding basic biological processes, as well as for potential therapeutic applications, stopping pathological processes from continuing or amplifying, and rescuing failing processes.

Keywords Luciferase · Photoreceptor · Optogenetics · Sensor · Bioluminescence

8.1 Introduction

One of the most powerful applications of BioLuminescent OptoGenetics (BL-OG) is as a tool for molecular integration, enabling all-molecular feedback regulation. Locally produced light can provide a means for cells to detect their own chemical fluctuations, with subcellular precision, and cause discrete responses. The lives of cells, organs, and organisms are crucially dependent on chemical fluctuations, and

U. Hochgeschwender (✉)
College of Medicine, Central Michigan University, Mt. Pleasant, MI, USA
e-mail: hochglu@cmich.edu

U. Hochgeschwender (ed.), *BioLuminescent OptoGenetics*,
https://doi.org/10.1007/978-3-032-18121-3_8

failure to regulate these ongoing dynamics is a common motif in disease. To integrate intracellular event sensing with molecular outcomes, two main methods have been used to couple bioluminescent light emission to downstream processes. In the first, upon administration of luciferin, an intact luciferase emits bioluminescent light that activates a photoreceptor, resulting in further downstream events. This interaction between the luciferase and the light-sensing protein is designed to be contingent on an intracellular event, such as protein-protein interaction or calcium (Ca^{2+}) increase. The second method uses a luciferase construct split by a sensing module. The intracellular event of interest supplies the molecule required to reconstitute the split luciferase into a functional light-emitting enzyme; bioluminescent light is produced in the presence of luciferin, and this light activates a photoreceptor. These two approaches of sensing and modulating, i.e., integrating, have been applied to drive diverse cellular outcomes, including transcription of fluorescent reporters to label cells that experienced particular intracellular events, the transcription of actuators to allow manipulation of cells, proximity labeling of nearby proteins, VAMP2 cleavage to enable neuronal silencing, and the activation of excitatory or inhibitory opsins for control of neuronal activity. Intracellular events that have been integrated with these target effects include intracellular Ca^{2+} increase, protein-protein interactions, G-protein-coupled receptor (GPCR) activation, pH change, and protein phosphorylation. Table 8.1 lists the light-emitting and sensing components employed in BL-OG applications for integration of cellular processes and the intracellular events and molecular outcomes targeted.

The following paragraphs describe a variety of cases where bioluminescence was used for integration of cellular processes.

8.2 Integration of Cellular Processes with Transcription

Reporting on protein-protein interaction (PPI) in living cells provides a stable 'memory' of a previous PPI event that can be used for cell-based selections or high-throughput screening. SPARK2 (Specific Protein Association tool giving transcriptional Readout with rapid Kinetics) is a molecular tool that integrates PPI with transcription [1] (Fig. 8.1). One protein is fused to a LOV domain that contains a protease cleavage site (TEVcs) hidden in its Jα helix and a tethered transcription factor. The partnering protein is fused to a luciferase on its N-terminus and to a protease (TEVp) on its C-terminus. Interaction of the two proteins brings the luciferase into FRET (Förster Resonance Energy Transfer, [2]) distance of the LOV protein, and light emission in the presence of the luciferin unhinges the Jα helix to expose the cleavage site that is cleaved by the nearby protease, allowing the transcription factor to enter the nucleus and initiate expression of the reporter. This was demonstrated in HEK293 cells expressing NanoLuc (NLuc) fused to arrestin-TEVp, co-expressed with GPCR-LOV-TEVcs-Gal4 and the UAS-Citrine reporter [1]. In the dark and in the absence of the GPCR's agonist, the TEVcs is caged by the LOV domain, and the NLuc-arrestin-TEVp is not in proximity to the TEVcs. In the

Table 8.1 Bioluminescence driving integration of cellular processes

Intracellular event	Target effect	Light emitter	Light sensor	Tool	In vivo	Ref.
PP interaction	Transcription	NLuc	LOV	SPARK2		[1]
PP interaction	Transcription	Split NLuc	FKF1/GI	LuCID		[3]
Ca^{2+} increase	Transcription	sbGLuc	LOV	FLARE		[6]
Ca^{2+} increase	Transcription	NLuc	LOV	FLiCRE		[7]
Ca^{2+} increase	Transcription	Split NLuc	FKF1/GI	LuCID		[3]
Ca^{2+} increase	Transcription	Split sbGLuc	EL222	BLADe	Fluorescent reporter expression in the mouse brain	[8]
P phosphorylation	Transcription	Split eKL9h	EL222	BlinKS		[10]
GPCR activation	Proximity labeling	NLuc	LOV	LOVTurbo		[11]
GPCR activation	VAMP2 cleavage	NLuc	LOV	LATeNT		[12]
Ca^{2+} increase	VAMP2 cleavage	Split SSLuc	LOV	LATeNT		[12]
pH change	Membrane potential change	RLuc8	NpHR	pHIL	Epilepsy treatment in mouse models	[15]
Ca^{2+} increase	Membrane potential change	Split sbGLuc	Opsins	BLADe	Control of network dynamics in the mouse brain	[8]

presence of both luciferin and agonist, NLuc-arrestin-TEVp is recruited to the GPCR, allowing NLuc to uncage LOV and the TEVp to cleave the exposed TEVcs. The released Gal4 then drives transcription of the UAS reporter gene, allowing to visualize that PPI occurred by imaging the fluorescent reporter Citrine (Fig. 8.1).

An alternative PPI indicator system is LuCID (Luciferin- and Calcium-Induced Dimerization; [3]), initially developed to convert calcium-generated bioluminescence into stable gene expression (see below). The system is based on reconstitution of the N- and C-terminal fragments of a split NanoLuc (LgBiT and SmBiT; [4]) emitting light that activates the blue-light induced dimerization of FKF1/GI fused to the Gal4 DNA binding domain and VP64 transcriptional activator, respectively (Fig. 8.2). Dimerization allows binding of the promoter complex to its cognate UAS sites upstream of a gene of interest, triggering transcription. The reconstitution of the split NLuc is dependent on Ca^{2+} when LgBiT and SmBiT are fused to calmodulin and calmodulin binding protein, respectively (see below). Adapted to detect PPI, the Ca^{2+}-dependent components were replaced with the chemically-induced protein dimerizing domains FRB and FKBP (Fig. 8.2) or GAI and GID. In HEK293 cells, addition of rapamycin together with the luciferin led to dimerization of FK506 binding protein (FKBP) and FKBP–rapamycin binding (FRB) domain, reconstituting NLuc for emission of bioluminescence and activating transcription of a FLuc

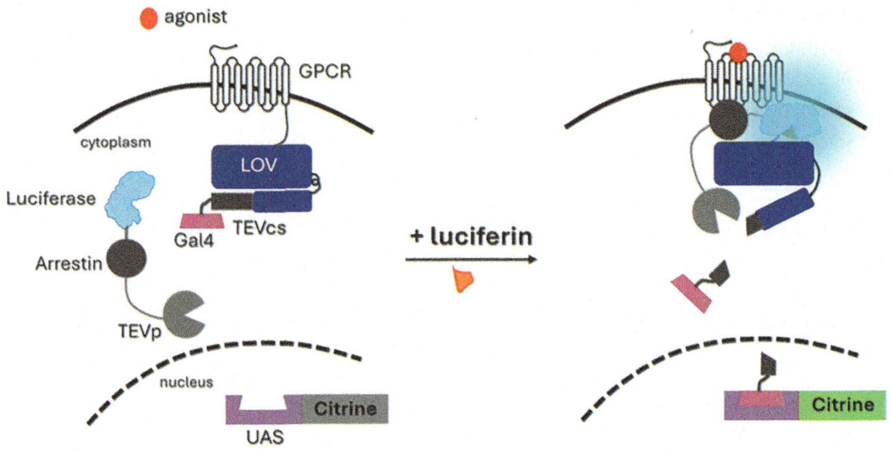

Fig. 8.1 Bioluminescence-mediated integration of protein-protein interaction with transcription: SPARK. The cell expresses SPARK2, where a G-protein-coupled receptor (GPCR) is fused to a LOV domain containing a protease cleavage site (TEVcs) and a tethered transcription factor (Gal4), as well as arrestin fused to NLuc and to a protease (TEVp), and the transcriptional reporter UAS-Citrine. Binding of an agonist triggers recruitment of arrestin to the GPCR, as well as fused NLuc and TEVp. In the presence of luciferin light emission from NLuc activates the LOV protein and exposes the protease cleavage site for TEVp to release the transcription factor Gal4. Gal4 – UAS interaction initiates Citrine expression

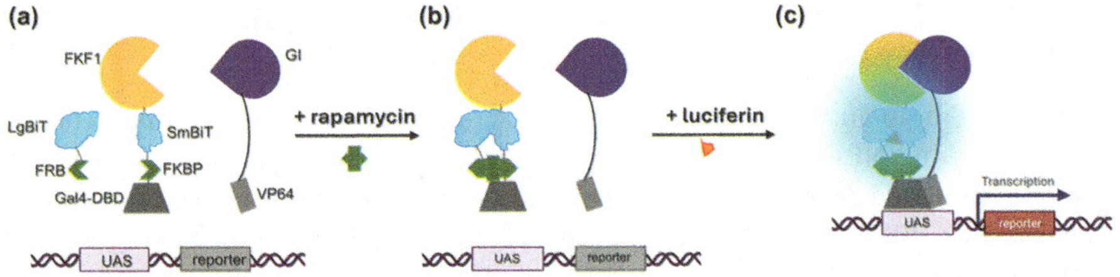

Fig. 8.2 Bioluminescence-mediated integration of protein-protein interaction with transcription: LuCID. (**a**) The plant photoreceptor proteins Flavin Kelch-repeat F-box 1 (FKF1) and its binding partner GIGANTEA (GI) are applied to a split Gal4 system, where the Gal4 DNA-binding domain (Gal4-DBD) is fused to FKF1 and the activation domain VP64 is fused to GI. The FKF1-Gal4-DBD fusion also contains the small part (SmBiT) of a split NLuc and the dimerizing domain FKBP. In a separate construct, the large part of the split NLuc (LgBiT) is fused to the dimerizing domain FRB. (**b**) The presence of rapamycin chemically induces FRB/FKBP interaction, resulting in reconstitution of the luciferase enzyme. (**c**) In the presence of luciferin, bioluminescence results in GI/FKF1 dimerization and reconstitution of the functional transcription factor. Gal4-DBD/VP64 binds to UAS, allowing reporter expression, reflecting FRB/FKBP protein interaction

reporter gene via the Gal4 system. This induced an 11.5-fold higher FLuc expression than rapamycin-only or furimazine-only treatments [3]. Similarly, treatment with gibberellic acid (GA) and the luciferin resulted in dimerization of GAI (Gibberellin-insensitive) and GID (GA-insensitive dwarf1), with the resulting light emission inducing 2.5-fold higher FLuc expression than GA-only or furimazine-only treatments.

An example of protease recruitment similar to that used in SPARK (see above) is FLARE (Fast Light- and Activity-Regulated Expression) [5]. In this case, protease recruitment is dependent on intracellular calcium (Ca^{2+}). Ca^{2+} flux is used as a proxy for neuronal activity in "Read & Write" or "Record & Play" approaches that link activity with the expression of a marker (e.g., a fluorescent protein) and/or an effector (e.g., the light-gated ion channel or pump) in cells active during a temporal window that overlaps with a behavior of interest. An increase in cytosolic Ca^{2+} levels leads to the binding of calmodulin and M13 required to bring the protease and its cleavage site in proximity (Fig. 8.3). The presence of bioluminescent light produced by the activity of a co-expressed luciferase activates the LOV domain, enabling cleavage and release of the transcription factor [6]. Crespo et al. [6] co-transfected HEK293 cells with the original FLARE components, a dual Firefly (FLuc)-dTomato reporter construct, and a membrane-anchored *Gaussia* luciferase variant, sbGLuc. In the presence of increased intracellular Ca^{2+} produced by exposure of cells to 2 μM ionomycin and 5 mM calcium chloride ($CaCl_2$), application of light from a blue LED led to robust expression of the fluorescence reporter compared with cells left in the dark. In addition, expression of FLuc was determined by measuring bioluminescence after adding the FLuc substrate D-luciferin. Similar levels of FLuc expression were achieved with bioluminescence emitted by sbGLuc upon application of the sbGLuc substrate (CTZ) in the presence of ionomycin and $CaCl_2$ used to increase intracellular calcium. This experiment showed that the

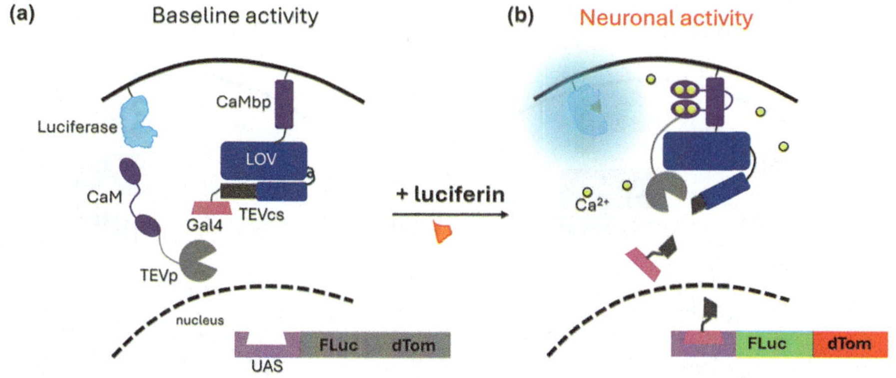

Fig. 8.3 Bioluminescence-mediated integration of calcium increase with transcription: FLARE. (**a**) The cytoplasm contains a membrane-anchored calmodulin binding peptide (CaMbp) fused to a LOV domain enclosing a protease cleavage site (TEVcs) and a tethered transcription factor (Gal4); calmodulin (CaM) fused to a protease (TEVp); and a membrane-anchored luciferase. The transcriptional reporter UAS-FLuc-dTomato is expressed in the nucleus. (**b**) Activity of a neuron causes Ca^{2+} influx, resulting in calmodulin binding to its peptide and bringing the protease in proximity to the LOV protein. In the presence of luciferin, bioluminescence will lead to unfolding of LOV, exposure of the cleavage site, and release of the transcription factor. Gal4 binds to UAS, allowing reporter expression reflecting neuronal activity

FLARE construct could be activated by either an external light source (blue LED) or bioluminescent light (sbGLuc with its luciferin CTZ) to drive transcription of a reporter (FLuc-dTomato) in the presence of calcium.

An improvement of several key design features of FLARE was introduced with FLiCRE (fast light and calcium-regulated expression) [7]. A variant of this tool family employs luciferin- and calcium-dependent gating of the LOV domain to address a key challenge to activity-dependent labeling tools, the background labeling of basally active neurons (Fig. 8.4). The luciferase NanoLuc (NLuc) is fused between the calmodulin and protease (TEVp) domains. In the presence of high calcium and luciferin, the calmodulin-NLuc-TEVp is brought into proximity of the membrane-bound component of FLiCRE, allowing NLuc-LOV FRET (Förster Resonance Energy Transfer) to occur. Following this energy transfer, the TEVp can access the cleavage site (TEVcs) and release the transcription factor to the nucleus. A key advantage of this luciferin-gated design versus the LED light-gated version of FLiCRE is a double dependency on calcium for transcription factor release: both the LOV domain uncaging via the recruited luciferase and the protease cleavage via the recruited protease require an increase in calcium. This feature reduces background labeling of basally active neurons in the dark. This was demonstrated in cultured primary hippocampal neurons, which exhibit high basal calcium activity, when the LED light-gated version of FLiCRE could not distinguish between the basal versus high calcium environments, while the luciferin-gated FLiCRE could do so robustly [7]. This luciferin-gated design reduces nonspecific labeling of basally active neurons, while also removing the requirement for optical fiber implant and light delivery into the brain.

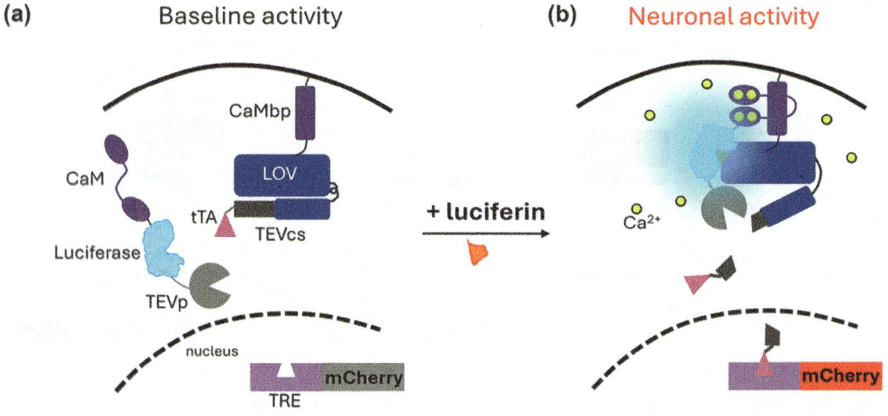

Fig. 8.4 Bioluminescence-mediated integration of calcium increase with transcription: FLiCRE. (**a**) At baseline (dark, low Ca^{2+}), the transcription factor (tTA) is tethered to a membrane-anchored calmodulin binding peptide (CaMbp) fused to a LOV domain enclosing a protease cleavage site (TEVcs). Calmodulin (CaM) is fused to a luciferase and the protease (TEVp). The transcriptional reporter TRE-mCherry is expressed in the nucleus. (**b**) Activity of a neuron causes Ca^{2+} influx and interaction of calmodulin with its binding protein, bringing TEVp into proximity of TEVcs, which is only accessible for cleavage when LOV is activated by bioluminescence from NLuc. Release of the transcription factor allows reporter expression reflecting neuronal activity

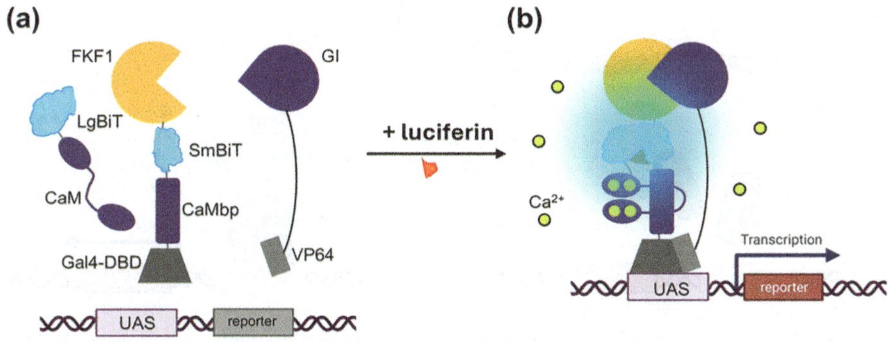

Fig. 8.5 Bioluminescence-mediated integration of calcium increase with transcription: LuCID. (**a**) The large part of a split NLuc (LgBiT) is fused to calmodulin (CaM). The small part (SmBiT) is fused to the light-sensing component FKF1 and to the calmodulin binding protein (CaMbp), to which is tethered the Gal4 DNA binding domain (DBD). The activation domain VP64 is fused to GI. The UAS-transcriptional reporter is expressed in the nucleus. (**b**) Ca^{2+} influx causes interaction of calmodulin with its binding protein, enabling reconstitution of the luciferase enzyme. In the presence of luciferin, bioluminescence results in GI/FKF1 dimerization and reconstitution of the functional transcription factor. Gal4-DBD/VP64 binds to UAS, allowing reporter expression reflecting Ca^{2+} increase

Rather than using Ca^{2+} for bringing together components that are then activated by an intact luciferase, a luciferase's bioluminescent light emission itself can be made dependent on the Ca^{2+}-driven reconstitution of a split luciferase enzyme. This method was realized in LuCID (luciferin- and calcium-induced dimerization; [3]), where calmodulin is fused to the N-terminal large fragment of NLuc (LgBiT) and calmodulin binding peptide is fused to the C-terminal small fragment (SmBiT) (Fig. 8.5). Calmodulin and its binding peptide interact in the presence of Ca^{2+}, thereby reconstituting the two separate parts of the luciferase into a functioning enzyme. In the presence of luciferin, this Ca^{2+}-dependent light emission then activates dimerization of fused light-sensitive FKF1 and its tethered Gal4 DNA binding domain with GI coupled to transcriptional activator VP64. Dimerization allows binding of the promoter complex Gal4-DBD/VP64 to the UAS site upstream of a reporter, triggering transcription. LuCID is a dual-purpose calcium sensor that provides a real-time bioluminescence readout of calcium dynamics in cells, while the GI/FKF1 split GAL4 component of LuCID converts calcium-generated bioluminescence into stable gene expression. In HEK293 cells, the presence of calcium and luciferin for 40 min resulted in robust expression of the FLuc reporter 12 h later [3]. However, LuCID has poor temporal resolution, and short-lived increases in cytosolic calcium induced by ATP or histamine did not give detectable bioluminescence or transcriptional signal. This is likely because GI/FKF1 requires several minutes of sustained NLuc bioluminescence to become activated.

The approaches described above for integrating intracellular Ca^{2+} events with transcription of a reporter are fairly complex, requiring expression and interaction of multiple components. The most straightforward approach would be to make light emission directly dependent on Ca^{2+} within a single-chain molecule that then

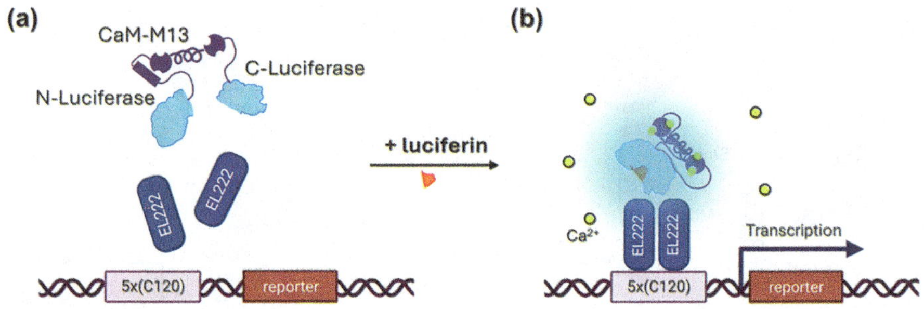

Fig. 8.6 Bioluminescence-mediated integration of calcium increase with transcription: BLADe. (**a**) In the dark, the light-sensing transcription factor EL222 remains a monomer in an "off" conformation. At low Ca^{2+} levels, the N- and C-terminal parts of the luciferase remain separated by the Calmodulin-M13 (CaM-M13) Ca^{2+} sensing moiety. (**b**) Ca^{2+} influx causes conformational change of the CaM-M13 sensor, reconstituting the luciferase enzyme. In the presence of luciferin, bioluminescence results in EL222 homodimerization and binding to the 5xC120 sequence, allowing reporter expression reflecting Ca^{2+} increase

directly activates a light-sensing transcription factor. This strategy was realized in the BLADe (BioLuminescence Activity Dependent) platform (Fig. 8.6) ([8]; see also Chap. 9). The *Gaussia* luciferase variant sbGLuc was split by the calmodulin-M13 Ca^{2+} sensing unit from GCaMP6f [9], generating lumicampsin (LMC). Increasing levels of Ca^{2+} led to reconstitution of the luciferase, and, in the presence of the luciferin coelenterazine, the bioluminescent light emitted was sufficient to activate the homodimerization of EL222. A bacterial light-dependent transcription factor, the EL222 dimer, binds to its promoter region 5 x C120, driving transcription of a gene-of-interest. Efficacy of the system was demonstrated in HEK293 cells, HeLa cells, and primary neurons in vitro as well as in vivo in the mouse cortex [8]. The critical aspect of this system's success lies in the unique characteristics of the split luciferase, having very low levels of light emission in the presence of luciferin at low levels of Ca^{2+}, resulting in a signal-to-noise ratio (SNR) that prevents activation of the light sensor at baseline but is sufficient for activation in the presence of higher levels of Ca^{2+}.

The principle of cofactor-dependent light emission can be expanded to a wide variety of critical chemical signals, essentially to any signal that bioluminescent enzymes can be engineered to detect, further expanding the impact of this integration platform. Chatterton et al. ([10], Chap. 10) developed a bioluminescent kinase sensor (BlinKS) that emits light in response to a specific kinase signaling event to then activate a light sensitive transcription factor (Fig. 8.7). The N- and C-terminal fragments of a luciferase are flanking two components: a phosphoamino acid-binding domain (PAABD), a molecular component that specifically recognizes and binds to a phosphorylated substrate after a kinase has acted on it, and a kinase-specific substrate, a modified protein that is phosphorylated by the kinase of interest. Phosphorylation of the peptide mimic due to increased kinase activity leads to binding of PAABD, causing a conformational change in the biosensor that

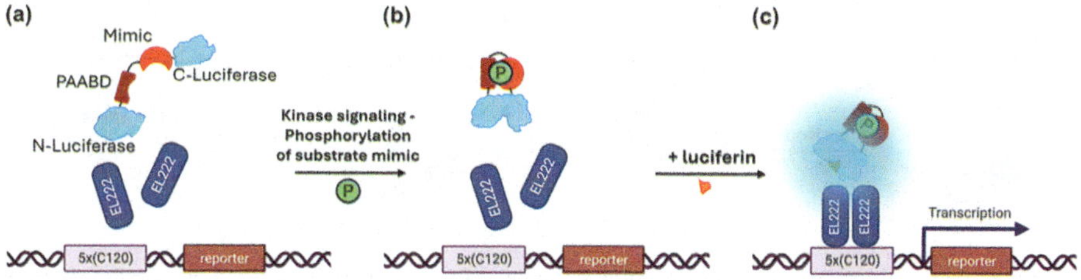

Fig. 8.7 Bioluminescence-mediated integration of kinase signaling with transcription: BlinKS. (**a**) The kinase sensor consists of the N- and C-terminal parts of a luciferase split by the phosphoamino acid-binding domain (PAABD) and a kinase-specific substrate mimic (Mimic). Co-expressed are the light-sensing transcription factor EL222 and the construct 5xC120-reporter. (**b**) When the kinase of interest is active, the kinase substrate mimic will be phosphorylated. PAABD recognizes and binds to this newly phosphorylated site, causing a conformational change that reconstitutes the split luciferase. (**c**) In the presence of luciferin, luciferase emits blue light that promotes the dimerization of El222 and binding to its promoter 5xC120, resulting in transcription of a gene of interest (reporter or therapeutic reagent)

reconstitutes the two luciferase halves into a functional enzyme. In the presence of luciferin, bioluminescent light emission activates a co-expressed light-sensing transcription factor, resulting in expression of a gene of interest. For example, a MEK1 mimic is used for sensing the phosphorylation state of MEK1, a crucial intermediate kinase in the MAPK/ERK pathway central for regulating cell processes like proliferation, differentiation, and survival. Its dysregulation is often implicated in various diseases, particularly cancer. Translating growth factor signaling events into light emission and activation of transcription can be leveraged to produce a negative regulator of the cell cycle, e.g., shRNA, to block the aberrant cell division. The availability of numerous kinase substrates enables a broadly useful all-molecular feedback system that links kinase signaling with a variety of downstream outcomes (e.g., reduced inflammation, blocking of cell proliferation) for potential therapeutic application (e.g., in inflammatory diseases and cancer).

8.3 Integration of Cellular Processes with Protein Labeling and Cleavage

GPCR (G protein-coupled receptor) activation is another intracellular event that would be valuable to couple to molecular modulation. Upon binding to an external signal (ligand), GPCRs undergo a conformational change that allows them to interact with and activate a heterotrimeric G protein, triggering a cascade of intracellular signaling events. GPCR activation has been utilized as the intracellular trigger to cause biotin labeling of proximal proteins [11] and to cleave the vesicle fusion protein VAMP2 for inhibiting synaptic transmission [12]. For both cases of intracellular integration, the following design was used: an intact luciferase (NLuc) was fused to a GPCR, and a protein that interacts with the GPCR (arrestin) was fused to a

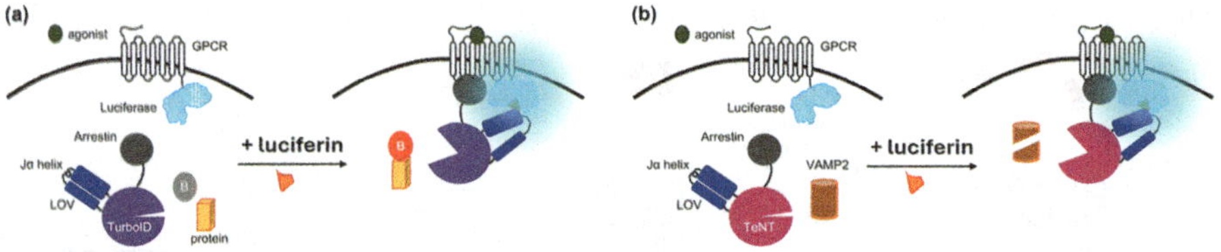

Fig. 8.8 Bioluminescence-mediated integration of GPCR activation with downstream enzyme activity. Agonist binding activates the GPCR that recruits arrestin, bringing the LOV protein in proximity to the luciferase. In the presence of luciferin, bioluminescence activates the LOV protein, thereby releasing its clamp on the synthetic enzyme. (**a**) GPCR activation enables LOVTurbo to biotin label proximal proteins. **b** GPCR activation enables LATeNT to cleave VAMP2

synthetic enzyme whose function is dependent on light-activated release of allosteric inhibition via a LOV domain (Fig. 8.8). In the presence of luciferin, activation of the GPCR through binding of its ligand recruits the GPCR-interacting protein, thus allowing light emission of the GPCR-bound luciferase to activate the enzyme to carry out its function.

Applied to the light-dependent synthetic enzyme TurboID (LOVTurbo) that labels nearby proteins via biotin covalent tagging, bioluminescence-mediated LOVTurbo activation and proximity labeling of proteins become contingent on GPCR activation (Fig. 8.8a). This was tested in HEK293 cells by expressing the GPCR CCR6 (C-C chemokine receptor type 6) fused to NanoLuc (NLuc) and arrestin-LOVTurbo, which is recruited to CCR6 on activation by its primary ligand, the chemokine CCL20. Biotinylation of nearby proteins by arrestin-LOV-Turbo was observed in the presence of luciferin and biotin in cells treated with the CCL20 agonist but not in untreated cells [11].

The same principle was used for a different outcome by engaging as integrator the synthetic enzyme LATeNT (light-activated light chain of tetanus neurotoxin) to drive the cleavage of the vesicular fusion protein VAMP2 [12] (Fig. 8.8b). In HEK293 cells co-expressing CCR6-NLuc and arrestin-LATeNT, stimulation with the GPCR agonist CCL20, together with luciferin, for 60 min resulted in 2.3-fold higher VAMP2 cleavage in the presence of CCL20 than in its absence, and no VAMP2 cleavage was detected in the absence of luciferin [12]. The efficacy of "GPCR-LATeNT" in converting GPCR activity into LATeNT activation was demonstrated in a similar experiment with a different GPCR, the kappa opioid receptor-based DREADD KORD, which is activated by the drug salvinorin B (SalB) [12]. Again, 2.1-fold higher VAMP2 cleavage was observed in the presence of SalB compared to its absence.

LATeNT was also incorporated in a synthetic circuit that converted an increase of intracellular Ca^{2+} into VAMP2 cleavage through fusion of LATeNT to the Ca^{2+}-activated luciferase CaBLAM [13] (Fig. 8.9). "Ca-LATeNT" expressed in HEK293 cells gave 6.7-fold greater cleavage of VAMP2 reporter in the presence of Ca^{2+} and luciferin than when treated with luciferin alone, with a 60-min exposure to Ca^{2+}/luciferin producing a similar extent of VAMP2 cleavage as 30 min treatment with

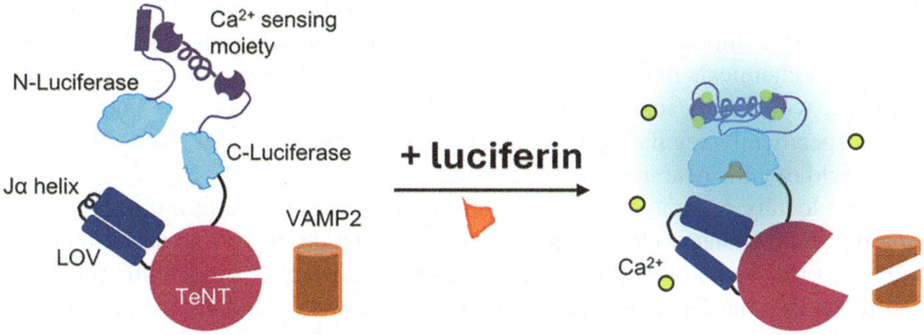

Fig. 8.9 Bioluminescence-mediated integration of calcium increase with downstream enzyme activity: Ca^{2+} LATeNT. Ca^{2+} influx causes conformational change of the Ca^{2+} sensor, reconstituting the split luciferase. In the presence of luciferin, bioluminescence activates LATeNT, allowing VAMP2 cleavage, reflecting Ca^{2+} increase

LED light [12]. These examples show that LATeNT can be gated by diverse inputs, including Ca^{2+} and GPCR receptor activity. In principle, any light-sensing actuator can be integrated with any intracellular process for which a bioluminescent sensor can be developed.

8.4 Integration of Cellular Processes with Membrane Potential Changes

Several of the above-described cases of molecular integration are amenable to applications in all-molecular feedback regulation. For example, local fluctuations in Ca^{2+} in pancreatic β cells can be employed in the regulation of insulin production, or the activity of kinases in tumor cells can be translated into the production of growth inhibitors. These molecular feedback processes work through bioluminescence-driven activation of transcription, involving timescales of hours and days. Many processes in the brain involve neural activity and are crucially dependent on chemical fluctuations on the time scale of milliseconds to seconds, such as the spreading of neural activity creating a seizure and neural activity-dependent cell death.

Current solutions to measuring and controlling cellular chemical fluctuations in neuroscience are based on implanted hardware and computer-based detection algorithms to conduct detection and deliver control, and there has not been the ability to sense subcellular processes as the trigger for regulation. For example, approaches to real-time detection and feedback control for altering maladaptive activity patterns rely on signals being tracked using electrodes and patterns detected by computer algorithms. These algorithms, in turn, trigger countermanding stimulation through implanted electrodes or fiber optics for electrical or optogenetic regulation. This strategy has shown success, for example, in stopping epileptic seizures dependent on thalamic bursting with optogenetic drive [14]. These closed-loop approaches for dynamic brain regulation are promising but have key limitations. These methods require chronic implants to detect patterns and deliver stimuli, and electrical stimuli or injected chemicals impact all adjacent cells. An ideal strategy would not require implants, would only regulate cells when they express maladaptive patterns, and would specifically regulate only those cells exhibiting aberrant patterns, not the entire field of cells near an electrode.

Using locally generated bioluminescent light created by luciferase enzymes that emit light in response to local signals, and optogenetic sensors that detect this light and effect the desired outcome, enables feedback control of intracellular events in real time. This has been applied to feedback regulation of membrane potential for controlling neuronal activity.

Merolla et al. [15] implemented an all-molecular negative feedback loop at the single neuron level by coupling a sensor for neuronal hyperactivity with a molecular actuator able to switch off neuronal activity (Fig. 8.10). Neuronal hyperactivity causes significant and rapid changes in intracellular pH towards acidosis. The ratiometric pH sensor E2GFP increases emission intensity in proportion to a drop in pH when excited at ~400–410 nm, but not at higher wavelengths, and emits fluorescence with a peak at 510 nm that increases upon acidification. By fusing the *Renilla* luciferase variant RLuc8 that emits at 405 nm when supplied with bisdeoxycoelenterazine (CTZ 400a) as luciferin substrate on the C-terminus and the chloride pump halorhodopsin variant eNpHR3.0, a hyperpolarizing actuator that is activated by 510 nm light, on the N-terminus, the authors generated the pH-sensitive inhibitory luminopsin pHIL. Through a resonant energy transfer cascade triggered by acidification in the presence of luciferin, from the luciferase to the fluorescent protein and from the fluorescent protein to the opsin, pHIL translates the hyperactivity-induced intracellular acidosis into silencing of neuronal activity. This was demonstrated in vitro in primary neurons, where hyperactivity evoked by the administration of convulsants

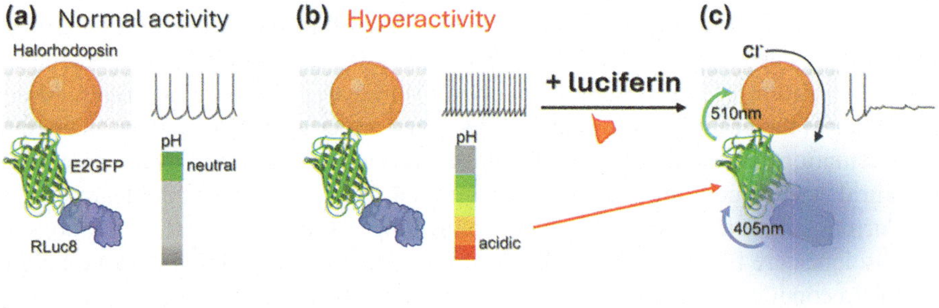

Fig. 8.10 Bioluminescence-mediated integration of pH with neuronal firing: pHIL (pH-sensitive inhibitory luminopsin). (**a**) The chloride pump halorhodopsin is fused to the pH sensor E2GFP and the *Renilla* luciferase variant RLuc8. Under normal neural activity, the intracellular pH is neutral. (**b**) Hyperactivity of neurons causes acidification. (**c**) In the presence of luciferin, RLuc8 emits light of 405 nm that excites E2GFP, which in turn emits light of 510 nm in proportion to the intracellular acidification, thus activating the chloride pump. Bioluminescence translates neuronal hyperactivity through hyperactivity-induced intracellular acidosis into the silencing of neuronal activity

was slowed, and neuronal firing rates returned to the physiological range [15]. The silencing of the neuronal firing activity occurred very rapidly, in ~2 s after the addition of CTZ 400a. In vivo expression of pHIL in the hippocampus of mice upon CTZ 400a administration mitigated acute tonic-clonic seizures induced by pilocarpine and counteracted audiogenic seizures in the PRoline-Rich Transmembrane protein 2 knockout (PRRT2 KO) mouse, a model of genetic epilepsy [15]. The integration tool pHIL is an all-molecular closed-loop system that detects incoming seizures and activates a negative feedback response to silence only hyperactive source neurons experiencing intracellular acidosis, leaving healthy bystander cells unaffected.

Instead of sensing intracellular pH shifts, Crespo et al. [8] developed a platform based on intracellular calcium (Ca^{2+}) level fluctuations. Ca^{2+} signals are ubiquitous in biology and essential for life, driving, for example, contraction in muscles, information processing in neurons, and insulin release in pancreatic cells. In this method, changes in Ca^{2+} are directly translated into bioluminescent light emission by making light production dependent on Ca^{2+} binding in a split luciferase construct, thereby establishing a BioLuminescence Activity Dependent platform (BLADe; see also Chap. 9). Molecular outcomes can cover a wide array of contexts, as Ca^{2+}-dependent light activation can be directed toward, in principle, any light-sensing actuator. BLADe was applied to activity-dependent transcription in neurons by combining it with the light-dependent transcription factor EL222 (see above). When paired with optogenetic actuators, BLADe provides an all-molecular feedback control of intracellular events in real time by allowing cells to sense and correct their own activity patterns (Fig. 8.11). This was demonstrated in vitro in single neurons

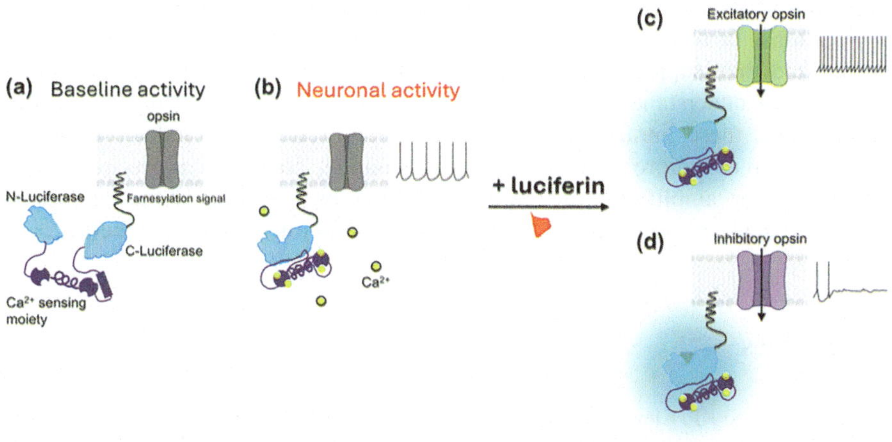

Fig. 8.11 Bioluminescence-mediated integration of Ca^{2+} increase with neuronal firing: BLADe. (**a**) Under baseline activity of neurons, a split Ca^{2+}-dependent luciferase is anchored in the cell membrane via a farnesylation signal, close to opsins. (**b**) Neuronal activity causes influx of Ca^{2+}, enabling reconstitution of the luciferase. In the presence of luciferin, (**c**) neurons expressing excitatory opsins will further increase firing, while (**d**) neurons expressing inhibitory opsins will decrease firing. Bioluminescence translates neuronal activation through intracellular Ca^{2+} increase into modulation of neuronal activity, depending on the biophysical properties of the opsin

through patch clamp recordings and in neuronal populations assessed in multielectrode recording experiments [8]. Co-expression of the integrator lumicampsin (LMC), sbGLuc split by the calmodulin-M13 Ca^{2+}-sensing unit from GCaMP6f, with an excitatory opsin, ChR2(C128S), or an inhibitory opsin, hGtACR2, resulted in neural activity-driven additional depolarization or hyperpolarization, respectively, of the individual cell and neural networks. These effects were only present with luciferin, not with the vehicle, and were not observed when either opsins or LMC were expressed by themselves. The efficacy of this real-time self-regulation of neural activity was demonstrated in vivo, where LMC converted spontaneous activity and sensory evoked changes in mouse sensory cortex into activity-dependent control of neocortical network dynamics [8]. In neurons of the mouse barrel cortex co-expressing LMC and the excitatory opsin CheRiff, changes in intracellular calcium caused the reconstitution of the luciferase and, in the presence of luciferin, bioluminescence photoactivated light-sensitive cation channels and enhanced spontaneous cortical neural activity. LMC also caused changes to neural activity via physiological sensory inputs, showing enhanced neural activity with luciferin versus vehicle as a function of sensory stimulation (vibrissae stimulus amplitude).

8.5 Conclusion

The designer forms of homeostatic control described here open up a wealth of new avenues for studying and repairing cells and cell networks. Further developed and improved versions of these tools are expected to have broad research and clinical applications. In basic research, this paradigm allows investigators to selectively detect, and then suppress or amplify, specific biological events; this is a unique approach to understanding the role of those processes. In potential therapeutic applications, these intracellular sensing tools allow detecting and discontinuing aberrant activity before it can cause harm, stopping pathological processes from continuing, or amplifying and rescuing failing processes, without the use of implanted devices and regulated by simple peripheral injection of a luciferin.

References

1. Kim CK, Cho KF, Kim MW, Ting AY. Luciferase-LOV BRET enables versatile and specific transcriptional readout of cellular protein-protein interactions. Elife. 2019;8:e43826.
2. Förster T. Zwischenmolekulare Energiewanderung Und Fluoreszenz. Ann Phys. 1948;2:55–75.
3. Erdenee E, Ting AY. A dual-purpose real-time indicator and transcriptional integrator for calcium detection in living cells. ACS Synth Biol. 2022;11(3):1086–95.
4. Dixon AS, Schwinn MK, Hall MP, Zimmerman K, Otto P, Lubben TH, et al. NanoLuc complementation reporter optimized for accurate measurement of protein interactions in cells. ACS Chem Biol. 2016;11(2):400–8.

5. Wang W, Wildes CP, Pattarabanjird T, Sanchez MI, Glober GF, Matthews GA, et al. A light- and calcium-gated transcription factor for imaging and manipulating activated neurons. Nat Biotechnol. 2017;35(9):864–71.
6. Crespo EL, Bjorefeldt A, Prakash M, Hochgeschwender U. Bioluminescent optogenetics 2.0: harnessing bioluminescence to activate photosensory proteins in vitro and in vivo. J Vis Exp. 2021;174 https://doi.org/10.3791/62850.
7. Kim CK, Sanchez MI, Hoerbelt P, Fenno LE, Malenka RC, Deisseroth K, et al. A molecular calcium integrator reveals a striatal cell type driving aversion. Cell. 2020;183(7):2003–2019.e16.
8. Crespo EL, Pal A, Prakash M, Silvagnoli AD, Zaidi Z, Gomez-Ramirez M, et al. A Bioluminescent Activity Dependent (BLADe) platform for converting intracellular activity to photoreceptor activation. bioRxiv. 2025:2023.06.25.546469.
9. Chen TW, Wardill TJ, Sun Y, Pulver SR, Renninger SL, Baohan A, et al. Ultrasensitive fluorescent proteins for imaging neuronal activity. Nature. 2013;499:295–300.
10. Chatterton M, Dubey O, VanDusen A, Robinette E, Abel J, Stewart A, Rossignol J, et al. Bioluminescent kinase sensors – BlinKS as optogenetic integrators. In: Hochgeschwender U, editor. BioLuminescent optogenetics. New York: Springer; 2026.
11. Lee SY, Cheah JS, Zhao B, Xu C, Roh H, Kim CK, et al. Engineered allostery in light-regulated LOV-Turbo enables precise spatiotemporal control of proximity labeling in living cells. Nat Methods. 2023;20(6):908–17.
12. Roh H, Kim D, Kim B, Jeon Y, Kim Y, Jacko M, et al. Light-activated tetanus neurotoxin for conditional proteolysis and inducible synaptic inhibition in vivo. bioRxiv. 2025:2025.01.27.635161.
13. Lambert GG, Crespo EL, Murphy J, Boassa D, Luong S, Celinskis D, et al. CaBLAM! A high-contrast bioluminescent Ca(2+) indicator derived from an engineered Oplophorus gracilirostris luciferase. Nat Methods. 2026;23(1):205–15.
14. Paz JT, Davidson TJ, Frechette ES, Delord B, Parada I, Peng K, et al. Closed-loop optogenetic control of thalamus as a tool for interrupting seizures after cortical injury. Nat Neurosci. 2013;16(1):64–70.
15. Merolla A, Michetti C, Moschetta M, Vacca F, Ciano L, Emionite L, et al. A pH-sensitive closed-loop nanomachine to control hyperexcitability at the single neuron level. Nat Commun. 2024;15(1):5609.

Chapter 9
Bioluminescent Activity-Dependent (BLADe) Platform

Nathan C. Shaner and Ute Hochgeschwender

Abstract Bioluminescence can be used for integrating intracellular activity states with user-defined changes in cellular function. Combining bioluminescent sensors, split luciferases whose light emission is dependent on the availability of an intracellular agent, with light-dependent actuators creates a Bioluminescent Activity-Dependent (BLADe) platform for converting biochemical signals within cells into photoreceptor activation and downstream cellular outcomes. To validate this experimental framework, a luciferase split by a calcium-sensing domain was used to convert neural activity into transcription of a reporter gene and into changes in membrane potential in vitro and in vivo. Facilitating the engineering of further BLADe sensors that couple biochemical events to light output motivated the molecular evolution of a new luciferase scaffold, SSLuc, that is highly amenable to insertion of sensor domains. This modular strategy of coupling an activity-dependent light emitter to a light-sensing actuator offers a generalizable framework for state dependent cell-autonomous control across biological systems.

Keywords Genetically encoded calcium indicator, GECI · Bioluminescent sensor · Photoreceptor · Optogenetics · Channelrhodopsin · Molecular integrator

---

N. C. Shaner
University of California San Diego, La Jolla, CA, USA
e-mail: ncshaner@health.ucsd.edu

U. Hochgeschwender (✉)
College of Medicine, Central Michigan University, Mt. Pleasant, MI, USA
e-mail: hochg1u@cmich.edu

U. Hochgeschwender (ed.), *BioLuminescent OptoGenetics*,
https://doi.org/10.1007/978-3-032-18121-3_9

9.1 BLADe Principle

A key advantage of employing a genetically encoded light source is that light emission can be made dependent on biochemical signals, such as intracellular calcium (Ca^{2+}) or cyclic adenosine monophosphate (cAMP) [1]. The most versatile approach to building a bioluminescent indicator is to "split" a luciferase by inserting a sensing moiety. In the presence of luciferin, light emission then becomes dependent on binding of a target co-factor. In the context of BioLuminescent OptoGenetics (BL-OG), bioluminescent indicator emission can not only be imaged but can also be used to activate nearby light-dependent proteins. This Bioluminescent Activity Dependent (BLADe) paradigm serves to integrate cellular activity states with downstream consequences from transcription to membrane potential changes, depending on the optogenetic actuator coupled to the conditional luciferase (Fig. 9.1). As proof-of-principle, Crespo et al. [2] developed a BLADe platform based on cellular calcium level fluctuations. Ca^{2+} level changes are directly translated into bioluminescence light emission by making light production dependent on Ca^{2+} binding in a split luciferase construct. The Ca^{2+} dependent light emitter was paired with a light-dependent transcription factor and a light-dependent ion channel that detected this light and affected transcription and neuronal firing, respectively. Beyond these proof-of-principle applications, molecular outcomes can cover a wide array of contexts, as Ca^{2+}-dependent light activation can be directed towards, in principle, any light-sensing actuator. Moreover, a luciferase construct can be split by a variety of sensing modules, providing a general mechanism that will allow cells to detect their own chemical fluctuations and then drive discrete outcomes (see also Chap. 10).

9.2 Requirements for BLADe Tools

As Ca^{2+} signals are ubiquitous in biology and essential for life, driving, for example, contraction in muscles, insulin release in pancreatic cells, and information processing in neurons, a Ca^{2+} dependent bioluminescent integrator was a logical choice for

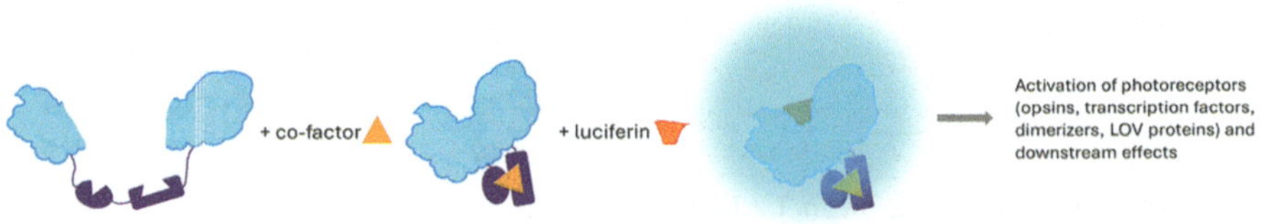

Fig. 9.1 Principle of Bioluminescent Activity Dependence (BLADe) A luciferase molecule is split for insertion of a co-factor sensing sequence. Presence of the co-factor is the condition for reconstituting a functional luciferase. Activation of the light emitter and subsequent initiation of photoreceptor activation and respective downstream effects are temporally regulated by administration of the luciferin. This principle can be designed for any intracellular co-factor for which a molecular sensing moiety can be engineered

testing BLADe. While fluorescent sensors are currently the predominant tools used for imaging Ca^{2+}, there is strong motivation for developing sensors based on bioluminescent light emission and several bioluminescent Ca^{2+} sensors have been developed based on luciferases split by Ca^{2+} sensing moieties [3–10]. Advantages are the absence of photobleaching and phototoxicity, the elimination of the need for external light excitation sources and accompanying hardware and of the limitation in the number of cells that can be simultaneously controlled.

Because bioluminescent probes have historically been far dimmer than fluorescent probes, increasing overall brightness has been the major focus of most efforts to produce bioluminescent (BL) Ca^{2+} indicators. As a result, most existing BL Ca^{2+} sensors are highly capable of reporting Ca^{2+} fluctuations, but with Ca^{2+} induced signal increases riding on top of a high baseline (Fig. 9.2). While the presence of a baseline BL signal in the absence of high Ca^{2+} is acceptable for many imaging and sensing applications of these probes, it is incompatible with the BLADe approach, as the background light emission by itself is already high enough to activate a photoreceptor. Thus, the split luciferase must produce very low levels of light emission in the presence of luciferin alone and at low levels of the specific co-factor needed for reconstitution of the functional enzyme (Fig. 9.2).

Crespo et al. [2] identified a *Gaussia* luciferase-based construct as fulfilling these criteria. The *Gaussia* luciferase variant sbGLuc [11] was split by the calmodulin sequence from GCaMP6f [12] and the M13 sequence from the Ca^{2+} sensor YC3.6 [3], generating LuMiCaMPsin (LMC; Fig. 9.3). In the presence of luciferin, this sensor had very low light emission at baseline levels of Ca^{2+}, while increase of Ca^{2+} led to a steep increase in light emission. Comparison to other bioluminescent Ca^{2+} sensors BlueCaMBI [7], GLICO [6], or GeNL_520 [4] revealed this critical difference in SNR [2] (Fig. 9.4). For BlueCaMBI and GeNL_520, the initial peak in light emission with application of luciferin is higher than the subsequent peak reflecting Ca^{2+} increase with histamine application, rendering these sensors unusable for BLADe applications (Fig. 9.4a, b). While GLICO shows a Ca^{2+}-driven increase in light emission exceeding the initial peak with luciferin alone, the difference in light emission reflecting Ca^{2+} flux is very small relative to the baseline (Fig. 9.4c). In contrast, LMC has very low levels of light emission at background conditions and displays a robust increase reflecting rising Ca^{2+}, making this sensor a suitable tool for BLADe, despite its lower overall brightness relative to the other indicators.

9.3 Applications of BLADe In Vitro and In Vivo

9.3.1 Transcription

By combining LMC with the bacterial light-dependent transcription factor EL222 [13], BLADe was applied to activity-dependent transcription in HeLa cells [2]. In the presence of the luciferin coelenterazine, increasing levels of Ca^{2+} led to

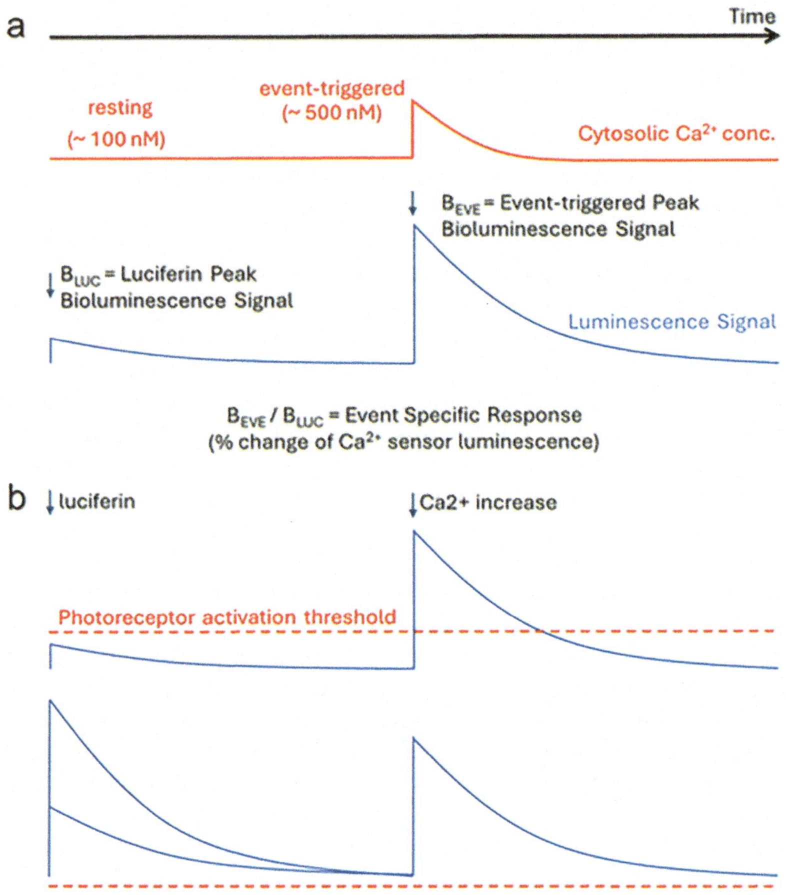

Fig. 9.2 Requirements for BLADe tools (**a**) Along a timeline, the resting cytosolic Ca^{2+} concentration in a cell acutely increases with an event, such as Ca^{2+} influx through opening of voltage-gated Ca^{2+} channels in activated neurons. A bioluminescent Ca^{2+} sensor will mirror this increase by an increase in luminescence that is significantly higher than the luminescence right before the event. (**b**) For a Ca^{2+} sensor to be usable in BLADe applications, luminescence upon application of luciferin in the absence of Ca^{2+} cannot exceed the threshold of light required for activation of a photoreceptor (upper panel) to allow the specific event to trigger photoreceptor activation. Bioluminescent Ca^{2+} sensors with high non-specific background (lower panel) can activate photoreceptors at baseline and are generally not suitable for BLADe

reconstitution of the luciferase and light emission sufficient to activate the homodimerization of EL222. The EL222 dimer binds to its promoter region (5 × C120), enabling transcription of a gene-of-interest (Fig. 9.5a). The performance of different Ca^{2+} dependent light emitters was compared in HeLa cells expressing the EL222 transcription factor and a reporter construct with the 5 × C120 promoter controlling transcription of firefly luciferase (FLuc). Treatment of HeLa cells with histamine reliably induced Ca^{2+} release from the ER. Reflecting their high level of baseline

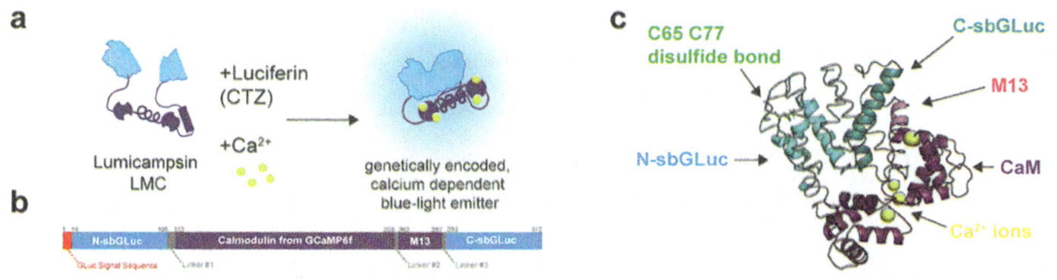

Fig. 9.3 LuMiCaMPsin (**a**) The *Gaussia* luciferase variant sbGLuc is split by the calmodulin sequence from GCaMP6f and the M13 sequence from YC3.6, making light emission dependent on the presence of Ca^{2+}. (**b**) Schematic detailing the various insertion sites and sequence elements used in LMC. (**c**) Structure model of LMC when bound to Ca^{2+}. (Reproduced from Ref. [2])

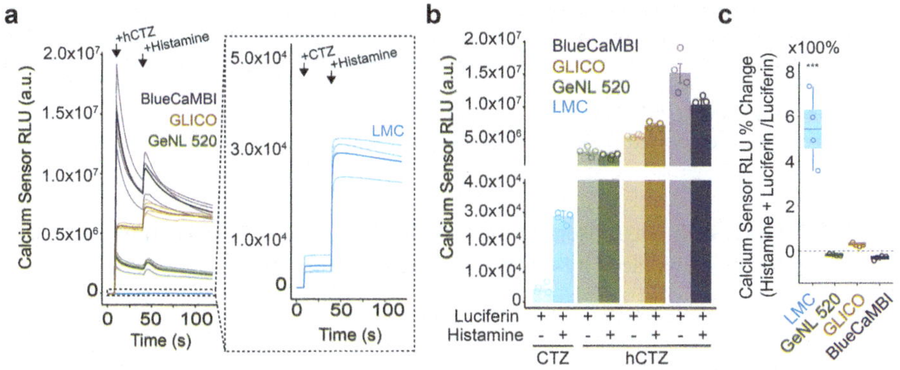

Fig. 9.4 Identifying a Ca^{2+} sensor suitable for BLADe (**a**) Luminometer time series from HeLa cells transiently expressing BlueCaMBI, GLICO, GeNL, or LMC following addition of luciferin and histamine. Inset shows LMC traces on an extended scale. (**b**) Peak bioluminescence measured after luciferin and histamine stimulation for each sensor from (**a**). (**c**) Histamine-specific percent change in bioluminescence normalized to luciferin baseline, expressed as percent change for each calcium sensor. Boxplots show median, 25th and 75th percentiles (box edges), whiskers to most extreme data points, and individual outliers. (Reproduced from Ref. [2])

light emission (see Fig. 9.4, above), BlueCAMBI, GLICO, and GeNL_520 each induced higher levels of FLuc transcription in the presence of luciferin alone than with luciferin and histamine (Fig. 9.5b, c). By contrast, LMC did not activate the transcription factor at baseline but was sufficient for its activation with histamine-induced Ca^{2+} release (Fig. 9.5b, c), functioning as a true integrator of intracellular activity with transcription as its readout.

The efficacy of the system was also demonstrated in vivo [2]. Mixtures of AAV encoding the Ca^{2+}-dependent light emitter LMC, the light-sensing transcription factor EL222, and the reporter construct 5 × C120-EYFP were injected into the mouse prefrontal cortex. Reporter expression was compared between groups receiving intraventricular injection of the luciferin, CTZ, together with the pan-excitatory reagent NMDA, CTZ alone, or NMDA together with vehicle. A significant increase

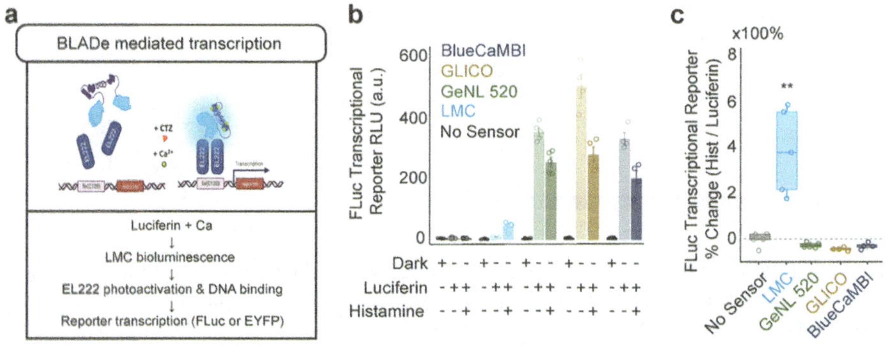

Fig. 9.5 BLADe control of transcription in vitro (**a**) Schematic of LMC leading to bioluminescence mediated dimerization of the light-sensing transcription factor EL222 in the presence of Ca^{2+} and luciferin (CTZ). The photoactivated EL222 homodimer then binds to the promoter 5 × C120 and leads to transcription of the reporter gene. (**b**) Firefly luciferase (FLuc) transcriptional reporter measured under dark, luciferin-only, or luciferin + histamine treatments in HeLa cells transiently expressing BlueCaMBI, GLICO, GeNL, or LMC. (**c**) FLuc transcriptional reporter normalized to luciferin baseline expressed as percent change for each sensor/condition. (Reproduced from Ref. [2])

in EYFP expression was observed only in the presence of both NMDA and CTZ, reflecting the neuronal activity-driven transcription (Fig. 9.6).

9.3.2 *Neuronal Activity*

When paired with optogenetic actuators, BLADe enables all-molecular feedback control of intracellular events in real time by allowing cells to sense and correct their own activity patterns. This was demonstrated in vitro in single neurons through patch clamp recordings [2]. In the presence of luciferin (CTZ), increasing current injections in neurons co-expressing LMC with an excitatory opsin ChR2(C128S) [14] resulted in a left-shift of the I/O curve due to bioluminescence activation of the opsin and additional depolarization of the cell, while co-expression with an inhibitory opsin hGtACR2 [15] resulted in a right-shift of the I/O curve due to BLADe-mediated hyperpolarization of the cell (Fig. 9.7). No shifts of the I/O curve were observed in the absence of luciferin.

BLADe-mediated, neural activity-driven change in firing activity was also applied successfully to neural networks in vivo [2]. LMC together with the excitatory opsin CheRiff [16] were targeted to excitatory pyramidal neurons within the mouse barrel cortex. Spiking behavior in the presence of CTZ led to enhanced multiunit activity as compared to vehicle application (Fig. 9.8). This LMC-mediated activity-dependent enhancement of neural dynamics could be driven through spontaneous neural activity (Fig. 9.8c) and through vibrissae sensory recruitment of LMC/CheRiff expressing neurons (Fig. 9.8d).

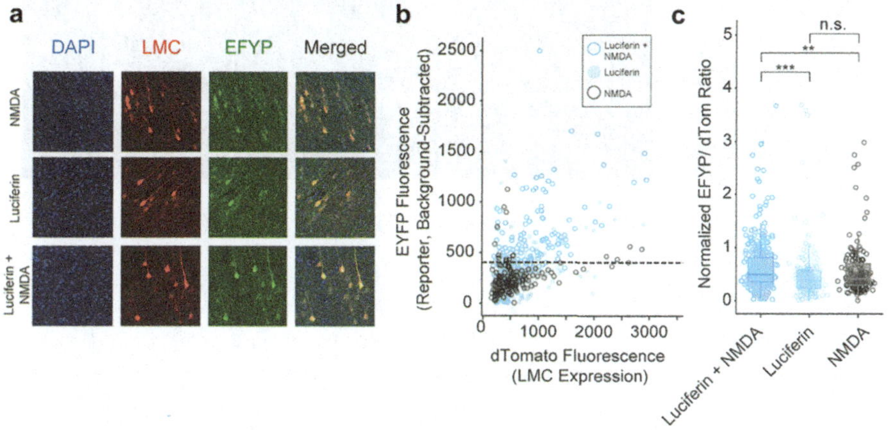

Fig. 9.6 BLADe control of transcription in vivo. (**a**) Representative confocal images showing EYFP transcriptional reporter (green), LMC P2A-dTomato viral expression (red), and nuclear DAPI stain (blue) from a field of view in the mouse prefrontal cortex following intraventricular injection of luciferin and/or NMDA. Mixtures of AAV9 preparations of the three constructs were stereotaxically injected into the medial prefrontal cortex of mice. Three or more weeks later mice received injections of NMDA alone, CTZ alone, or CTZ and NMDA. Brains were collected 18 h later and sections imaged under a confocal microscope. (**b**) EYFP reporter expression quantified as per-cell EYFP/dTomato ratio across treatment groups. Each point represents a single cell derived from three treatment groups: Luciferin-NMDA ($N = 350$ cells/2 mice), Luciferin-ONLY $N = 197$ cells/2 mice), NMDA-ONLY ($N = 147$ cells/2 mice). Dotted line indicates the 90th percentile EYFP/dTomato ratio observed in the NMDA-ONLY group, used as a threshold to define transcriptional activation above baseline. (**c**) EYFP reporter intensity plotted against dTomato fluorescence for each cell and condition. Boxplots show median, 25th and 75th percentiles (box edges), whiskers to most extreme data points, and individual outliers. (Reproduced from Ref. [2])

9.4 Evolution of BLADe Tools

BLADe depends on conditional light emission that stays *low* at baseline and *rises steeply* when the target co-factor is present. LMC, a split sbGLuc gated by the CaM–M13 sensing unit from YC3.6 and GCaMP6f, satisfied this criterion but was too dim for broad use, which prompted a focused program in molecular evolution resulting in a new luciferase scaffold named SSLuc [17]. SSLuc is derived from *Oplophorus gracilirostris* luciferase [18–21] and is several-fold more active than NanoLuc [22] with furimazine in vitro. It also folds well in protein fusions and is highly amenable to insertion of sensor domains, making it a useful starting point for a next-generation BL Ca^{2+} indicator specifically and bioluminescent sensors generally.

To build the sensor, SSLuc was split in the loop that joins the final two β-strands for insertion of Ca^{2+}-sensing domains. Circular permutations and direct fusions that used the RS20-calmodulin pair or troponin C were tested (Fig. 9.9a). Prototypes all responded *in vitro*, but the RS20–CaM topology gave the largest change in bioluminescence. This layout echoes a NanoBit–GCaMP hybrid design (GLICO) while

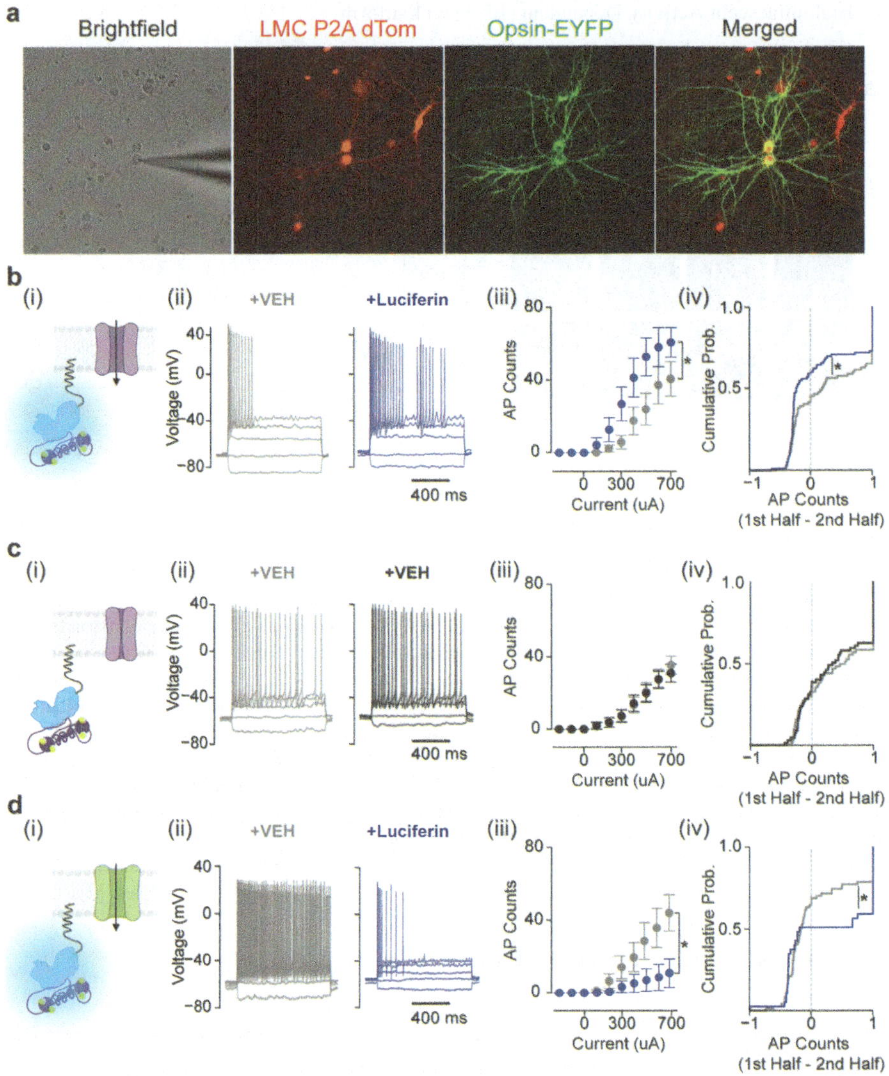

Fig. 9.7 BLADe couples depolarization-induced calcium influx to activation of light-sensing ion channels. (**a**) Brightfield and fluorescence images of cultured cortical neurons nucleofected with opsin (ChR2(C128S))-EYFP and LMC-P2A-dTomato and used for whole-cell patch-clamp recordings. The patch electrode is visible in the brightfield image. (**b**) (i) Schematic of farnesylated LMC anchored to the inner cell membrane near the excitatory opsin ChR2(C128S) enabling light-mediated activation in the presence of depolarization-induced Ca^{2+} flux and the luciferin (CTZ). (ii) Representative membrane voltage responses to step current injections before (gray) and after (blue) bath perfusion of luciferin in a neuron co-expressing LMC and ChR2(C128S). (iii) Action potential counts under vehicle (gray) and luciferin (blue) conditions. Data expressed as mean ± SEM. (iv) Empirical cumulative distribution of spike fraction changes across voltage traces for vehicle and luciferin conditions, represented as the difference in spike counts between the first and second halves of the recording. (**c**) (i) Schematic of a control condition, where LMC is co-expressed with ChR2(C128S) in the presence of Ca^{2+} but without luciferin, preventing light-mediated activation. (ii–iv) Same analyses as in **b**. (**d**) (i) Schematic of LMC co-expressed with the inhibitory opsin hGtACR2, enabling light-mediated suppression of activity in the presence of Ca^{2+} and luciferin. (ii–iv) Same analyses as in **b**, showing inhibition rather than activation. Data are presented as mean ± SEM in panels. (Reproduced from Ref. [2])

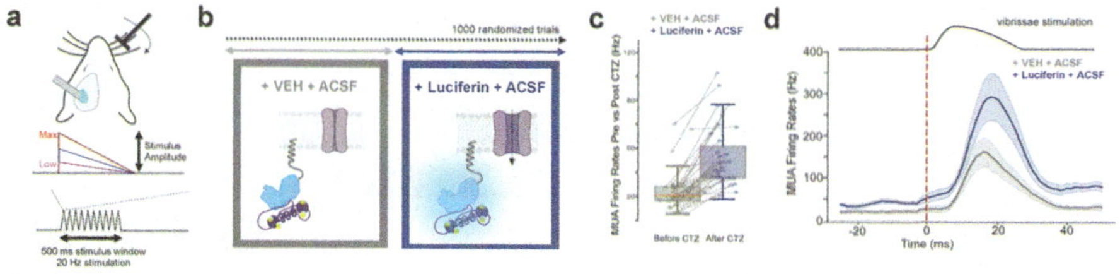

Fig. 9.8 BLADe converts neural activity into change in firing rate in vivo. (**a**) Schematic of the experimental design. Extracellular electrophysiology recordings from SI barrel cortex during varied vibrissal deflections (zero, low, mid, max intensities) for 500 ms at 20 Hz. (**b**) Schematic of the experimental timeline. Each recording consisted of 1000 trials (~2.5-h sessions) where ACSF containing vehicle or CTZ was constantly perfused over the craniotomy. During each treatment epoch, vibrissal deflections were randomly presented. In the presence of the sensory stimulus and vehicle (grey box), there is no activity-dependent light emission of LMC. In contrast, in the presence of CTZ and the sensory stimulus (blue box), sensory recruitment leads to activity-dependent bioluminescence and subsequently photoactivation of opsins in neurons. (**c**) MUA firing rate before and after the bath application of CTZ over the craniotomy. (**d**) Peristimulus time histograms from two distinct recording sites reflect changes in spiking activity evoked by contralateral sensory driven recruitment during the vehicle (grey) and CTZ (blue) treatment periods during low amplitude vibrissae stimulation. All data are plotted as mean +/− SEM. (Reproduced from Ref. [2])

preserving SSLuc's fixed N-terminal fluorescent partner for FRET enhancement and restricting Ca^{2+} readout to the BL channel. Affinity tuning was an essential part of the engineering process, as early clones bound Ca^{2+} much too tightly. Thus, Ca^{2+} affinity was reduced and contrast increased through a set of changes that included swapping out the RS20 and CaM sequences with those used in various other fluorescent indicators, changing linker lengths and compositions, and EF-hand mutations in CaM (Fig. 9.9b). Ultimately, affinity variants were generated covering much of the physiological range of Ca^{2+} concentrations.

The C-terminal peptide of SSLuc also proved to be a powerful control point. An initial screen of previously identified low-affinity peptides identified the lowest-affinity sequence, VTGYRLFEEIL (pep114), as producing the largest increase in emission between low and high Ca^{2+}. Redesigning the optimized peptide to VTGYRLLEEISN maintained the Lys-to-Glu charge-swap required for low peptide-enzyme affinity while enhancing the brightness of the sensor in the Ca^{2+}-bound state. This final architecture was named CaBLAM (Ca^{2+} BL Activity Monitor, [17]) (Fig. 9.9c). Purified CaBLAM showed a K_D near 439 nM and >80-fold absolute contrast in vitro. Two EF-hand variants of CaBLAM, CaBLAM_294W (K_D ~ 3 μM) and CaBLAM_332W (K_D ~ 280 nM) maintained its high brightness and contrast while shifting affinity, together giving a small family of indicators that covers a useful range.

CaBLAM's low baseline signal and high peak brightness position it as the Ca^{2+} indicator of choice for future iterations, while the SSLuc scaffold is a flexible starting point for engineering an array of new BLADe indicators that couple other biochemical events to light output. Each new BL indicator presents an opportunity to apply the BLADe strategy to transduce its specific signal into tailored optogenetic

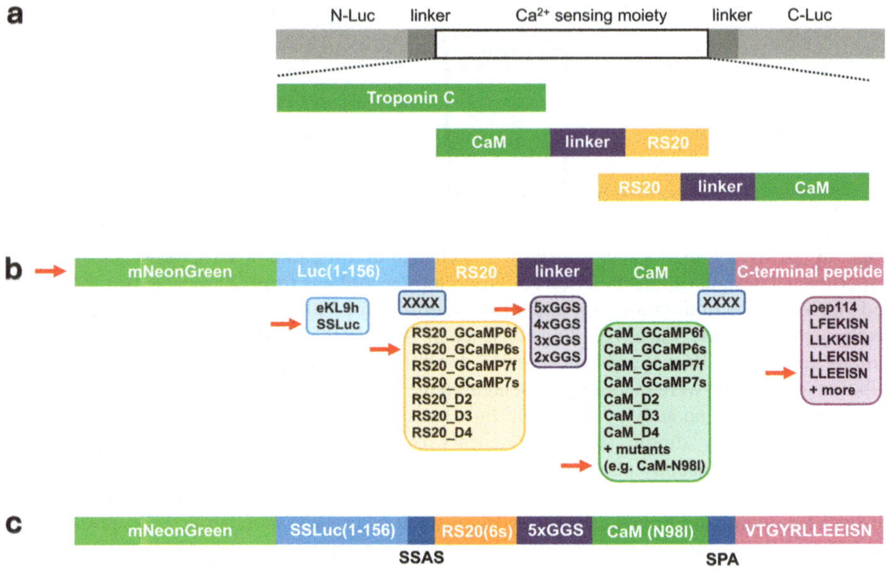

Fig. 9.9 Iterative refinement of bioluminescent Ca^{2+} sensor. (**a**) Three Ca^{2+} sensing moieties were initially tested, with RS20-CaM performing best. (**b**) Red arrows point at best-performing variants within several sequence elements tested: N-terminal fluorescent partner mNeonGreen for FRET enhancement, SSLuc ("Sensor Scaffold Luciferase") as the brightest and most soluble luciferase, and affinity tuning by swapping out the RS20 sequences with those used in various other fluorescent indicators, changing linker lengths and compositions, and introducing EF-hand mutations in CaM as well as exploring C-terminal peptide sequence variations. (**c**) Final architecture of CaBLAM. (Modified from Ref. [17])

output. Future optimization of substrate delivery via engineering modified and caged luciferins will support deeper imaging and stronger pairing with optogenetic actuators in BLADe workflows.

9.5 Summary

The Bioluminescent Activity-Dependent (BLADe) platform is a highly versatile approach for converting intracellular activity to photoreceptor activation. By using the concept of a luciferase split by a sensing moiety, any bioluminescent sensor can be made into an integrator by combining it with a photoreceptor. The proof-of-principle data show that a Ca^{2+} dependent luciferase can integrate Ca^{2+} fluctuations in activated neurons to drive transcription and change membrane potential in vitro and in vivo. The BLADe platform will be furthered by the recent development of a new luciferase scaffold, SSLuc, that enabled generation of the bioluminescent Ca^{2+} sensor, CaBLAM, with low baseline signal and high peak brightness, positioning these tools as flexible starting points for engineering an array of new BLADe indicators. The modularity of the BLADe framework allows light activation to be coupled to a wide range of light-sensing actuators, providing experimental control over diverse molecular outcomes depending on the user's goal. Moreover, the luciferase can be split by different sensing domains, providing a general mechanism for cells to detect their own biochemical state and then drive discrete user-defined outcomes.

References

1. Yeh HW, Ai HW. Development and applications of bioluminescent and chemiluminescent reporters and biosensors. Annu Rev Anal Chem. 2019;12(1):129–50.
2. Crespo EL, Pal A, Prakash M, Silvagnoli AD, Zaidi Z, Gomez-Ramirez M, et al. A bioluminescent activity dependent (BLADe) platform for converting neuronal activity to photoreceptor activation. bioRxiv. 2025; 2023.06.25.546469.
3. Saito K, Chang YF, Horikawa K, Hatsugai N, Higuchi Y, Hashida M, et al. Luminescent proteins for high-speed single-cell and whole-body imaging. Nat Commun. 2012;3:1262.
4. Suzuki K, Kimura T, Shinoda H, Bai G, Daniels MJ, Arai Y, et al. Five colour variants of bright luminescent protein for real-time multicolour bioimaging. Nat Commun. 2016;7:13718.
5. Yang J, Cumberbatch D, Centanni S, Shi SQ, Winder D, Webb D, et al. Coupling optogenetic stimulation with NanoLuc-based luminescence (BRET) Ca(++) sensing. Nat Commun. 2016;7:13268.
6. Farhana I, Hossain MN, Suzuki K, Matsuda T, Nagai T. Genetically encoded fluorescence/bioluminescence bimodal indicators for Ca(2+) imaging. ACS Sens. 2019;4(7):1825–34.
7. Oh Y, Park Y, Cho JH, Wu H, Paulk NK, Liu LX, et al. An orange calcium-modulated bioluminescent indicator for non-invasive activity imaging. Nat Chem Biol. 2019;15(5):433–6.
8. Qian Y, Rancic V, Wu J, Ballanyi K, Campbell RE. A bioluminescent Ca(2+) indicator based on a topological variant of GCaMP6s. Chembiochem. 2019;20(4):516–20.
9. Tian X, Zhang Y, Li X, Xiong Y, Wu T, Ai HW. A luciferase prosubstrate and a red bioluminescent calcium indicator for imaging neuronal activity in mice. Nat Commun. 2022;13(1):3967.
10. Tian X, Zhang Y, Du H, Huang W, Bizimana LA, Nishimura N, et al. An enhanced red bioluminescent indicator for responsive detection of physiological calcium dynamics in cells and mice. ACS Sens. 2025;10(8):5826–33.
11. Welsh JP, Patel KG, Manthiram K, Swartz JR. Multiply mutated Gaussia luciferases provide prolonged and intense bioluminescence. Biochem Biophys Res Commun. 2009;389(4):563–8.
12. Chen TW, Wardill TJ, Sun Y, Pulver SR, Renninger SL, Baohan A, et al. Ultrasensitive fluorescent proteins for imaging neuronal activity. Nature. 2013;499:295–300.
13. Motta-Mena LB, Reade A, Mallory MJ, Glantz S, Weiner OD, Lynch KW, et al. An optogenetic gene expression system with rapid activation and deactivation kinetics. Nat Chem Biol. 2014;10(3):196–202.
14. Berndt A, Yizhar O, Gunaydin LA, Hegemann P, Deisseroth K. Bi-stable neural state switches. Nat Neurosci. 2009;12(2):229–34.
15. Govorunova EG, Sineshchekov OA, Janz R, Liu X, Spudich JL. Natural light-gated anion channels: a family of microbial rhodopsins for advanced optogenetics. Science. 2015;349:647–50.
16. Hochbaum DR, Zhao Y, Farhi SL, Klapoetke N, Werley CA, Kapoor V, et al. All-optical electrophysiology in mammalian neurons using engineered microbial rhodopsins. Nat Methods. 2014;11(8):825–33.
17. Lambert GG, Crespo EL, Murphy J, Boassa D, Luong S, Celinskis D, et al. CaBLAM - a high-contrast bioluminescent Ca(2+) indicator derived from an engineered Oplophorus gracilirostris luciferase. Nat Methods. 2026;23(1):205-215.

18. Shimomura O, Masugi T, Johnson FH, Haneda Y. Properties and reaction mechanism of the bioluminescence system of the deep-sea shrimp Oplophorus gracilorostris. Biochemistry. 1978;17(6):994–8.
19. Inouye S, Watanabe K, Nakamura H, Shimomura O. Secretional luciferase of the luminous shrimp Oplophorus gracilirostris: cDNA cloning of a novel imidazopyrazinone luciferase. FEBS Lett. 2000;481(1):19–25.
20. Inouye S, Sasaki S. Overexpression, purification and characterization of the catalytic component of Oplophorus luciferase in the deep-sea shrimp, Oplophorus gracilirostris. Protein Expr Purif. 2007;56(2):261–8.
21. Inouye S, Sato JI, Sahara-Miura Y, Yoshida S, Hosoya T. Luminescence enhancement of the catalytic 19 kDa protein (KAZ) of Oplophorus luciferase by three amino acid substitutions. Biochem Biophys Res Commun. 2014;445(1):157–62.
22. Hall MP, Unch J, Binkowski BF, Valley MP, Butler BL, Wood MG, et al. Engineered luciferase reporter from a deep sea shrimp utilizing a novel imidazopyrazinone substrate. ACS Chem Biol. 2012;7(11):1848–57.

Chapter 10
Bioluminescent Kinase Sensors (BlinKS) as Optogenetic Integrators

Michael Chatterton, Osheen Dubey, Amy VanDusen, Emily Robinette, Jackson Abel, Andrew Stewart, Julien Rossignol, Jesse Bakke, and Eric D. Petersen

Abstract The ability to monitor and manipulate cell signaling events is critical for understanding cellular functions in health and pathology. Our lab has been working to expand on the tools available to noninvasively monitor and manipulate cellular functions both in cell models and model organisms. To address the need for more advanced methods to study and manipulate cell signaling events, our lab developed *B*io*l*um*in*escent *K*inase *S*ensors (BlinKS) that increase in brightness in response to specific kinases and pairing them with optogenetic transcription systems to create kinase dependent *G*enetic *C*ircuits, BlinKS-GC. The use of bioluminescent-based sensors paired with optogenetic actuators to create cell signaling-based integrators offers high flexibility in that manipulations are dependent on a specific kinase activity and the presence of luciferin, where brightness can be tuned based on luciferin concentration and the on/off kinetics of the sensor can be tuned by engineering. BlinKS can be used for noninvasive reporting of cell signaling and dynamic control over signaling events in genetically defined cell types for basic research relevant to a variety of different diseases and biological processes. The use of these sensors can not only enhance our understanding of cell signaling processes, but also open new avenues for therapeutic intervention either through mechanistic understanding of cellular processes or by adapting this system to correct aberrant cell signaling events.

Keywords Kinase sensor · Cell signaling · Biosensor · Synthetic biology · Bioluminescent optogenetics

---

M. Chatterton · O. Dubey · A. VanDusen · E. Robinette · J. Abel · J. Rossignol · J. Bakke
E. D. Petersen (✉)
College of Medicine, Central Michigan University, Mt. Pleasant, MI, USA
e-mail: chatt1mp@cmich.edu; dubey1o@cmich.edu; vandu1as@cmich.edu; robin4ej@cmich.edu; abel2jd@cmich.edu; Rossi1j@cmich.edu; bakke2jl@cmich.edu; Peter1ed@cmich.edu

A. Stewart
University of Kentucky, Lexington, KY, USA
e-mail: Anst265@uky.edu

U. Hochgeschwender (ed.), *BioLuminescent OptoGenetics*,
https://doi.org/10.1007/978-3-032-18121-3_10

10.1 Introduction

There is a technological gap between whole organ imaging and cellular imaging with microscopy using single wavelength fluorescent kinase sensors, limiting our ability to study and control dynamic cell signaling events over time. Bioluminescent indicators of cell signaling activity can fill this gap by offering a technology for imaging with cell type specificity (genetically encoded), covering large volumes of tissue (meso- to macroscopic scale) for whole organ imaging, and allowing continuous recording over extended time scales (hours-days) while remaining financially and technically accessible to researchers.

With bioluminescence, light is produced by an enzymatic reaction between luciferases and their unique substrates (luciferins). Compared to fluorescent imaging, this negates the need for an excitation light source, thereby eliminating major limitations of fluorescent imaging such as phototoxicity, autofluorescence, and excitation/emission light scattering to overcome depth and volume barriers. Critically, the use of bioluminescence enables noninvasive imaging of deep tissue structures throughout the body and subcortical structures in the brain such as the hippocampus [1, 2], basal nuclei (ganglia) [3–5], and hypothalamus [6]. Examples are the imaging of neuronal activity with a bioluminescent calcium sensor 5 mm deep through the mouse skull at 10 Hz [7] or bioluminescence signal at 6 mm through brain tissue at 10 Hz temporal resolution [8]. A notable example of recent progress toward moving bioluminescent cellular activity imaging beyond the brain is whole liver imaging of calcium fluctuations with whole body imaging in the intact mouse [9].

An attractive feature of bioluminescent sensors for molecular and cellular activity imaging is the absence of photobleaching, instead the timescale for continuous imaging is determined by the availability of the substrate (luciferins) and their delivery method. A single intraperitoneal (IP) delivery can provide multiple hours of continuous imaging, the longest reported being *6 h* of continuous bioluminescent voltage imaging in mice, *14 h* of continuous calcium imaging in drosophila, and *24 h* in cells [10–14]. The marine-based and synthetic luciferins that can be used with our sensors, coelenterazine, furimazine, fluorofurimazine (FFZ), and cephalofurimazine (CFz) are not toxic at physiological concentrations and can efficiently cross the blood-brain barrier when delivered IP [15–17]. Imaging sessions can also be sustained indefinitely by continuous delivery of the luciferin.

Beyond applications for optically reporting cell signaling activity, one of the most significant applications for these light emitting sensors is their potential use to control a variety of optogenetic actuators. Other approaches have been developed to achieve logical manipulation of cell signaling pathways. For example, synthetic biological logic gate systems were developed to integrate activation of multiple inflammation signaling pathways leading to the initiation of transcription of an anti-inflammatory cytokine [18]. The drawback of these systems is a lack of external control, tuneability, and flexibility. These drawbacks are alleviated with BlinKS as delivery of the luciferin allows external time gating of when the cell signaling modulation can occur. Furthermore, the gating is adjustable both in magnitude and time from minutes to hours by the luciferin variety, delivery route, and dose chosen. For example, BlinKS emitting light can be used in response to pro-growth signaling to activate an optogenetic transcription factor, which transcribes a protein that inhibits cell division. In this example, by adjusting both the dose and exposure time of luciferin, researchers can control transcription levels to express a growth inhibitor selectively in cells with abnormal growth factor signaling, such as cancer cells driven by pro-growth mutations.

10.2 Engineering of Bioluminescent Kinase Sensors (BlinKS)

Our previously designed bioluminescent glutamate sensor, BLING, was the basis for our initial design of BlinKS, created by swapping the bacterial glutamate binding protein from our library of BLINGs with a kinase substrate as well as the Phospho-Amino Acid Binding Domains (PAABDs) commonly used for fluorescent kinase sensors (Fig. 10.1). We initially used a kinase substrate domain that responded to growth factors and then expanded to kinases within inflammation, insulin, and cell cycle signaling pathways by swapping out the kinase substrate domain with a peptide sequence involved in the different signaling pathways. We designed our initial prototypes with three different PAABDs including 14-3-3, FHA1 and WW domains, which bind the substrate. When the kinase substrate is phosphorylated, the binding of PAABD to the phosphorylated substrate located between the halves of the split luciferase will cause the split luciferase to reconstitute and allow it to produce light in the presence of luciferin, serving as a reporter or controlling transcriptional systems. Our initial designs for sensing domains were based on the fluorescent FRET based PKA sensors AKAR1 and AKAR2 [19–22], where we replaced each fluorescent protein with a luciferase half and the substrate sequence with one derived from MEK1 with the goal of creating a BlinKS that responds to growth factor signaling pathways.

To create new sensor variants that respond to different cell signaling pathways, we synthesized and cloned a rationally designed library containing combinations of different luciferases, split sites, affinity mutations, PAABDs, and flexible or rigid

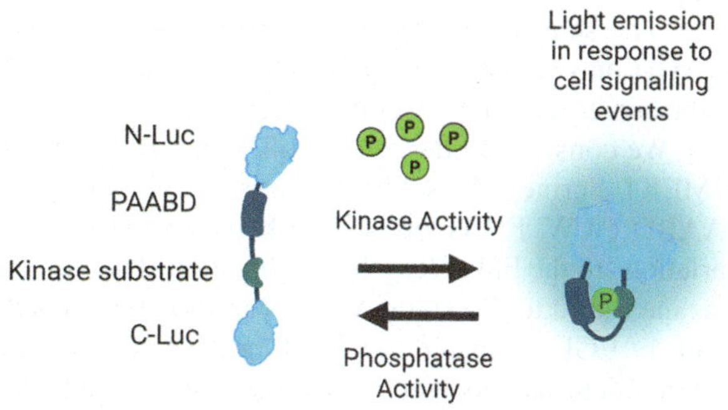

Fig. 10.1 General design of BlinKS where an N and C terminal half of a luciferase are split and, in between, the kinase sensing module consists of a PAABD and kinase substrate that will be phosphorylated by the kinase of interest, with each component separated by a linker

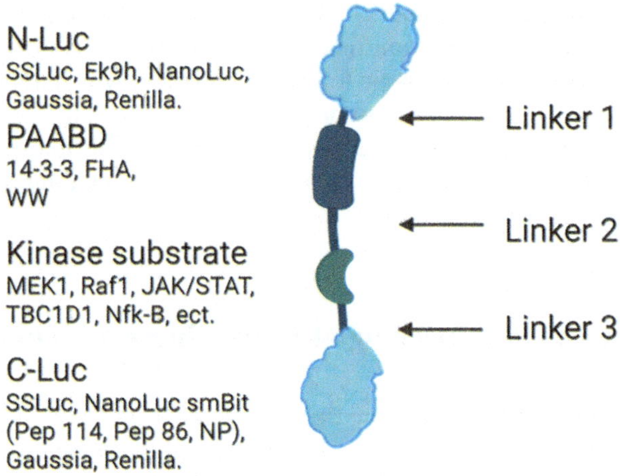

Fig. 10.2 Modular components that make up BlinKS including various luciferases, split sites and pairings, kinase substrates, various PAABDs, and multiple linkers that can be modified

linkers, fusing these components taking advantage of the modular design of BlinKS (Fig. 10.2). These are transfected into HEK 293 or a cell line specific to the signaling pathway of interest and screened for luminescence using a plate reader to detect change in brightness of the BlinKS following activation or inhibition of the pathway of interest. Next, we selected the top variants with the best response and low background signal with the goal of identifying variants with high signal-to-noise-ratio (SNR). We repeated this with iterative rounds of directed evolution including further rational modifications to the top sensors, site-saturated mutagenesis, or random mutations.

10.3 Variants and Applications of Bioluminescent Kinase Sensors (BLinKS)

10.3.1 Growth Factor Signaling

As our first design for BlinKS, we aimed to create bioluminescent sensors to detect alterations associated with the epidermal growth factor (EGF) pathway in cancer cells. The sensors are designed with peptide sequences that correspond to phosphorylation sites on kinase substrates associated with the EGF pathway. The design of the construct utilizes a split luciferase from the *Oplophorus gacrilirostris* eKL9h [23]. We constructed the sensor following the general design described above with a MEK1 Serine 222 substrate inserted, including flanking amino acids, with a PAABD (WW, FHA1 or 14-3-3) between the split halves of the luciferase, with variations to the linker lengths and location. The initial sensor library's performance was measured to determine which variation responded best when treated with high levels of EGF compared to inhibiting RAS. The sensors were transfected into pancreatic ductal adenocarcinoma cancer cells (PANC-1) that are serum-starved and

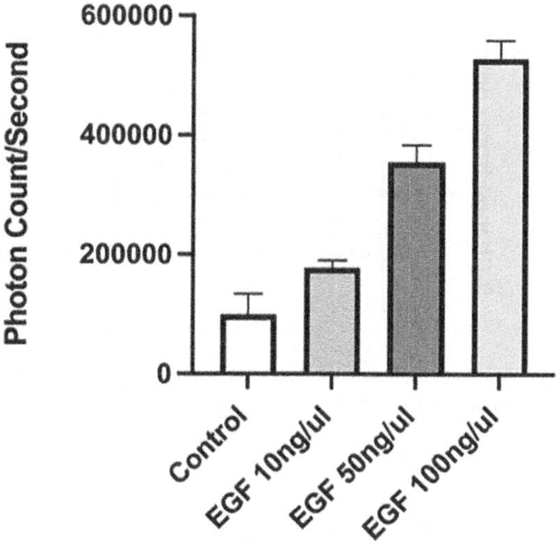

Fig. 10.3 Dose-dependent response of BlinKS with a MEK1 mimic responding to increasing concentrations of EGF

treated with high levels of EGF+luciferin. The sensors are then read through a kinetic assay to measure photons per second given off by the luciferase. The dose-dependent response of the top variant to increasing levels of EGF is shown in Fig. 10.3.

10.3.2 Insulin Signaling

We used the BlinKS design to create bioluminescence sensors to detect intracellular insulin signaling, using peptide sequences corresponding to phosphorylation sites on either AS160 or TBC1D1 as kinase substrates. AS160 and TBC1D1 act downstream of the insulin receptor and play an essential role in glucose uptake by regulating glucose transporter 4 (GLUT4) translocation to the plasma membrane. We tested multiple insulin-specific phosphorylation sites on AS160 and TBC1D1, known to modulate GLUT4 activation. To construct the sensor, for mimics, two kinase substrates, each from AS160 and TBC1D1, were paired with the PAABDs 14-3-3 or FHA1 and inserted between the halves of a split luciferase with a variation of rigid and flexible linkers. These were tested in C2C12 mouse myoblast cells, where they exhibited a dose-dependent emission response to insulin treatment and inhibition via wortmannin, a PI3K inhibitor (Fig. 10.4). Unlike traditional methods such as Western blotting or fixed-cell imaging, these sensors enable dynamic, real-time, and noninvasive monitoring of insulin signaling in living cells. This technology therefore provides a versatile tool to investigate physiological insulin signaling, mechanisms of insulin resistance, and supports therapeutic development for type 2 diabetes mellitus and other metabolic diseases.

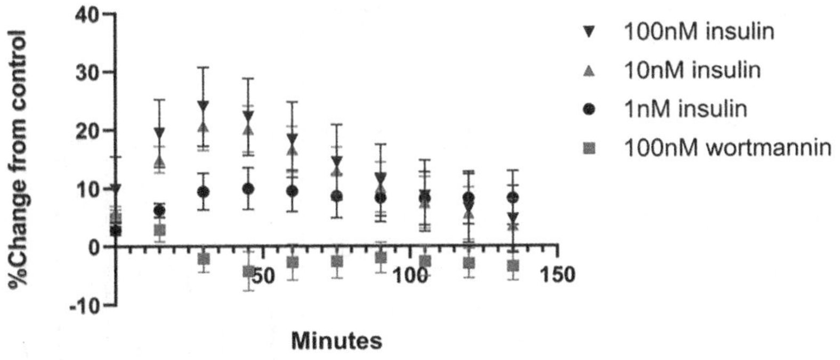

Fig. 10.4 The dose-dependent kinetic response of sensor with AS160 Thr642 showed the greatest response at 100 nM insulin concentration in C2C12 cells and a negative response to wortmannin, an inhibitor of PI3K

10.3.3 Inflammation Signaling

We designed BlinKS for inflammation sensing that could detect kinase activity, specifically JAK2 and STAT1, involved in IL-6 signaling. For these constructs, we attached the FHA1 PAABD to peptide substrates derived from endogenous phosphorylation motifs of JAK2 and STAT1 to provide selective recognition for the cytokine-activated signaling pathway. The substrate-PAABD module was placed between split Ek9h fragments in such a way that, upon phosphorylation by a targeted kinase, the FHA1 domain would bind to the phosphorylated version of the substrate and allow for reconstitution of luciferase activity to generate a change in light output. The performance of the sensors was evaluated by transfecting HEK293 cells with the FHA1-based BlinKS variants and then treating them with lipopolysaccharide (LPS) which induces IL-6 expression and thereby activates downstream JAK/STAT signaling. Each sensor targeted at either JAK2 or STAT1 signaled a significant increase in bioluminescence, exhibiting phosphorylation with the mimic substrate from LPS stimulation compared to the unstimulated controls (Fig. 10.5). These results demonstrate the capability of FHA1-based BlinKS to dynamically report inflammatory signaling events in living cells and serve as a platform for expansion to other kinases and eventual coupling to genetic circuit designs as feedback mechanisms for control of inflammation.

10.3.4 Mitochondrial Quality Control Signaling

We designed BlinKS to create sensors that could detect kinase activity associated with the PINK1/Parkin pathway, which governs mitochondrial quality control via mitophagy induction during stress. To facilitate PINK1 reporter activity at the mitochondrial membrane, the BlinKS reporter was fused to the 3′-amino acid sequence

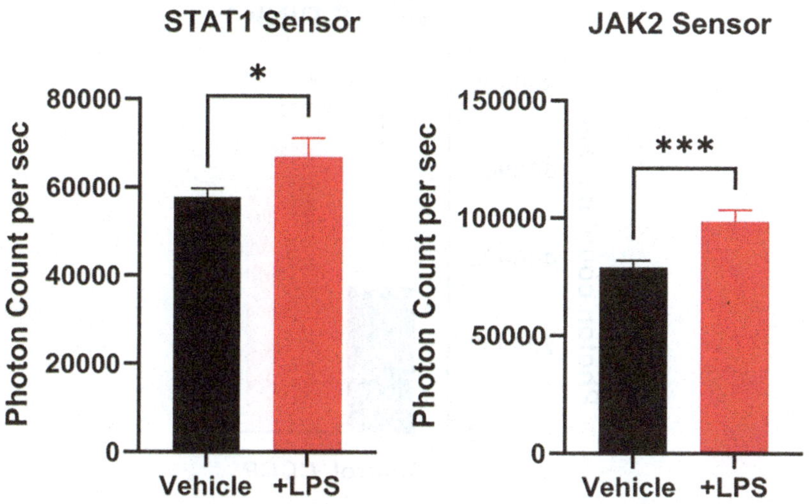

Fig. 10.5 BlinKS sensors based on either a STAT1 or JAK2 substrate sequence respond to inflammation after 12 h of LPS treatment in HEK293 cells. Data represent mean ± SEM from $n = 8$ technical replicates derived from a single transfection. Statistical significance was determined by two-tailed paired *t*-test (STAT1: *, $p = 0.0196$; JAK2: ***, $p = 0.0006$)

of the outer mitochondrial membrane protein 25 (OMP25) to facilitate a specific subcellular localization onto the cytosolic side of the outer mitochondrial membrane [24]. The PINK1/Parkin reporter BlinKS was engineered with the WW phospho-amino acid binding domain to peptide substrates derived from endogenous phosphorylation motifs surrounding serine-65 Parkin to enable selective recognition of PINK1-mediated signaling [25]. The substrate-PAABD module was positioned between split fragments such that, upon mitochondrial depolarization, PINK1 translocates to the outer mitochondrial membrane and phosphorylates the target peptide. The WW domain binds to the phosphorylated substrate and allows reconstitution of luciferase activity to generate a measurable change in light output.

The performance of the PINK1/PARK sensor was evaluated by transfecting HEK293 cells with WW-based BlinKS variants and treating them with 10 μM carbonyl cyanide m-chlorophenyl hydrazone (CCCP), a mitochondrial uncoupler known to activate the PINK1/Parkin pathway. A delayed but distinct increase in bioluminescence was observed approximately 60 min after CCCP addition (Fig. 10.6), consistent with previous reports showing Parkin phosphorylation around 80 min following mitochondrial depolarization [26]. These results demonstrate the capability of WW-based BlinKS to dynamically report mitochondrial stress signaling events in living cells and serve as a foundation for expanding BlinKS to other pathways involved in mitophagy and neuroprotection.

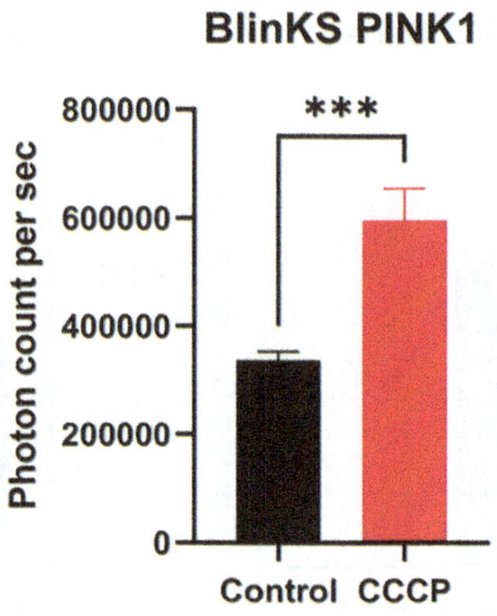

Fig. 10.6 HEK293 cells expressing the WW-based PARK BlinKS sensor were treated with 10 μM CCCP. After 60 min, treated wells showed a significant increase in bioluminescence compared to untreated controls, indicating phosphorylation-dependent activation of the PINK1/Parkin pathway. Data represent mean ± SEM from $n = 8$ replicates per group derived from a single pooled transfection. Significance was determined using an unpaired two-tailed *t*-test ($p = 0.0008$)

10.4 BlinKS as Integrators of Cellular Processes

Building on previous work using bioluminescence to drive transcription via intact luciferases or bioluminescent calcium sensors [8–10, 12–14] we are using our kinase sensors to drive gene expression. Specifically, we have used an EL222-based transcription system, an optogenetic gene circuit that is efficiently activated by optical methods, including a luciferase-luciferin system that emits blue light. We have used our lab's bioluminescent kinase sensor that emits light in response to growth factor signaling, described above, to activate this optogenetic gene circuit, demonstrating that BlinKS can efficiently drive transcription in response to a cell signaling event (Fig. 10.7).

10.4.1 Example Applications for BlinKS-GC: Tumor Growth Control of Glioblastoma (GB) and Pancreatic Ductal Adenocarcinoma (PDAC)

Glioblastoma and pancreatic cancer cells are robust models of unregulated cell growth in which the cell cycle does not turn off due to mutations within growth factor signaling pathways. We will use the optogenetic transcription system BlinKS-GC to respond to dysregulated growth factor signaling for inhibiting growth by

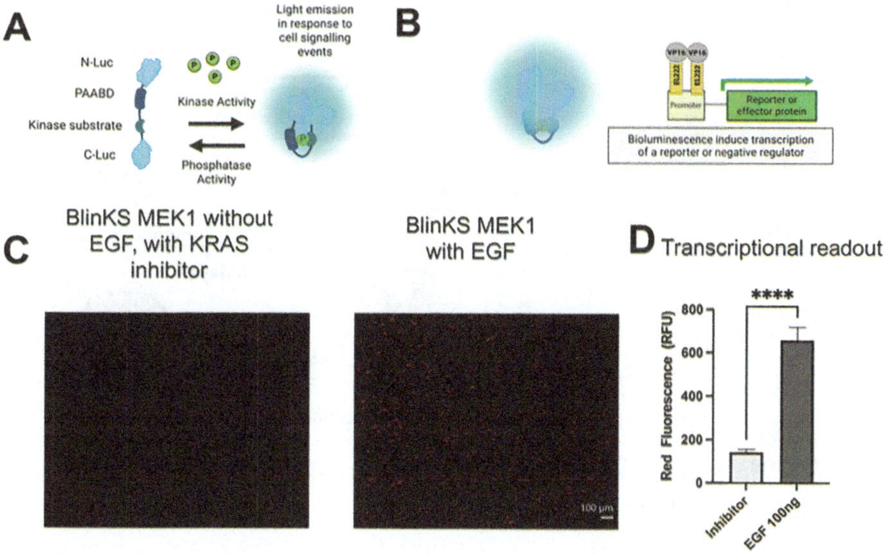

Fig. 10.7 BlinKS as integrator. (**a**) Schematic of the Bioluminescent Kinase Sensor (BlinKS). A split luciferase (blue) reconstitutes when kinase substrate is phosphorylated, allowing binding to phospho-amino acid binding domain (PAABD). (**b**) Light emission induces transcription by dimerizing the light-responsive transcription factor EL222. (**c**) Expression of mScarlet-3I in PANC1 cells expressing growth factor-responsive BlinKS with the gene circuit in the presence of luciferin and KRAS inhibitor (left) compared to cells expressing the sensor and gene circuit with high levels of EGF, showing a significantly higher level of reporter expression (right). (**d**) Quantification of reporter expression in cells expressing the growth factor-responsive gene circuit with KRAS inhibitor vs EGF treatment. ($n = 5$ cultures)

inducing dynamic expression of a suppressor of cell cycle (p53 or p21) (Fig. 10.8). Our goal is to reprogram cancer cells to respond to their dysregulated growth factor signaling by expressing a suppressor of cellular division. This should result in selective over expression of these tumor suppressor proteins in cells with excessive growth factor signaling. In contrast, healthy cells should not express these tumor suppressor proteins since our growth factor sensing BlinKS does not activate the gene circuit in cells with healthy, low levels of growth factor signaling.

In future experiments, we will utilize spheroid assays, where BlinKS-GC (p53) expressing cancer cells will form a well-rounded spheroid. We will add luciferin to induce production of growth factor signaling-dependent light from BlinKS and subsequently p53 expression via the optogenetic transcription system to arrest cell cycle. We will maintain the luciferin in the cultures, recording spheroid size via imaging. The expected outcome would be that p53 induction leads to G1/S and/or G2/M cell cycle arrest and inhibition of the spheroid growth in response to the aberrantly high growth factor signaling. Control spheroids (BlinKS-GC + vehicle) would be expected to continue to grow, whereas the BlinKS-GC + luciferin cells would have cell cycle arrest, not continue to grow, and potentially undergo apoptosis.

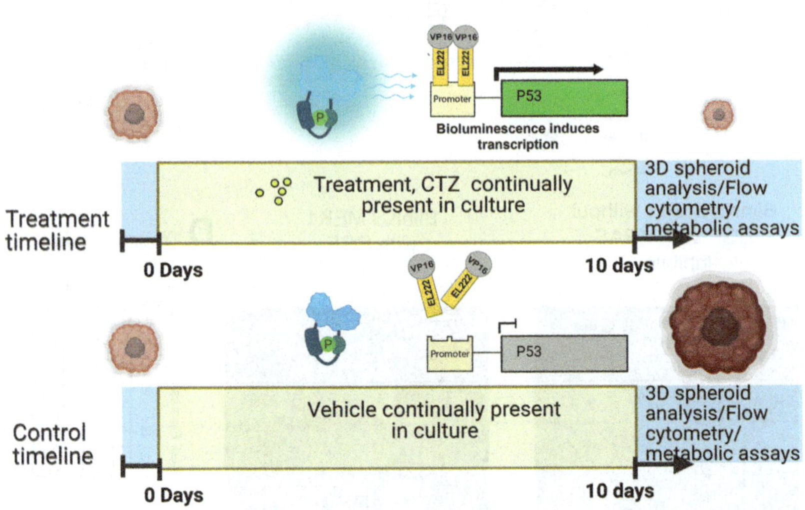

Fig. 10.8 Experimental timeline for in vitro testing in cancer spheroids expressing BlinKS-GC that are treated with luciferin (hCTZ) continuously are expected to halt uncontrolled cellular division and even reduce spheroid size. Spheroids only expressing BlinKS, without receiving luciferin, are expected to continue growing at the same rate as non-expressing spheroids

10.5 Summary

Over the last several years, our lab has demonstrated that we can adapt existing bioluminescent sensor design schemes to respond to kinase activity with a positive increase in light in response to phosphorylation events. We have also demonstrated that these sensors are bright enough to drive an optogenetic gene circuit with a low level of noise when upstream kinases within the pathway are inhibited. We expect that soon we will have applied this approach to detect activity and respond with a transcriptional readout in multiple signaling pathways (Table 10.1). We anticipate that once these tools are further optimized, they will be utilized for basic research in cell signaling and regulation. These systems also have a high potential for future translational applications like regulating cancer growth as in the above example, where some of the most difficult to treat cancers with dysregulated growth signaling can have that dysregulation leveraged to selectively be killed off with a genetic switch, leaving healthy cells unaffected. Other examples could be the use of our insulin-responsive BlinKS to induce the transcription of excess GLUT4 glucose transporter to increase glucose uptake in individuals with insulin resistance, in essence, amplifying the response of cells to insulin signaling. In another example, we can apply our BlinKS that responds to inflammation to combat chronic inflammation by transcribing an anti-inflammatory cytokine in response to high levels of inflammation. In this case, BlinKS is able to locally mitigate inflammation and turn itself off once the inflammatory state has reached a homeostatic level while being

Table 10.1 Example signaling pathways, kinase substrates used for each pathway, imaging applications, reporter protein examples, and example effector proteins that can be transcribed in response to activity within each signaling pathway

Cell signaling process	Kinase substrate	Induction of BL emission/imaging	Effector gene transcription	Integration with cellular processes
Cell cycle	B-Myb, RB, CDK2,4,6	S to G phase kinase activity	Suppressors of cell cycle (p53 or p21)	Cell cycle arrest, growth inhibition
Growth factor	MEK1, Raf1	Growth factor receptor activation	Suppressors of cell cycle (p53 or p21)	Cell cycle arrest, growth inhibition
Insulin	TBC1D1, AS160	Insulin receptor activation	GLUT4 glucose transporter	Increase glucose uptake
Mitophagy	PINK1/ Parkin	Activation of mitophagy	PGC1α	Mitogenesis/ metabolic augmentation
Inflammation	JAK/STAT, Nfk-B	Inflammation kinase signaling	Anti-inflammatory cytokines	Turn off inflammatory state

able to turn back on if the inflammatory state returns. In each of these examples, the BlinKS-GC concept has several key advantages that are not achieved by other synthetic biology approaches to self-regulated systems. Importantly, rather than being constitutively active, the sensor component utilizes the luciferin to be activated, so the system is inherently in the off state. It can be activated by delivery of the luciferin; in addition, the magnitude of response can be adjusted on a day-to-day basis by increasing or decreasing dosage. The time window of activation can also be tuned by using different designer luciferins that have been modified to have short, medium, or long windows of activity.

Acknowledgments This work was supported by grants from the National Institutes of Health (R35GM156968, R21EB034494, P20GM148326), the Elsa U Pardee Foundation, the American Heart Association (25AIREA1360438) and Diabetes Research Connection.

References

1. Dunn-Meynell AA, Dowling P, Marchese M, Rodriguez E, Blumberg B, Choi YB, et al. In vivo bioluminescence imaging used to monitor disease activity and therapeutic response in a mouse model of Tauopathy. Front Aging Neurosci. 2019;11:252.
2. Fricke IB, Schelhaas S, Zinnhardt B, Viel T, Hermann S, Couillard-Despres S, et al. In vivo bioluminescence imaging of neurogenesis - the role of the blood brain barrier in an experimental model of Parkinson's disease. Eur J Neurosci. 2017;45(7):975–86.
3. Aswendt M, Vogel S, Schafer C, Jathoul A, Pule M, Hoehn M. Quantitative in vivo dual-color bioluminescence imaging in the mouse brain. Neurophotonics. 2019;6(2):025006.
4. Kim DE, Tsuji K, Kim YR, Mueller FJ, Eom HS, Snyder EY, et al. Neural stem cell transplant survival in brains of mice: assessing the effect of immunity and ischemia by using real-time bioluminescent imaging. Radiology. 2006;241(3):822–30.

5. Tung JK, Gutekunst CA, Gross RE. Inhibitory luminopsins: genetically-encoded bioluminescent opsins for versatile, scalable, and hardware-independent optogenetic inhibition. Sci Rep. 2015;5:14366.
6. Cordonier EL, Liu T, Saito K, Chen SS, Xu Y, Fukuda M. Luciferase reporter mice for in vivo monitoring and ex vivo assessment of hypothalamic signaling of Socs3 expression. J Endocr Soc. 2019;3(7):1246–60.
7. Tian X, Zhang Y, Li X, Xiong Y, Wu T, Ai H-W. A luciferase prosubstrate and a red bioluminescent calcium indicator for imaging neuronal activity in mice. Nat Commun. 2022;13(1):3967.
8. Silvagnoli A, Taylor A, Slaviero A, Petersen E. A method for optimizing imaging parameters to record neuronal and cellular activity at depth with bioluminescence. Neurophotonics. 2024;11(2):024206.
9. Oh Y, Park Y, Cho JH, Wu H, Paulk NK, Liu LX, et al. An orange calcium-modulated bioluminescent indicator for non-invasive activity imaging. Nat Chem Biol. 2019;15(5):433–6.
10. Gomez-Ramirez M, More AI, Friedman NG, Hochgeschwender U, Moore CI. The BioLuminescent-OptoGenetic in vivo response to coelenterazine is proportional, sensitive, and specific in neocortex. J Neurosci Res. 2020;98(3):471–80.
11. Inagaki S, Agetsuma M, Ohara S, Iijima T, Yokota H, Wazawa T, et al. Imaging local brain activity of multiple freely moving mice sharing the same environment. Sci Rep. 2019;9(1):7460.
12. Minocci D, Carbognin E, Murmu MS, Martin JR. In vivo functional calcium imaging of induced or spontaneous activity in the fly brain using a GFP-apoaequorin-based bioluminescent approach. Biochim Biophys Acta. 2013;1833(7):1632–40.
13. Orioka M, Eguchi M, Mizui Y, Ikeda Y, Sakama A, Li Q, et al. A series of furimazine derivatives for sustained live-cell bioluminescence imaging and application to the monitoring of myogenesis at the single-cell level. Bioconjug Chem. 2022;33(3):496–504.
14. Petersen ED, Sharkey ED, Pal A, Shafau LO, Zenchak-Petersen J, Peña AJ, et al. Restoring function after severe spinal cord injury through BioLuminescent-OptoGenetics. Front Neurol. 2022;12:792643.
15. Otto-Duessel M, Khankaldyyan V, Gonzalez-Gomez I, Jensen MC, Laug WE, Rosol M. In vivo testing of Renilla luciferase substrate analogs in an orthotopic murine model of human glioblastoma. Mol Imaging. 2006;5(2):57–64.
16. Su Y, Walker JR, Park Y, Smith TP, Liu LX, Hall MP, et al. Novel NanoLuc substrates enable bright two-population bioluminescence imaging in animals. Nat Methods. 2020;17(8):852–60.
17. Gaspar N, Walker JR, Zambito G, Marella-Panth K, Lowik C, Kirkland TA, et al. Evaluation of NanoLuc substrates for bioluminescence imaging of transferred cells in mice. J Photochem Photobiol B Biol. 2021;216:112128.
18. Schukur L, Geering B, Hamri GC-E, Fussenegger M. Implantable synthetic cytokine converter cells with AND-gate logic treat experimental psoriasis. Sci Transl Med. 2015;7(318):318ra201.
19. Zhang J, Ma Y, Taylor SS, Tsien RY. Genetically encoded reporters of protein kinase A activity reveal impact of substrate tethering. Proc Natl Acad Sci USA. 2001;98(26):14997–5002.
20. Zhang J, Hupfeld CJ, Taylor SS, Olefsky JM, Tsien RY. Insulin disrupts beta-adrenergic signalling to protein kinase A in adipocytes. Nature. 2005;437(7058):569–73.
21. Allen MD, Zhang J. Subcellular dynamics of protein kinase A activity visualized by FRET-based reporters. Biochem Biophys Res Commun. 2006;348(2):716–21.
22. Depry C, Allen MD, Zhang J. Visualization of PKA activity in plasma membrane microdomains. Mol BioSyst. 2011;7(1):52–8.
23. Lambert G, Crespo E, Murphey J, Turner K, Gershowitz E, et al. CaBLAM - a high-contrast bioluminescent Ca2+ indicator derived from an engineered *Oplophorus gracilirostris* luciferase. bioRxiv. 2025:2023.06.25.546478.
24. Fecher C, Trovò L, Müller SA, Snaidero N, Wettmarshausen J, Heink S, et al. Cell-type-specific profiling of brain mitochondria reveals functional and molecular diversity. Nat Neurosci. 2019;22(10):1731–42.

25. McWilliams TG, Barini E, Pohjolan-Pirhonen R, Brooks SP, Singh F, Burel S, et al. Phosphorylation of Parkin at serine 65 is essential for its activation in vivo. Open Biol. 2018;8(11):180108.
26. Kazlauskaite A, Kondapalli C, Gourlay R, Campbell DG, Ritorto MS, Hofmann K, et al. Parkin is activated by PINK1-dependent phosphorylation of ubiquitin at Ser65. Biochem J. 2014;460(1):127–39.

Chapter 11
Application of BL-OG Tools in Animal Studies

Ute Hochgeschwender

Abstract BioLuminescent OptoGenetics (BL-OG) tools, molecular reagents that utilize light emitted by a luciferase when interacting with its luciferin to activate light-sensing molecules, have been employed in animal models including *C. elegans*, zebrafish, mice, rats, and marmosets. This brief overview chapter highlights important parameters to be considered for in vivo applications of BL-OG. Routes of luciferin administration can be as simple as adding it to the food source for worms to applying it via several different routes in mammals, including intracranial, intravenous, and intraperitoneal injections. The choice of a particular route of administration in mammals will be affected by the luciferase–luciferin combination used and the kinetics desired. The choice of luciferin is determined by the luciferase employed in a BL-OG tool, with some luciferases offering a variety of luciferins optimized for in vivo applications and with varying signal stability. This overview chapter is followed by a chapter reviewing reports of BL-OG applications in therapeutic and basic science contexts in animal models. The second chapter highlights results from systematic studies of the relationships between bioluminescent light production, neural spiking activity, luciferin concentrations and injection parameters in the mammalian brain.

Keywords In vivo · Luciferin · Luciferase · Application route · Mammals

BioLuminescent OptoGenetics (BL-OG) tools have been employed in animal models, including *C. elegans*, zebrafish, mice, rats, and marmosets. Particular attributes of a given species determine how the genetically encoded components are introduced and expressed and how the luciferin is administered. For example, it is straightforward to generate transgenic strains of *C. elegans* expressing BL-OG tools, while viral transduction with adeno-associated virus (AAV) is the most practical approach for rodents and nonhuman primates. Luciferin can be administered to

---

U. Hochgeschwender (✉)
College of Medicine, Central Michigan University, Mt. Pleasant, MI, USA
e-mail: hochg1u@cmich.edu

U. Hochgeschwender (ed.), *BioLuminescent OptoGenetics*,
https://doi.org/10.1007/978-3-032-18121-3_11

C. elegans by simply adding it to the animal's food source or to zebrafish larvae by immersion in luciferin containing water. For mammals, a variety of routes have been explored (Table 11.1).

The choice of a particular route of administration in mammals is affected by the luciferase–luciferin combination used and the kinetics desired. The choice of luciferin is determined by the luciferase employed in a BL-OG tool (Table 11.2). For example, tools using *Gaussia* luciferase variants have to use native coelenterazine (CTZ), as this is the only substrate for this group of luciferases. Tools employing NanoLuc variants can use CTZ formulations, but higher efficacy in vivo can be achieved with furimazine derivatives specifically developed for use with NanoLuc in animals. These are fluorofurimazine (FFz, [16]) and cephalofurimazine (CFz, [17]), with the latter showing improved access to the brain compared to FFz and CTZ analogs. Most recently, an optimized formulation, cephalofurimazine-9 (CFz9, [18]), was developed for improved performance both inside and outside the brain while reducing toxicity. For long-term experiments, furimazine variants with signal stability for hours to days have been developed (Endurazine, Vivazine). In addition, specific luciferin preparations are available to meet unique experimental demands. For example, the pH-sensitive inhibitory luminopsin pHIL needs light emission from RLuc at ~400 nm for excitation of the tethered fluorescent pH sensor to activate a fused inhibitory opsin [19]. The use of CTZ400a enabled emission of the desired wavelength.

Intuitively, intravenous administration of luciferin results in faster onset and offset of bioluminescence and its effects compared with intraperitoneal administration. Several studies compared routes of administration and/or luciferin concentrations for a given route by measuring the bioluminescent light created at the target region. Klein et al. [7] imaged bioluminescent light production following intracerebroventricular, intravenous, and intraperitoneal injections of coelenterazine in mice. Dose-response curves were generated for intracranial [3], intraperitoneal [20], and

Table 11.1 Routes of luciferin administration in mammalian experimental animals

Route of luciferin	Experimental animal	References (examples)
Intracranial	Rat	[1, 2]
	Mouse	[3]
Cerebral ventricle	Rat	[4, 5]
	Mouse	[6, 7]
Intranasal	Mouse	[8, 9]
Intravenous (IV)		
IV - Jugular vein	Rat	[1, 10]
IV - Tail vein	Rat	[2]
	Mouse	[7, 11]
IV - Saphenous vein	Marmoset	[12]
Intraperitoneal	Rat	[13]
	Mouse	[7, 14]
Intragastric	Mouse	[15]

Table 11.2 Luciferase–luciferin combinations for use in experimental animals

Luciferase	Luciferin	Comments
Gaussia luciferase (GLuc, sbGLuc)	Coelenterazine, CTZ (native)	GLuc and GLuc variants can only use native CTZ
Renilla luciferase (RLuc, RLuc variants)	Coelenterazine, CTZ (native)	Rluc substrate
	h-Coelenterazine, hCTZ (a 2-deoxy derivative of native Coelenterazine)	Longer lasting kinetic with RLuc
	Coelenterazine 400a (1bisdeoxycoelenterazine, di-dehydrocoelenterazine, DeepBlueC™)	Changes light emission to 400 nm (from 480 nm with native CTZ)
Oplophorus luciferase (NanoLuc and variants, e.g., eKL9h, SSLuc)	h-Coelenterazine, hCTZ	hCTZ can be used, however, with less efficacy compared to furimazine analogs (below)
	Fluorofurimazine, FFz (fluorination of the phenyl and benzyl rings of furimazine)	Improved solubility and bioavailability
	Cephalofurimazine, CFz (difluorinated analog of furimazine)	Enhances bioluminescence in the brain
	Endurazine, Vivazine	Signal stability over longer durations (>2 h compared to <2 h for FFz and CFz)

intravenous [11] coelenterazine administration in mice. Table 11.1 lists example references for the various routes of luciferin application utilized in BL-OG studies in mammals. In addition to the listed routes, luciferin can be directly injected into target tissues. For example, in a study designed to probe the role for proprioceptors in sensing acidosis (sngception) in a mouse model of chronic hyperalgesia induced by repeated intramuscular acid injection, Lee et al. [21] injected CTZ directly into the muscle of luminopsin 3 expressing mice. In another example in mice, FFz was directly injected into subcutaneous tumors expressing LiPOP-NLuc (light-induced non-apoptotic tool fused to NanoLuc), where bioluminescence drives necroptosis by directly permeabilizing the plasma membrane, inducing cell death and significant reduction in tumor size [22].

As for any in vivo studies, the toxicity of administered reagents is important. A number of the in vivo studies reviewed in Chap. 12 include assessments of substrate toxicity by evaluating major organs, including the heart, liver, spleen, lung, kidney, and muscles, and the circulatory system. Across different studies and luciferins, toxic effects were observed only at substrate levels exceeding the effective doses, not at the concentrations needed to drive BL-OG tools.

As for any tool development, an important step in the generation of BL-OG tools is the proof-of-concept demonstration in animals. Table 2.3 in Chap. 2 and Table 3.1 in Chap. 3 list the publications that applied and validated, in vivo, a given bioluminescence-driven opsin or non-opsin photoreceptor, respectively. In this section, Chap. 12 provides a brief overview of the published reports of BL-OG tools applied in animal models that use the tools in therapeutic contexts to improve disease outcomes or in basic science contexts to test specific hypotheses about biological mechanisms. Chapter 13 highlights the results from systematic studies that tested the relationships between bioluminescent light production and neural spiking activity in the brain and between luciferin concentrations and injection parameters. These studies showed that the dynamics of neuromodulation effects of BL-OG tools can be systemically controlled by the experimenter.

References

1. Tung JK, Gutekunst CA, Gross RE. Inhibitory luminopsins: genetically-encoded bioluminescent opsins for versatile, scalable, and hardware-independent optogenetic inhibition. Sci Rep. 2015;5:14366.
2. Tung JK, Shiu FH, Ding K, Gross RE. Chemically activated luminopsins allow optogenetic inhibition of distributed nodes in an epileptic network for non-invasive and multi-site suppression of seizure activity. Neurobiol Dis. 2018;109:1–10.
3. Gomez-Ramirez M, More AI, Friedman NG, Hochgeschwender U, Moore CI. The BioLuminescent-OptoGenetic in vivo response to coelenterazine is proportional, sensitive and specific in neocortex. J Neurosci Res. 2020;98(3):471–80.
4. Petersen ED, Sharkey ED, Pal A, Shafau LO, Zenchak-Petersen J, Peña AJ, et al. Restoring function after severe spinal cord injury through BioLuminescent-OptoGenetics. Front Neurol. 2022;12:792643.
5. Anderson KA, Whitehead BJ, Petersen ED, Kemme MR, Wedster A, Hochgeschwender U, et al. Behavioral context improves optogenetic stimulation of transplanted dopaminergic cells in unilateral 6-OHDA rats. Behav Brain Res. 2023;441:114279.
6. Zenchak JR, Palmateer B, Dorka N, Brown TM, Wagner LM, Medendorp WE, et al. Bioluminescence-driven optogenetic activation of transplanted neural precursor cells improves motor deficits in a Parkinson's disease mouse model. J Neurosci Res. 2020;98:458–68.
7. Klein E, Marsh S, Becker J, Andermann M, Lehtinen M, Moore CI. BioLuminescent OptoGenetics in the choroid plexus: integrated opto- and chemogenetic control in vivo. Neurophotonics. 2024;11(2):24210.
8. Yu SP, Tung JK, Wei ZZ, Chen D, Berglund K, Zhong W, et al. Optochemogenetic stimulation of transplanted iPS-NPCs enhances neuronal repair and functional recovery after ischemic stroke. J Neurosci. 2019;39(33):6571–94.
9. Zhang JY, Tung JK, Wang Z, Yu SP, Gross RE, Wei L, et al. Improved trafficking and expression of luminopsins for more efficient optical and pharmacological control of neuronal activity. J Neurosci Res. 2020;98:481–90.
10. Berglund K, Fernandez AM, Gutekunst CAN, Hochgeschwender U, Gross RE. Step-function luminopsins for bimodal prolonged neuromodulation. J Neurosci Res. 2020;98:422–36.
11. Murphy EF, Means A, Li C, Baez H, Gomez-Ramirez M. Strength of activation and temporal dynamics of bioluminescent-optogenetics in response to systemic injections of the luciferin. Neuroimage. 2024;301:120882.
12. Murphy EF, Gomez-Ramirez M. Application of BL-OG tools in the mammalian brain. In: Hochgeschwender U, editor. BioLuminescent OptoGenetics. New York: Springer Nature; 2026.
13. Ikefuama EC, Kendziorski GE, Anderson K, Shafau L, Prakash M, Hochgeschwender U, et al. Improved locomotor recovery in a rat model of spinal cord injury by BioLuminescent-OptoGenetic (BL-OG) stimulation with an enhanced luminopsin. Int J Mol Sci. 2022;23(21):12994.

14. Ikefuama EC, Slaviero AN, Silvagnoli AD, Crespo EL, Schalau R, Gott M, et al. Presymptomatic targeted circuit manipulation for ameliorating Huntington's disease pathogenesis. iScience. 2025;28(3):112022.
15. Li T, Chen X, Qian Y, Shao J, Li X, Liu S, et al. A synthetic BRET-based optogenetic device for pulsatile transgene expression enabling glucose homeostasis in mice. Nat Commun. 2021;12(1):615.
16. Su Y, Walker JR, Park Y, Smith TP, Liu LX, Hall MP, et al. Novel NanoLuc substrates enable bright two-population bioluminescence imaging in animals. Nat Methods. 2020;17(8):852–60.
17. Su Y, Walker JR, Hall MP, Klein MA, Wu X, Encell LP, et al. An optimized bioluminescent substrate for non-invasive imaging in the brain. Nat Chem Biol. 2023;19:731–9.
18. Gao C, Wu Y, Fitzgerald C, Wang H, Ugo T, Uyeda T, et al. An optimized luciferin formulation for NanoLuc-based in vivo bioluminescence imaging. Sci Rep. 2025;15(1):12884.
19. Merolla A, Michetti C, Moschetta M, Vacca F, Ciano L, Emionite L, et al. A pH-sensitive closed-loop nanomachine to control hyperexcitability at the single neuron level. Nat Commun. 2024;15(1):5609.
20. English AW, Berglund K, Carrasco D, Goebel K, Gross RE, Isaacson R, et al. Bioluminescent optogenetics: a novel experimental therapy to promote axon regeneration after peripheral nerve injury. Int J Mol Sci. 2021;22(13):7217.
21. Lee CH, Lin JH, Lin SH, Chang CT, Wu YW, Bewick G, et al. A role for proprioceptors in sngception. Sci Adv. 2025;11(5):eabc5219.
22. He L, Huang Z, Huang K, Chen R, Nguyen NT, Wang R, et al. Optogenetic control of non-apoptotic cell death. Adv Sci. 2021;8(13):2100424.

Chapter 12
Application of BL-OG Tools in Translational and Basic Science

Ute Hochgeschwender

Abstract The ultimate goal of tool development is application in contexts of disease therapies and/or to address basic biological questions. Bioluminescent optogenetics (BL-OG) tools, molecular reagents that utilize light emitted by a luciferase when interacting with its luciferin to activate light-sensing molecules, have been employed in both areas. BL-OG was successfully used in preclinical, translational studies in animal models of epilepsy, peripheral nerve and spinal cord injury, stroke, neurodegenerative diseases, sleep disorders, diabetes, and various forms of cancer. Basic science studies were conducted in areas, including neurodevelopment, memory and learning, sngception, fear and anxiety, and the dissection of motor circuitry. All applications benefited from the advantages of BL-OG over standard optogenetic and chemogenetic approaches, particularly its noninvasiveness, spatial access, and fast kinetics and absence of cross-reactivity of luciferins. Moreover, BL-OG applications will increasingly enable unique capabilities not afforded by any currently available technology, such as bimodal opto- and chemogenetic activation, genetically targeted cell-to-cell communication, and integration of intracellular states with cellular outcomes. The studies reviewed in this chapter highlight the potential of BL-OG for therapeutic and fundamental basic science applications.

Keywords In vivo · Chemogenetics · Optogenetics · Preclinical · Therapeutic

12.1 Introduction

As with any molecular tool development, bioluminescent optogenetics (BL-OG) tools are first tested in cultured cells, usually proceeding from cell lines (e.g., HEK293 or HeLa) to primary cells (e.g., cultured embryonic neurons). The next step in the tool development pipeline is to assess functionality in vivo in a suitable

---

U. Hochgeschwender (✉)
College of Medicine, Central Michigan University, Mt. Pleasant, MI, USA
e-mail: hochg1u@cmich.edu

U. Hochgeschwender (ed.), *BioLuminescent OptoGenetics*,
https://doi.org/10.1007/978-3-032-18121-3_12

experimental animal system, such as mice, rats, or *C. elegans*. These proof-of-principle studies aim to demonstrate the function of the tool in live animal systems. This includes multichannel recordings in vivo of triggered neural activity, fluorescent imaging of induced transcription of reporter constructs, and eliciting behavioral sequences that allow quantification of in vivo effects of the tool. These in vivo applications are described for each tool in the context of the chapter that introduces the tool. The focus of this chapter is to highlight instances where BL-OG tools have been applied in animal studies in a translational or basic research context that aimed to utilize BL-OG for therapy of disease conditions or to address basic biological mechanistic questions. Table 12.1 lists the studies conducted grouped under the disease conditions and basic science subjects they are addressing, together with the BL-OG tool employed, the animal system used, and a brief description of the study. The subsequent sections go into more detail about the experiments conducted.

12.2 Translational Studies Involving the Nervous System

The majority of studies in the translational arena thus far involve disorders of the nervous system. This reflects the earlier availability of luminopsins relative to other BL-OG tools and the prevalence of standard optogenetics applications in neuroscience over other fields.

12.2.1 Epilepsy

One of the earliest therapeutic applications of BL-OG was in epilepsy. Tung et al. [26] developed the inhibitory luminopsin iLMO2 from a motivational background of epilepsy research and applied it in a rat model of epilepsy [1]. They showed that cell-specific multi-focal control of pathological circuits enabled by BL-OG was more effective at suppressing behavioral seizures than inhibition of the individual structures [1]. Simultaneous inhibition of four different nodes in the brain, including the dentate gyrus and anterior nucleus of the thalamus, had an additive effect of significantly reducing seizure duration and severity induced by pentylenetetrazol (PTZ) compared to inhibition of the individual nodes.

The same optogenetic actuator used in iLMO2 for silencing neuronal activity, halorhodopsin (NpHR), was exploited in developing pHIL, a pH-sensitive inhibitory luminopsin that uses a change in pH as a proxy for hyperexcitation [2]. Here, a fluorescent sensor of intracellular pH (E2GFP) allows light from a fused RLuc8 luciferase to activate the tethered NpHR in acidic conditions. The authors showed that pHIL is able to sense acidosis caused by an incoming seizure and reestablish a physiological excitation/inhibition balance by silencing the activity of hyperexcited principal neurons in the hippocampus. They demonstrated the therapeutic potential of this BL-OG integrator in an acute seizure model evoked by administration of the

Table 12.1 Translational and basic research studies employing BL-OG tools in vivo

Translational				
	BL-OG tool	Animal	Main findings	Reference
Epilepsy				
	iLMO2	Rat	Multi-nodal inhibition of seizures in chemical model	[1]
	pHIL	Mouse	Effective in acute chemical and genetic models	[2]
Peripheral nerve injury				
	iLMO2	Mouse	Inhibition of motoneuron activity prevents exercise-induced functional recovery	[3]
	LMO3	Mouse	Motoneuron-specific excitation promotes axon regeneration	[4]
	LMO3	Mouse	Regeneration of sensory axons; effective with LMO expression before and after the nerve injury	[5]
	LMO3	Mouse	Impaired sympathetic neuron regeneration in contrast to moto- and sensory neurons	[6]
Spinal cord injury				
	LMO3	Rat	Activation of spinal neurons results in accelerated and enhanced locomotor recovery	[7]
	LMO3.2	Rat	Using LMO with higher light sensitivity allows peripheral application of luciferin	[8]
Stroke				
	LMO3	Mouse	Activation of transplanted neural stem cells promotes neural network repair and functional recovery	[9]
Neurodegenerative disorders				
	LMO3	Mouse	Activation of transplanted neural stem cells improves motor performance in genetic model of Parkinson's disease	[10]
	LMO3	Rat	Activation of transplanted neural stem cells improves motor performance in 6-OHDA model of Parkinson's disease	[11]
	GeNL_SS-ChRmine	Mouse	Circuit-specific activation during pre-symptomatic stages alleviates disease manifestation in later stages of Huntington's disease	[12]
	GeNL_SS-hGtACR1			
Sleep disorders				
	LMO3	Mouse	Activation of hypoglossal motor neurons alleviates sleep-disordered breathing	[13]

(continued)

Table 12.1 (continued)

Translational				
Diabetes				
	luminGAVPO	Mouse	Highly tunable control of glucose homeostasis in STZ diabetes model	[14]
Cancer				
	NLuc-CRY2	Mouse	Killing of human renal cell carcinoma cells through necroptosis	[15]
	NLuc-CRY2/CIB	Mouse	Inhibition of human retinoblastoma cell growth through apoptosis	[16]
	NLuc-CoChR	Mouse	Inhibition of glioblastoma and breast cancer tumor growth through mitochondrial-targeted membrane depolarization	[17]
	RLuc-KillerRed	Mouse	Inhibition of breast cancer tumor growth through production of reactive oxygen species	[18]
	NLuc-MiniSOG	Mouse	Inhibition of localized and deep-seated widespread tumor growth through production of reactive oxygen species	[19]
	NLuc-SOPP3	Mouse		[20]
Basic research				
	BL-OG tool	Animal	Description	Reference
Neurodevelopment	LMO3	Mouse	Developmental hyperexcitation results in key behavioral and neurophysiological signatures of ASD in adults	[21]
Memory & learning	LMO3 iLMO	Mouse	Activation or inhibition of CA3 neurons affects working and short term spatial and episodic memory	[22]
Sngception	LMO3	Mouse	Acidosis perception from muscle (sngception) is transmitted by proprioceptors	[23]
Fear & anxiety	LMO3	Mouse	Activation of ASIC4-expressing cells induces anxiogenic behavior	[24]
Motor circuits	PhAST	*C. elegans*	Deciphering motor circuit for ovipositioning	[25]

chemo-convulsant pilocarpine and in a genetic mouse model of human epileptic disorders, where seizures can be easily triggered by environmental stimuli such as sound [2]. They observed significant delays in the appearance of the most severe seizure behaviors and reduction in the duration of and the percentage of mice developing generalized tonic-clonic seizures for the acute seizure model and suppression or significant weakening of epileptic outbreaks evoked by sound in the genetic epilepsy model, leading the authors to suggest that it could be a seizure-preventive treatment in chronic epilepsy forms, where seizures are triggered by sensory stimuli.

In addition to showing the potential for therapeutic applications of BL-OG, these two studies also reflect the evolution in BL-OG tool development. The initial application of iLMO2 takes advantage of the strengths of the opto-chemical approach, specifically the access to brain-wide treatment without the need for any hardware and the use of ion channels for direct membrane potential changes compared to chemogenetic approaches utilizing G-protein coupled receptors (GPCRs) as used in DREADDs [27]. The most recent application of a bioluminescent optogenetic integrator, pHIL, exemplifies the unique power provided by coupling biological light locally generated in response to intracellular conditions with optogenetic actuators for self-regulation of disease states in living systems.

12.2.2 Peripheral Nerve Injury

There are several studies utilizing BL-OG tools in therapeutic approaches for the injured nervous system. One area of application is peripheral nerve injury (PNI). Results from preclinical and clinical studies indicate that neuronal activity enhances nerve regeneration and regain of function after PNI. To probe underlying mechanisms, sciatic nerve transection-repair in mice was employed as a model of PNI [3–6]. In these studies, the sciatic nerve is cut, and its ends are immediately secured together by application of fibrin glue. In mice, daily exercise by interval treadmill-training following this injury results in enhanced motor axon regeneration and muscle fiber reinnervation. To probe the motoneuron-specific contributions to axon regeneration and recovery of function, Jaiswal et al. [3] injected Cre-dependent inhibitory iLMO2 into the gastrocnemius and tibialis anterior muscles of ChAT-IRES-Cre mice, followed by sciatic nerve transection and repair. Inhibiting motoneuron activity during post-injury exercise by injection of luciferin, administered 30 min prior to exercise, completely blocked the exercise-induced enhancements in functional recovery.

In a subsequent study, the authors explored whether motoneuron-specific excitation will promote axon regeneration after PNI [4]. They performed the sciatic nerve transection-repair in double transgenic ChAT-Cre::lox-STOP-lox-LMO3 mice, restricting expression of the excitatory LMO to motoneurons. A single intraperitoneal administration of luciferin immediately after sciatic nerve transection and repair markedly enhanced motor axon regeneration when tested 4 weeks later. The magnitude of evoked compound muscle action potentials in the reinnervated tibialis anterior and gastrocnemius muscles was significantly larger than found in controls. Retrograde labeling of motoneurons revealed that BL-OG treatment significantly increased the number of motoneurons whose axons regenerated successfully. Both metrics of enhanced axon regeneration after PNI indicated that increasing the excitability of motoneurons by BL-OG treatment promotes motor axon regeneration. The results observed with transgenic mice were comparable to those of mice in which the LMO was expressed using viral vectors.

Another study evaluated whether directly stimulating the injured neurons via BL-OG treatment would have a similar effect on the regeneration of muscle sensory axons [5]. Using the same experimental paradigm, no significant difference in the number of fluorescent, retrogradely labeled sensory neurons was found between BL-OG-treated PNI mice and intact controls, suggesting complete sensory reinnervation in those animals.

The results presented above are based on experiments where the excitatory luminopsin was expressed in neurons prior to the injury of the peripheral nerve containing these neurons' axons. Using a delayed reinnervation model, the authors evaluated the efficacy of BL-OG when LMOs were expressed after the nerve injury [5]. In this case, the common fibular nerve (CFn) was cut, and AAV encoding LMO3 was injected into the proximal stump of the cut nerve. Four weeks later, the unligated proximal stump of the CFn was attached to the distal segment of a freshly cut tibial nerve, and animals were treated with luciferin. Four weeks later, compound muscle action potentials elicited in the reinnervated gastrocnemius muscle by electrical stimulation of the sciatic nerve proximal to the injury site were significantly larger in BL-OG-treated mice compared to controls, indicating the efficacy of BL-OG treatment in this clinically relevant scenario.

In addition to motor and sensory deficits, peripheral nerve injuries also cause autonomic deficits that impede functional recovery after nerve injury, manifesting in muscle atrophy and deficits in muscle cellular energy charge. Tian et al. [6] compared motor and sensory axon regeneration to sympathetic regeneration in the acute sciatic nerve transection-repair model. Expression of LMO3 was directed to lumbar sympathetic neurons in ThCre:lox-STOP-lox-LMO3 mice. Neither BL-OG nor electrical stimulation of sympathetic neurons enhanced their regeneration, with the cell-specific BL-OG treatment actually resulting in impaired sympathetic regeneration, indicated by a decrease in the number of retrograde-traced sympathetic neurons. These findings suggest that sympathetic neurons do not follow the same principles when responding to regeneration-enhancing neural stimulation that is effective in motor and sensory axon regeneration. The distinct molecular signature underlying sympathetic neuron regeneration needs to be elucidated to develop therapeutic options that will enhance the regeneration of all axons.

12.2.3 Spinal Cord Injury

Many spinal cord injuries leading to paralysis are crush injuries that leave neuronal connections intact and amenable to being maintained and strengthened by stimulating their activity. Two studies explored BL-OG to reengage functionally ineffective, dormant populations of neurons below an injury to activate spinal circuitry for improved functional output [7, 8]. Neurons of the rat lumbar enlargement were transduced with AAV vectors to express LMOs, followed by a contusion injury of the thoracic spinal cord 3 weeks later, at which time BL-OG treatment and testing of locomotor behavior were carried out for 2 weeks. In the first study, the excitatory LMO3 (sbGLuc-VChR1) was expressed under control of a pan-neuronal or motor neuron-specific promoter, and the luciferin was applied through a lateral ventricle cannula [7]. Activation of either neuron population below the level of injury significantly improved locomotor recovery measured against a standard assessment scale. The second study employed LMO3.2, a fusion of sbGLuc to the fast and sensitive opsin CheRiff [8]. The higher light sensitivity of CheRiff required lower levels of luciferin for optimal activation and thus allowed for peripheral delivery. BL-OG treatment with intraperitoneal luciferin injections achieved results in locomotor recovery comparable to the use of a lateral ventricular cannula while avoiding its invasiveness and bringing this approach closer to clinical application.

12.2.4 Stroke

In the previous examples of application of BL-OG tools in the brain, spinal cord, and peripheral nervous system, neuronal excitation or inhibition was carried out in affected neurons still able to fire. In the case of stroke, neurons in the affected area are too damaged to be directly manipulated. Yu et al. [9] explored creating an activity-enriched microenvironment that confers regenerative benefits through the excitation of transplanted induced pluripotent stem cell-derived neural progenitor cells (iPS-NPCs) for the treatment of stroke. Stable LMO3-expressing iPS-NPCs were grafted into the core and peri-infarct region of the ischemic sensorimotor cortex of mice, where they differentiated into mature neurons and formed synapses with host neurons. Stimulation of transplanted cells with daily intranasal administration of luciferin promoted neuronal differentiation, axonal remyelination, and long-term survival in the ischemic brain. *In vivo* single-unit recordings provided evidence for the reestablishment of cortical connections and neural network repair. Motor coordination was restored to normal levels and anxiety was decreased, demonstrating functional and behavioral recovery after stroke. These significant improvements were achieved with BL-OG treatment of transplanted cells, not with cell transplantation alone. This application illustrates the advantage of the BL-OG approach using locally produced light for neural excitation of transplanted cells across brain regions because transplanted cells are mobile and capable of migrating long distances.

12.2.5 Neurodegenerative Disorders

Neurodegenerative diseases, where the affected neurons are too damaged or are gone altogether, is another area benefitting from the excitation of transplanted cells. Two studies explored bioluminescence-driven excitation of transplanted cells in models of Parkinson's disease (PD). Zenchak et al. [10] generated a self-renewing neural stem cell line from mouse embryonic stem cells carrying a knock-in allele of LMO3. Bilateral transplantation into the striatum of homozygous aphakia mice (Pitx3 mice), which carry a spontaneous mutation leading to a lack of dopaminergic neurons and symptoms of PD, followed by stimulation of transplanted cells with luciferin delivered via a lateral ventricle cannula, resulted in significantly improved motor skills in an accelerating rotarod test. In contrast, transplantation of cells without luciferin application or luciferin application without cell transplants had no effects on motor deficits.

Anderson et al. [11] used a rat PD model with unilateral 6-OHDA injection of the medial forebrain bundle. Adipose-derived mesenchymal stem cells stably expressing LMO3 were pre-differentiated in vitro into dopaminergic cells, then transplanted into the striatum of the same site that previously received the 6-OHDA injection. Stimulation of transplants by administration of luciferin through a lateral ventricle cannula caused lasting reduction of motor deficits, especially the asymmetric limb usage characteristic of the unilateral 6-OHDA lesioned rat model of PD.

Huntington's disease (HD) is an inherited neurodegenerative disorder caused by a specific mutation, and thus incidence of the disease can be determined through genetic testing. Early diagnosis opens the door for presymptomatic interventions that might slow down or significantly ameliorate disease manifestation. In both patients and animal models, imbalanced output from the cortex to striatal and subcortical structures has been observed in early stages of HD before the onset of motor and cognitive symptoms. Ikefuama et al. [12] used BL-OG for cell-type-specific modulation to impede disease progression in R6/2 HD mice. Excitatory and inhibitory luminopsins combining a fusion protein of a molecularly evolved shrimp luciferase, SSLuc, and the fluorescent protein mNeonGreen (GeNL_SS [28]), with highly light sensitive opsins, ChRmine and hGtACR1, were expressed in specific populations of the motor cortex to either inhibit pyramidal neuron activity or to strengthen inhibitory activity of parvalbumin interneurons. Directly decreasing the activity of pyramidal neurons or increasing the activity of interneurons to dampen pyramidal neurons through daily intraperitoneal luciferin application during the presymptomatic period decreased the overall drive from the M1 motor cortex to the striatum. This resulted in attenuated locomotor impairments determined by two behavioral paradigms: motor coordination on a rotarod and gait parameters in a Catwalk. This study revealed that bioluminescence-driven optogenetic modulation of neuronal activities before the onset of symptoms attenuates motoric phenotypes seen in later stages of HD.

12.2.6 Sleep Disorders

Wang et al. [13] applied BL-OG as a gene therapy approach to obstructive sleep apnea. LMO3 was targeted to hypoglossal motor neurons through peripheral injection into the tongue of retrogradely transported AAV in diet-induced obese mice as a model of sleep-disordered breathing. Luciferin administered via intraperitoneal injection caused increased firing of hypoglossal motor neurons with an increase in phasic and tonic activity of the genioglossus muscle measured with electromyographic recordings. BL-OG-mediated amplification of genioglossus muscle discharge activity increased airflow, while the inspiratory flow limitation was substantially alleviated after luciferin treatment during both NREM and REM sleep. This study demonstrated that bioluminescent stimulation-induced activation of hypoglossal motor neurons can alleviate airway obstruction and hypoventilation in sleep-disordered breathing mice and could be an effective means to treat obstructive sleep apnea.

12.3 Translational Studies Outside the Nervous System

With more BL-OG tools that utilize bioluminescence to activate photoreceptors other than opsins becoming available, translational applications will expand beyond modulating neural activity. Two examples are the treatment of diabetes and of cancer.

12.3.1 Diabetes

Under normal conditions, insulin secretion is maintained at low levels before quickly rising in response to food intake. Type 1 diabetic patients do not produce enough insulin and rely on insulin administration from external sources, with the risk of hypoglycemia if too much insulin is used. Li et al. [14] used a BL-OG-based system for maintenance of glucose homeostasis that is highly tunable to different dietary habits. Light emission from NanoLuc catalyzes dimerization of a fused Vivid-based transcription construct (luminGAVPO), with the dimerized luminGAVPO binding its promoter region in front of the insulin sequence, leading to expression of insulin (see also Sect. 3.2, Fig. 3.1). HEK293 cells genetically engineered to carry within their genome luminGAVPO and the promoter-insulin expression cassette were microencapsulated and implanted intraperitoneally into type 1 diabetic mice (TD1 mice, a streptozotocin-based mouse model for diabetes). Instead of the traditional intramuscular injection of insulin before and after meals, oral administration of different dosages of luciferin at different time intervals before feeding enabled a highly flexible management of glucose homeostasis within narrow fluctuations in glucose levels before and after feeding. The optimal dosage and time for luciferin administration and resulting insulin production were closely correlated with the feeding intervals, with increased feeding intervals managed by increased doses of luciferin administered at longer time periods before feeding [14]. Taking advantage of the unique kinetics of bioluminescent light emission that allows pulsatile and quantitative activation of transgene expression via luciferin administration, the luminGAVPO system can be used to precisely control the dynamics of a key signaling protein in a pulsatile fashion, providing a highly tunable pharmacological intervention for sustainable restoration of blood glucose.

12.3.2 Cancer

The ultimate goal of any cancer therapy is the specific killing of tumor cells without detrimental effects on other cells. Genetically targetable local light emission combined with light activation of mediators of cell death can achieve this goal. Below are examples of diverse strategies developed for the application of BL-OG in cancer treatment.

He et al. [15] used NanoLuc to photoactivate LiPOP1, an optogenetic tool based on CRY2 oligomerization, triggering necroptosis and causing cell death by directly permeabilizing the plasma membrane of tumor cells (see also Sect. 3.7, Fig. 3.8a). These authors subcutaneously inoculated human renal cell carcinoma 786-O cells engineered to stably express LiPOP1-NanoLuc into the dorsal flanks of immunodeficient SCID mice. Intratumoral injection of luciferin over 3 weeks resulted in a significant reduction in the tumor size compared to controls, as determined through in vivo bioluminescence imaging and direct measurement of tumor diameters at endpoint.

Ding et al. [16] employed NanoLuc to trigger activation of the apoptotic pathway based on CRY-CIB dimerization (see also Sect. 3.7, Fig. 3.8b) for BL-OG-mediated therapy of retinoblastoma. Y79 human retinoblastoma cells were injected into the vitreous body of the right eye of nude mice and allowed to form tumors for 16 days. To test efficacy in preventing further tumor growth, the authors used vitreous injection of a nanoparticle-based optogenetic system that delivered the light emitter NanoLuc, the light sensor CRY2, and all plasmids associated with the optogenetic activation of the apoptotic pathway. The optogenetic system was activated the next day by either intraperitoneal injection of luciferin or by direct exposure of the eye to blue light. Tumor sizes and the general health of the eyes were assessed after three rounds of these 2-day treatments, applied 8 days apart. The in situ bioluminescence-treated group showed significant tumor cell apoptosis, resulting in effective inhibition of further tumor growth observed by in vivo imaging and histological analyses as well as lack of elevated intraocular pressure. Importantly, BL-OG-treated mice had clear corneas without neovascularization. In contrast, blue LED-treated mice showed effective inhibition of tumor growth but also exhibited light irradiation-based damage to the eye, such as corneal vascular hyperplasia.

The specific targeting of mitochondria in cancer cells is an approach applicable to various cancer types. Proper mitochondrial function relies on maintaining the electrical potential gradient across the inner mitochondrial membrane; disruption of this gradient is a crucial trigger for cell death. Chen et al. [17] targeted this property to develop mitochondrial-targeted BL-OG-based therapy for treating cancer. Their mLumiOpto system consists of the cationic channelrhodopsin CoChR targeted to the inner mitochondrial membrane and NanoLuc expressed in the cytosol. Application of luciferin induces mitochondrial depolarization and cytotoxicity, leading to cancer cell death. They applied this system in various xenograft mouse models of different types of glioblastoma (GBM) and triple-negative breast cancer (TNBC) tumors. Growth of intracranially grafted human GBM U87 and PDX tumor cells was monitored by MRI or IVIS imaging. Established tumors were treated by intracerebroventricular injection of AAV-mLumiOpto, followed by intravenous administration of luciferin. In both GBM models, depolarizing the mitochondria achieved dramatic reductions in tumor burden without damage to other organs and significantly extended survival compared with controls. Two preclinical TNBC models also showed effective inhibition of tumor growth. Tumor growth was induced by subcutaneously injecting human MDA-MB-231 or mouse 4 T1 cells into the mammary fat pads of mice. The mLumiOpto genes were delivered to tumor-bearing mice by intravenous injection of exosome-encapsulated AAV surface tagged with an EGFR mAB, focusing AAV transduction to the EGFR-positive tumor cells. Tumor growth stopped shortly after the start of intravenous luciferin administration, with tumor volume shrinking during the treatment. As with GBM, for TNBC mLumiOpto had high antitumor efficacy and minimal off-target toxicity.

Bioluminescent light emission has been harnessed to cause cancer cell death by activating genetically encoded photosensitizers that induce the production of reactive oxygen species (ROS), a form of bioluminescent optogenetics photodynamic therapy (PDT). Excessive ROS production directly damages mitochondria and cellular molecules, including DNA and proteins, thereby disrupting cell functions and triggering programmed cell death pathways.

Kim et al. [18] generated a fusion protein of RLuc8.6 and KillerRed that produces ROS in the presence of luciferin. They injected a protein solution of this fusion construct directly into tumors established by subcutaneous injection of triple-negative MDA-MB-231 breast cancer cells into mice. Luciferin was subcutaneously injected in the tumor area the following day. This sequential injection of protein and luciferin at 1-day intervals performed three times, 8 days apart, significantly inhibited the growth of the tumors, compared to untreated controls and to protein-injected laser light–irradiated mice.

A fusion construct of NanoLuc and the phototoxic flavoprotein MiniSOG was packaged into lentiviral vectors for intratumoral injection in mice carrying human ductal carcinoma BT-474 cells xenografted subcutaneously [19]. Generation of ROS by intravenous administration of luciferin resulted in significant tumor-growth inhibition. Further improvements of this BL-OG PDT system swapped MiniSOG for the improved version SOPP3 that is characterized by high photoinduced phototoxicity through increased singlet oxygen production [20]. In addition, a delivery method closer to clinical use was employed by loading the NLuc-SOPP3 protein into liposomes. The therapeutic effect of this system was tested on subcutaneous and intraperitoneal xenograft models. In mice bearing subcutaneous BT-474 tumor xenografts, intravenous application of liposomes loaded with proteinaceous NLuc-SOPP3 and luciferin achieved a substantial 78% in tumor reduction [20]. To demonstrate efficacy in a deep-seated and widespread tumor model, the authors generated mice bearing SKOV3ip-Kat intraperitoneal tumor xenografts, human ovarian cancer cells that form multiple small tumor nodules in the abdominal cavity upon intraperitoneal implantation. Intravenous application of loaded liposomes together with intraperitoneal administration of luciferin resulted in fewer and smaller intraperitoneal tumor nodules [20].

12.4 Basic Science Studies

As the application of newly generated tools lags behind their development, BL-OG tools applied to basic research questions are limited, thus far, to the use of luminopsins (LMOs) for changing membrane potential in neurons. Compared to standard optogenetics that uses implanted light devices, the strength of the BL-OG approach is its noninvasiveness. The concern about invasive hardware is less of a priority in basic research compared to translational work. Compared to chemogenetic approaches, such as DREADDs, LMOs provide much better kinetics with faster on/off times, and several studies have taken advantage of this. For a list of advantages of using bioluminescence compared to external light or chemicals, see Chap. 1, Table 1.1.

An investigation by Medendorp et al. [21] made use of the unique opportunity LMOs provide for bimodal, opto- and chemogenetic control of neural activity. The study investigated whether early enhancement of neural activity within the developing neocortical network can induce adult behavioral phenotypes and related circuit changes characteristic of neurodevelopmental disorders. Pan-neocortical pyramidal activity levels were systematically enhanced in early development in healthy mice by chemogenetically activating LMO3 expressed in Emx1-positive neurons through daily intraperitoneal injections of luciferin during postnatal days 4–14. Adult animals were evaluated for changes in behavior and for related circuit changes by taking advantage of optogenetic access to the same neurons initially chemogenetically hyperexcited during development. Developmental pyramidal overdrive selectively led to decreased social interaction and increased grooming activity, two key symptoms of autism spectrum disorder (ASD). Optogenetic interrogation of cortical pyramidal neurons identified enhanced stimulus-evoked firing, decreased output connectivity, and decreased cortico-striatal communication, all key neurophysiological signatures of ASD. These results show that brief and selective increases in pyramidal activity over a limited developmental window can generate key behavioral and neurophysiological signatures of ASD, implicating aberrant neocortical pyramidal activity during early development as a potential common driver of ASD-like phenotypes. The combined chemogenetic and optogenetic access through the same genetically targeted molecule enabled by BL-OG made this study possible by allowing developmental perturbation and adult functional probing of system connectivity and excitability.

To study the neurobiology of memory, specifically how CA3 functions in spatial and episodic memory in short versus long time scales, another investigation employed LMO3 and iLMO [22]. AAV injections delivered excitatory LMO3 and inhibitory iLMO into CA3 neurons, and luciferin was applied intraperitoneally to interfere with CA3 neural networks. Executing an array of behavioral tests revealed that activation or inhibition of neurons in CA3 does not influence long-term spatial or episodic memory. Instead, controlling these neurons had a dramatic effect on working and short-term spatial memory as well as short-term episodic memory, underscoring the importance of CA3 on neuronal circuits of short-term memory.

Proprioceptors are non-nociceptive mechanoreceptors that monitor the status of muscle contraction and/or body position and express acid-sensing ion channels. This property led Lee et al. [23] to propose a new biological concept that segregates the perception of muscle acidosis, coined sng (muscle soreness), from the perception of pain. According to their evidence, sngception is detected and transmitted by mechanosensory neurons, while nociception is detected and transmitted by specific somatosensory neurons. To disambiguate the causal role of proprioceptors and nociceptors in acid-induced pain chronicity, the authors crossed Cre-dependent LMO3 mice with Cre-driver lines restricting expression of LMO3 to proprioceptors and nociceptors, respectively. Taking advantage of the unique kinetics of BL-OG for fast- and short-term chemogenetic effects, they selectively primed each target by injecting luciferin and acid into muscle, followed by a second acid injection, to test the development of chronic hyperalgesia. Their results indicated that activated proprioceptors, but not nociceptors, contribute to hyperalgesic priming. This finding is evidence of a neurobiological basis for sngception to be distinguished from nociception.

The same group applied BL-OG in studies of the neurobiology underlying the regulation of anxiety and fear responses [24]. To clarify whether activating ASIC4 (acid-sensing ion channel 4)-expressing cells could modulate fear and anxiety responses, mice Cre-dependently expressing LMO3 in ASIC4-positive cells underwent behavioral tests assessing fear responses after intraperitoneal application of luciferin. BL-OG activation of ASIC4-expressing cells induced anxiogenic behavioral responses in mice.

The *C. elegans* egg-laying circuit is a seemingly simple, yet functionally not well-understood, neuronal circuit. Porta-de-la-Riva et al. [25] took advantage of their photon-assisted synaptic transmission (PhAST) approach to express the Ca^{2+}-dependent luciferase in two different presynaptic neuron populations (the motoneurons HSB and VC) and the channelrhodopsin in their shared postsynaptic target population, vm2, in *C. elegans* vulval muscles. After selective inhibition of synaptic vesicle release in motoneurons by tetanus toxin expression, PhAST was used to reinstate excitatory transmission at the neuromuscular junction. When both motoneurons lacked endogenous transmission, PhAST could not rescue activity in muscle regardless of presynaptic luciferase expression in HSN, VC, or both. However, when VC transmission remained intact, PhAST-mediated restoration of HSN activity successfully rescued muscle cell activity. These findings support a proposed disinhibitory circuit, where VC activity suppresses the inhibitory influence of neuroendocrine cells on HSN. These studies illustrate the potential of synaptic engineering for dissecting the wiring pattern of a neuronal circuit by systematic cell-specific expression of light-emitting pre- and light-sensing postsynaptic components.

As exemplified in this last example and given the aforementioned lag between tool development and application, one can anticipate the impact of BL-OG tools enabling cell-to-cell communication and cellular integration on discovery of biological mechanisms in basic science studies using unique experimental approaches not afforded by any other molecular platform.

12.5 Summary

This chapter provides an overview of reports that used BL-OG tools for preclinical, translational studies in animal models. Areas of disease model applications included epilepsy, peripheral nerve and spinal cord injury, stroke, neurodegenerative diseases, sleep disorders, diabetes, and various forms of cancer. Across applications, the BL-OG approach proved to be efficient while profiting from distinct benefits. Advantages compared to standard optogenetic approaches include the following: noninvasiveness and the spatial reach across all actuator-expressing cells in any anatomical location. Advantages relative to other chemogenetic approaches are as follows: the fast kinetics afforded by luciferins and the absence of cross-reactivity with endogenous molecules, reducing the likelihood of unintended cellular effects. None of the studies reported any toxic or undesirable side effects, highlighting the positive safety profile of BL-OG for therapeutic use. Nevertheless, more extensive studies of dose and dosing of luciferins will be required in clinically relevant models.

The advantages of the BL-OG approach extend to basic science studies in areas including neurodevelopment, memory and learning, sngception, fear and anxiety, and the dissection of motor circuitry. Moreover, taking advantage of aspects uniquely provided by BL-OG, such as bimodal opto- and chemogenetic activation, genetically targeted cell-to-cell communication, and integration of intracellular states with the cell's response, will enable the discovery and dissection of biological processes not afforded by any currently available technology.

Taken together, the studies reviewed in this chapter underscore the potential of BL-OG for therapeutic and fundamental basic science applications.

References

1. Tung JK, Shiu FH, Ding K, Gross RE. Chemically activated luminopsins allow optogenetic inhibition of distributed nodes in an epileptic network for non-invasive and multi-site suppression of seizure activity. Neurobiol Dis. 2018;109(Pt A):1–10.
2. Merolla A, Michetti C, Moschetta M, Vacca F, Ciano L, Emionite L, et al. A pH-sensitive closed-loop nanomachine to control hyperexcitability at the single neuron level. Nat Commun. 2024;15(1):5609.
3. Jaiswal PB, Tung JK, Gross RE, English AW. Motoneuron activity is required for enhancements in functional recovery after peripheral nerve injury in exercised female mice. J Neurosci Res. 2020;98:448–57.
4. English AW, Berglund K, Carrasco D, Goebel K, Gross RE, Isaacson R, et al. Bioluminescent optogenetics: a novel experimental therapy to promote axon regeneration after peripheral nerve injury. Int J Mol Sci. 2021;22(13):7217.
5. Ecanow A, Berglund K, Carrasco D, Isaacson R, English AW. Enhancing motor and sensory axon regeneration after peripheral nerve injury using bioluminescent optogenetics. Int J Mol Sci. 2022;23(24):16084.
6. Tian T, Kim D, Yu K, Hartzell HC, Ward PJ. Regenerative failure of sympathetic axons contributes to deficits in functional recovery after nerve injury. Neurobiol Dis. 2025;209:106893.

7. Petersen ED, Sharkey ED, Pal A, Shafau LO, Zenchak-Petersen J, Peña AJ, et al. Restoring function after severe spinal cord injury through BioLuminescent-OptoGenetics. Front Neurol. 2022;12:792643.
8. Ikefuama EC, Kendziorski GE, Anderson K, Shafau L, Prakash M, Hochgeschwender U, et al. Improved locomotor recovery in a rat model of spinal cord injury by BioLuminescent-OptoGenetic (BL-OG) stimulation with an enhanced luminopsin. Int J Mol Sci. 2022;23(21):12994.
9. Yu SP, Tung JK, Wei ZZ, Chen D, Berglund K, Zhong W, et al. Optochemogenetic stimulation of transplanted iPS-NPCs enhances neuronal repair and functional recovery after ischemic stroke. J Neurosci. 2019;39(33):6571–94.
10. Zenchak JR, Palmateer B, Dorka N, Brown TM, Wagner LM, Medendorp WE, et al. Bioluminescence-driven optogenetic activation of transplanted neural precursor cells improves motor deficits in a Parkinson's disease mouse model. J Neurosci Res. 2020;98:458–68.
11. Anderson KA, Whitehead BJ, Petersen ED, Kemme MR, Wedster A, Hochgeschwender U, et al. Behavioral context improves optogenetic stimulation of transplanted dopaminergic cells in unilateral 6-OHDA rats. Behav Brain Res. 2023;441:114279.
12. Ikefuama EC, Slaviero AN, Silvagnoli AD, Crespo EL, Schalau R, Gott M, et al. Presymptomatic targeted circuit manipulation for ameliorating Huntington's disease pathogenesis. iScience. 2025;28(3):112022.
13. Wang Y, Liu X, Zhang Q, Zhao D, Zhou B, Pan Z, et al. Bioluminescence-optogenetics-mediated gene therapy in a sleep-disordered breathing mouse model. Biomed Pharmacother. 2024;178:117159.
14. Li T, Chen X, Qian Y, Shao J, Li X, Liu S, et al. A synthetic BRET-based optogenetic device for pulsatile transgene expression enabling glucose homeostasis in mice. Nat Commun. 2021;12(1):615.
15. He L, Huang Z, Huang K, Chen R, Nguyen NT, Wang R, et al. Optogenetic control of non-apoptotic cell death. Adv Sci. 2021;8(13):2100424.
16. Ding J, Lu J, Zhang Q, Xu Y, Song B, Wu Y, et al. Camouflage nanoparticles enable in situ bioluminescence-driven optogenetic therapy of retinoblastoma. ACS Nano. 2023;17(8):7750–64.
17. Chen K, Ernst P, Sarkar A, Kim S, Si Y, Varadkar T, et al. mLumiOpto is a mitochondrial-targeted gene therapy for treating cancer. Cancer Res. 2024;84(23):4049–65.
18. Kim EH, Park S, Kim YK, Moon M, Park J, Lee KJ, et al. Self-luminescent photodynamic therapy using breast cancer targeted proteins. Sci Adv. 2020;6(37):eaba3009.
19. Shramova EI, Chumakov SP, Shipunova VO, Ryabova AV, Telegin GB, Kabashin AV, et al. Genetically encoded BRET-activated photodynamic therapy for the treatment of deep-seated tumors. Light Sci Appl. 2022;11(1):38.
20. Shramova EI, Frolova AY, Filimonova VP, Deyev SM, Proshkina GM. System for self-excited targeted photodynamic therapy based on the multimodal protein DARP-NanoLuc-SOPP3. Acta Nat. 2023;15(4):100–10.
21. Medendorp WE, Bjorefeldt A, Crespo EL, Prakash M, Pal A, Waddell ML, et al. Selective postnatal excitation of neocortical pyramidal neurons results in distinctive behavioral and circuit deficits in adulthood. iScience. 2021;24(3):102157.
22. Song D, Yang Q, Lang Y, Wen Z, Xie Z, Zheng D, et al. Manipulation of hippocampal CA3 firing via luminopsins modulates spatial and episodic short-term memory, especially working memory, but not long-term memory. Neurobiol Learn Mem. 2018;155:435–45.
23. Lee CH, Lin JH, Lin SH, Chang CT, Wu YW, Bewick G, et al. A role for proprioceptors in sngception. Sci Adv. 2025;11(5):eabc5219.
24. Chien YC, Lin SH, Lien CC, Wood JN, Chen CC. Lacking ASIC1a in ASIC4-positive amygdala/bed nucleus of the stria terminalis (BNST) neurons reduces anxiety and innate fear in mice. J Biomed Sci. 2025;32(1):43.
25. Porta-de-la-Riva M, Gonzalez AC, Sanfeliu-Cerdán N, Karimi S, Malaiwong N, Pidde A, et al. Neural engineering with photons as synaptic transmitters. Nat Methods. 2023;20(5):761–9.

26. Tung JK, Gutekunst CA, Gross RE. Inhibitory luminopsins: genetically-encoded bioluminescent opsins for versatile, scalable, and hardware-independent optogenetic inhibition. Sci Rep. 2015;5:14366.
27. Armbruster BN, Li X, Pausch MH, Herlitze S, Roth BL. Evolving the lock to fit the key to create a family of G protein-coupled receptors potently activated by an inert ligand. Proc Natl Acad Sci USA. 2007;104(12):5163–8.
28. Lambert GG, Crespo EL, Murphy J, Boassa D, Luong S, Celinskis D, et al. CaBLAM - a high-contrast bioluminescent Ca(2+) indicator derived from an engineered Oplophorus gracilirostris luciferase. Nat Methods. 2026;23(1):205–15.

Chapter 13
Application of BL-OG Tools in the Mammalian Brain

Emily F. Murphy and Manuel Gomez-Ramirez

Abstract Bioluminescent optogenetics (BL-OG) is a neuromodulation platform that combines the genetic precision of optogenetics with the noninvasive delivery of chemogenetics. BL-OG employs luminopsins (LMOs), fusion proteins that tether a light-emitting luciferase to a light-sensitive opsin, enabling modulation of neural activity through either external illumination or internally generated bioluminescence following systemic delivery of a luciferin such as coelenterazine. The bioluminescent light produced by LMOs robustly drives neural activity, confirming that photon emission is sufficient to activate the opsin and modulate targeted neuronal populations in vivo. The emitted light also serves as a built-in optical readout of neuromodulation in real time. BL-OG effects can be tuned by adjusting luciferin dose or delivery parameters, and coupling BL-OG with focused ultrasound (FUS) enables spatially selective blood-brain barrier (BBB) opening for region-specific activation. Extending to larger-brained species such as marmosets, BL-OG offers a powerful, noninvasive tool for causal circuit dissection and translational neuromodulation.

Keywords In vivo · Chemogenetics · Optogenetics · Focused ultrasound · Marmoset

13.1 Introduction

The ability to manipulate neuronal activity is a cornerstone of neuroscience research, enabling researchers to causally probe the functional architecture of the brain and develop therapeutic strategies for neurological and psychiatric disorders. Various techniques have been developed for this purpose, including electrical, magnetic,

E. F. Murphy · M. Gomez-Ramirez (✉)
University of Rochester, Rochester, NY, USA
e-mail: mgomezra@ur.rochester.edu

U. Hochgeschwender (ed.), *BioLuminescent OptoGenetics*,
https://doi.org/10.1007/978-3-032-18121-3_13

pharmacological, and/or advanced genetic interventions. Among the earliest and most influential methods is intracortical microstimulation (ICMS), which has been widely used to explore the roles of specific brain areas for many decades, particularly in sensory and motor systems, by injecting electrical current directly into the brain [1–4]. Beyond basic research, ICMS has also contributed significantly to clinical applications, particularly in the development of brain-computer interfaces (BCIs) and neuroprosthetics [5–8]. Despite its utility, ICMS has notable limitations: It is highly invasive, requiring extensive surgical procedures to implant chronic devices (e.g., electrode arrays) that carry risks of infection and other complications. ICMS effects can degrade across weeks, months, and years due to biological responses (e.g., gliosis) to the chronic implants. Functionally, ICMS is restricted to excitatory modulation and cannot selectively inhibit neural activity. It also activates broad and heterogeneous neuronal populations, providing little control over cell-type specificity. These shortcomings highlight the need for more refined tools that can both advance our understanding of neural circuit function in vivo and provide safer, more effective strategies for clinical neuromodulation. Accordingly, more recent advances in genetic-based neuromodulation strategies have emerged that provide bidirectional and cell-type-specific control of neural populations and afford greater spatial resolution than ICMS (e.g., optogenetics and chemogenetics). Below, we describe a versatile method, bioluminescent optogenetics (BL-OG), a hybrid technique that combines the benefits of optogenetic and chemogenetic systems.

13.2 BioLuminescent OptoGenetics (BL-OG): A Versatile Method to Study Cell Class-Specific Ensembles Noninvasively

BL-OG is a neuromodulation method that uses a genetically engineered molecule (termed luminopsin; LMO) that tethers a light-emitting enzyme (luciferase) to an opsin that is activated by bioluminescent activity generated when a luciferin (e.g., coelenterazine, CTZ) is catalyzed by a luciferase. Newer LMOs (e.g., LMO7) further improve the efficiency of BL-OG effects by relying on bioluminescence resonance energy transfer (BRET), wherein the luciferase is tethered to a fluorescent protein that enhances photonic activity to boost optogenetic drive (Fig. 13.1). Importantly, the luciferin (e.g., CTZ, h-CTZ, or furimazine) can be delivered systemically, enabling optogenetic drive without chronically implanted optical hardware (e.g., fiber optics), thus reducing the risk of infections and other surgical complications. CTZ also does not produce off-target effects, and it is not harmful to body functions. As a hybrid method, BL-OG also allows for opsins to be activated using conventional light sources (e.g., implanted fiber optic cables), providing neuromodulation effects across broad spatiotemporal scales [9, 10]. Figure 13.2 shows optogenetic effects generated via an external light source (Fig. 13.2 *left image*) vs. bioluminescence (Fig. 13.2 *right image*) of a neuron in the substantia nigra pars reticulata (SNr) that expressed the BL-OG molecule LMO3. External illumination

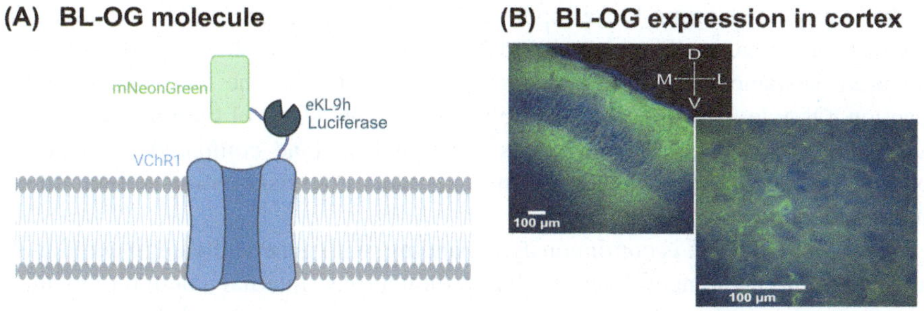

Fig. 13.1 Molecular and histological expression of the BL-OG molecule. (**a**) Schematic illustration of the luminopsin-7 (LMO7) molecule. The LMO7 molecule comprises a fused fluorescent and bioluminescent protein, known as NCS2 (mNeonGreen-eKL9h, a modified Oplophorus-based luciferase), that is tethered to the N-terminus of Volvox channelrhodopsin 1 (VChR1). The LMO7 molecule is anchored to the channelrhodopsin embedded in the cell membrane, with the NCS2 molecule residing in the extracellular space. The opsin can be activated using conventional light-emitting sources (e.g., fiber optics) or via bioluminescence when a luciferin (e.g., h-CTZ, yellow spheres) binds to the luciferase (purple cylinder object). (**b**) Representative histological images showing fluorescence expression in the left primary somatosensory (SI) cortex of a mouse. The lower panel shows a zoomed image highlighting the expression of LM07 around the cell membrane. The blue oval-like shapes represent DAPI. (Figure adapted with permission from Ref. [16])

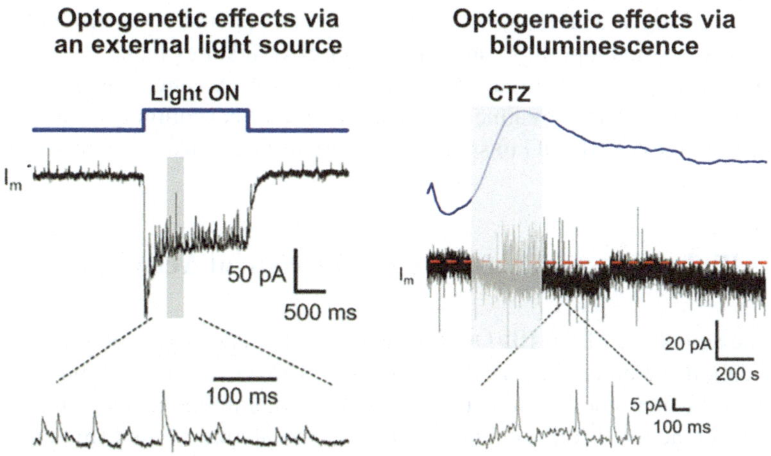

Fig. 13.2 BL-OG effects generated by different forms of optical stimulation. The left figure shows the response of a neuron in the SNr to a 2 s light pulse. Fast outward currents that resemble IPSCs are superimposed on an inward photocurrent due to LMO3 expressed by the cell. The inset (image below) shows the IPSCs on an expanded time scale. The right figure shows the response of the same SNr neuron to bioluminescence generated by the injection of CTZ. CTZ induced a slow inward current (red dashed line) and superimposed fast outward currents that resemble IPSCs. The holding potential was −60 mV. The inset (image below) shows CTZ-induced IPSCs on an expanded timescale. (Figure adapted with permission from Ref. [9])

evoked transient and time-locked spiking that followed the ON/OFF light pattern. In contrast, optogenetic effects produced by bioluminescence showed slower onsets but more sustained modulation, reflecting the kinetics of the luciferase/luciferin reaction. Of note, bioluminescence produces very low levels of thermal energy [11, 12], so prolonged activation imposes minimal heat load compared to extended external light stimulation, a mechanism that reduces the risks associated with thermal tissue damage.

The BL-OG toolkit is continuously expanding, with genetically engineered molecules that modulate neural activity based on neurons' internal calcium dynamics, across trans-synaptic networks [13], and with bidirectional control. For excitation, LMOs commonly pair genetically engineered variants of Gaussia with fast-acting opsins (e.g., ChR2 or VChR1) [9, 10, 14]. For inhibition, LMOs use luciferase and opsins that work on Mac proton pumps (iLMO) [9] or chloride channels (e.g., iChloC; iLMO4) [15]. Because the wavelength palette of bioluminescent enzymes is broad [11, 12], BL-OG constructs can be mixed and matched with different types of opsins to create multiplexed systems with excitatory and/or inhibitory control that are active to populations' intrinsic dynamics. Consequently, BL-OG has great potential for closed-loop neuromodulation effects that self-adjust to the ongoing patterns of neural activity, enhancing or suppressing firing depending on the mechanisms of the construct.

In sum, BL-OG uniquely combines hardware-free, luciferin-driven modulation with optional light activation to deliver cell-type-specific, bidirectional control across broad spatiotemporal scales. These collective features make BL-OG a powerful platform for causal circuit dissection, with a promising route toward more precise neuromodulation treatments in the clinic. In what follows, we describe the major advantages of BL-OG to study neural circuit mechanisms noninvasively.

13.3 Bioluminescence as a Proxy for Neural Activity

A unique advantage of the BL-OG system over other chemogenetic (or most) neuromodulation methods (e.g., DREADDs and PSAMs) is that it enables real-time readout of the neuromodulation effect through the bioluminescent signal created by the BL-OG reaction. From a chemogenetic standpoint, optical confirmation that the systemically delivered luciferin reaches the intended brain target is key for validating whether neuromodulation occurred in response to the BL-OG reaction. Of note, the optical readout tracks the neuromodulation effect with good temporal resolution [10]. The bioluminescent signal also tracks the neuromodulatory effect with temporal precision [10]. The left image in Fig. 13.3a shows the bioluminescence map elicited after coelenterazine (CTZ) injection in a mouse expressing LMO3 in the primary somatosensory (S1) cortex. The right plots of Fig. 13.3a show that multi-unit spiking activity covaries with the bioluminescence activity following CTZ

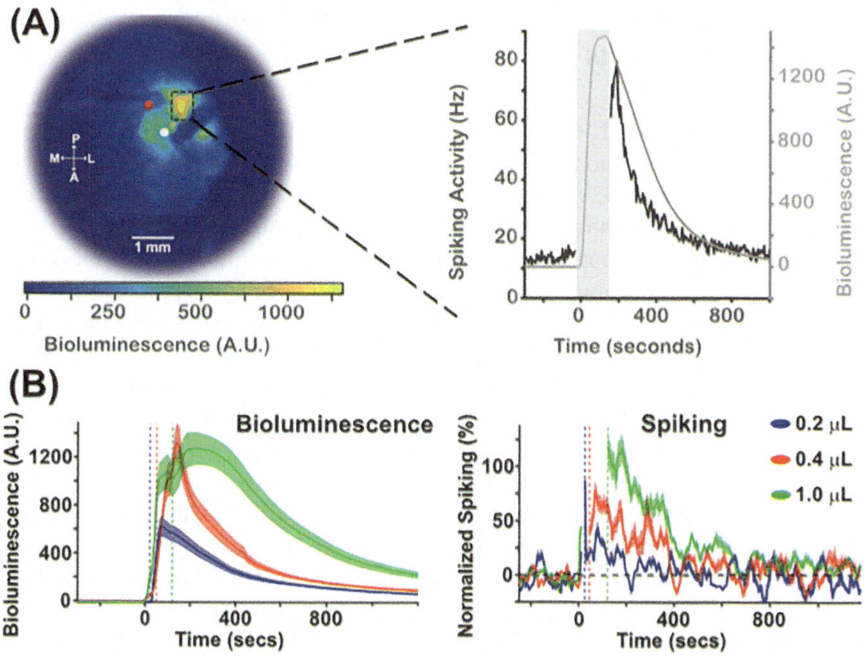

Fig. 13.3 Relationship between bioluminescence and spiking activity produced by BL-OG. (**a**) The left image shows the bioluminescence map in response to a 1.0 μL injection of CTZ directly in S1 cortex. The white and red circles indicate the surface locations of the recording electrode and syringe delivering CTZ, respectively. A = anterior, P = posterior, M = medial, L = lateral. The right image shows bioluminescence (gray trace) and multiunit spiking activity (black trace) time courses within the ROI of the bioluminescence map (dashed box). The gray rectangle represents the onset/offset time of the CTZ injection. (**b**) The left and right panels show the averaged normalized bioluminescence and multiunit spiking activity as a function of CTZ dose conditions, respectively (N = 10 mice). The dotted vertical lines indicate the offset of the injection for each CTZ dose condition. Shaded error bars denote the standard error of the mean values (SEM). The discontinuity in the multiunit spiking in both (**a**) and (**b**) traces stemmed from an electrical artifact in the electrophysiological recordings created by the mechanical injector. (Figure adapted with permission from Ref. [10])

delivery. Importantly, the relationship between BL-OG-evoked bioluminescence and spiking activity is consistent across CTZ dosage (Fig. 13.3b) [10], indicating that bioluminescence provides a reliable proxy for BL-OG-mediated neuromodulation across luciferin dosing conditions. Furthermore, the higher photon output of next-generation luminopsins (e.g., LMO7) allows bioluminescence to be captured with short exposures and less invasive procedures (e.g., without a chronically implanted window) [16], enabling higher-temporal-resolution neuromodulation readouts with reduced damage to the body.

13.4 Time Dynamics of the Noninvasively Induced BL-OG Effects

The temporal profile of BL-OG effects is highly dependent on the mode of luciferin injection, with fast activation and "flash" kinetic patterns for luciferin doses injected intracerebroventricular (ICV) and/or intravenously (IV) that diminished in seconds to minutes [16, 17] and intraperitoneal (IP) administration having slow-rising activations that can last for hours [17, 18]. Yet, the temporal profile of the BL-OG effect offers a significant advantage over conventional chemogenetic approaches whose activation profiles via IV injections have been shown to last many hours (e.g., DREADDs) [19–21]. The faster dynamics of BL-OG is likely due to the fast catalytic reaction of the luciferin. Importantly, the time dynamics of the BL-OG effects can be systematically controlled based on the injection parameters. Using bioluminescence activity evoked from the LMO7 molecule as a proxy for BL-OG effects, it was shown that the infusion rate of luciferin shapes the early kinetic response, with faster injection rates leading to earlier responses (Fig. 13.4a, b). However, the injection rate does not change the decay time of BL-OG effects [16]. Notably, despite these temporal dynamic shifts, the overall magnitude of BL-OG effects is similar across injection rates (Fig. 13.4c). Similarly, the onset and peak times of BL-OG effects are insensitive to dosage concentration [16]. Collectively, these data indicate that the temporal dynamics of BL-OG effects can be parametrically tuned by adjusting the parameters of the luciferin's injection rate, enabling precise experimental control over the timing and duration of neuromodulation.

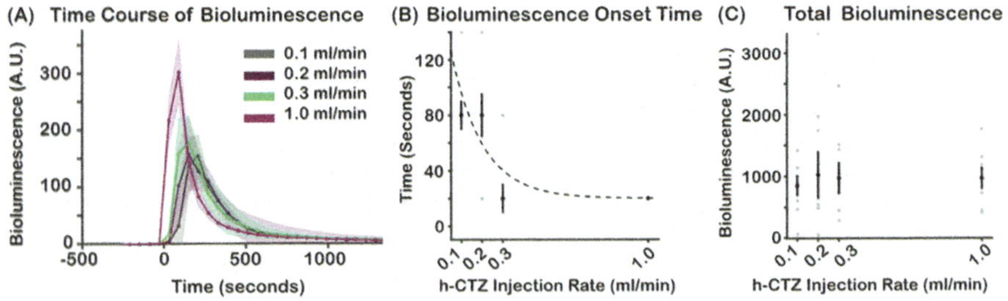

Fig. 13.4 Time dynamics of BL-OG effects: (**a**) Time course of bioluminescence across the rate of injection of h-CTZ, with the dosage in each rate set at 3.93 mM. Time point zero indicates the onset of the h-CTZ injection. Each dot is plotted between consecutive time frames. The first point after the zero time point in the 1.0 mL/min condition corresponds to 30 s (i.e., the middle point between 0 and 60 s). (**b**) Onset time of the bioluminescence response across the rate of injection of h-CTZ. The dashed lines represent an exponential curve fit. (**c**) Total bioluminescence response, estimated as the area under the curve (AUC) between the start and end points of bioluminescent activity. The black dots indicate the median bioluminescence values across animals, whereas the gray dots indicate each individual mouse's response. Note that the AUC of the bioluminescence response was constant across rate conditions; thus, the fits were not reliable. Error bars indicate SEM. Number of mice = 6. (Figure adapted with permission from Ref. [16])

13.5 Dissecting Large-Scale Neural Circuits In Vivo Using *BL-OG*

Complex behaviors that highlight evolved species (e.g., humans and nonhuman primates, NHPs) emerge from coordinated dynamics between functional ensembles distributed across the brain. Neuromodulation tools can provide key insight into the role(s) of a specific area (or areas) within the large-scale network by measuring how behavior or the network's dynamics is affected by inhibiting (or upregulating) that particular area. Genetically targeted approaches (e.g., optogenetics and chemogenetics) are especially powerful because they enable selective opsin expression in pre- and postsynaptic partners, allowing causal mapping of specific connections between network nodes. However, using optogenetics to study network-level interactions is challenging because it requires implanting several optical probes, increasing tissue damage and infection risk. Chemogenetics are also challenging because a systemic administration of the substrate leads to activation of all brain regions expressing the chemogenetic construct, resulting in all-or-none effects. This lack of spatial targeting makes it difficult to assess the function of individual nodes within a circuit. Thus, what is needed is a noninvasive approach that preserves the genetic tractability of optogenetics/chemogenetic methods while selectively targeting individual nodes within a network in the same organism. Ongoing work aims to address this limitation by pairing BL-OG with transcranial focused ultrasound (FUS) to noninvasively modulate the activity of individual nodes within a network. Because the BL-OG substrate, CTZ, has limited permeability to the blood-brain barrier (BBB) [22], it can be exploited for spatial specificity. Typically, low systemic doses of CTZ fail to provide enough bioavailability to the brain, but the bioavailability can be increased by using FUS to disrupt the BBB in the areas that express the BL-OG molecule. Under this approach, BL-OG activation would be localized to the sonicated region(s) only. Importantly, BBB opening using FUS is reversible (with repairs completed within hours after disruption) and does not require surgical procedures, providing selective and repeatable access to deep or distributed targets within the same organism. Figure 13.5 shows the results of BL-OG + FUS approach in a mouse expressing LMO7 in the S1 to a low-dose injection (0.74 mM of h-CTZ) via the tail vein. As shown in Fig. 13.5a, the low-dose injection of h-CTZ failed to elicit bioluminescence activity when the BBB was intact. However, the same dose of h-CTZ elicited bioluminescence after applying FUS in S1 (right image in Fig. 13.5a). The right trace in Fig. 13.5a shows the time course of bioluminescence increasing after the h-CTZ injection (black trace), but only when FUS was applied. The FUS-guided BBB disruption mechanism can also be used to amplify BL-OG effects. As Fig. 13.5b shows, the bioluminescence response to a high-dose h-CTZ injection (2.46 mM) is approximately 5 times greater when the BBB is disrupted. Importantly, Fig. 13.5c shows that BL-OG effects before and after FUS are reflected in neuronal activity, as measured by calcium responses in a mouse co-expressing the LMO7 molecule and the calcium indicator jRGECO in S1. Together, BL-OG and FUS provide a noninvasive way to modulate cell-class–specific activity in selected

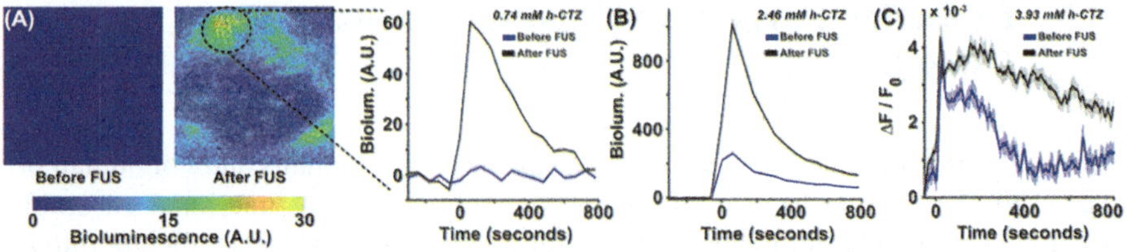

Fig. 13.5 FUS + BL-OG neuromodulation method. (**a**) Bioluminescence maps in mouse S1 in response to a low dosage of h-CTZ (0.74 mM) before (left image, BBB intact) and after applying FUS (right image). The graph on the right shows the bioluminescence response to 0.74 mM h-CTZ injection delivered via the tail vein when the BBB is intact (blue trace) vs. when it has been disrupted by FUS (black trace). (**b**) Bioluminescence response to a high dosage concentration of h-CTZ (2.46 mM) before and after FUS (blue and black traces, respectively). $N = 3$. (**c**) Widefield calcium fluorescence in response to a high dosage of h-CTZ injection (3.93 mM) before and after FUS application

nodes across distributed networks, a highly useful function for studying complex behavioral functions in large-brain animals (e.g., NHPs), which are typically supported by multi-area interactions.

13.6 Noninvasive Neuromodulation of Large-Brain Mammals

NHPs are indispensable animal models for studying behavior and neural functions that mirror those of humans (e.g., mechanisms of executive control, social cognition, hand-guided behaviors, and many others). A detailed understanding of the brain mechanisms that support these complex behaviors requires selective manipulation of the different nodes within the network underlying these functions. BL-OG offers a powerful way to study these mechanisms in large-brain animals (i.e., NHPs) for a variety of reasons. Because bioluminescent light is generated near the opsin itself, BL-OG delivers more uniform effects across neurons expressing the optogenetic elements, a mechanism that likely yields stronger and more homogeneous effects at the population level. Systemic delivery of the luciferin also removes the need to co-register a fiber tip with the neural area(s) that express the BL-OG molecule. This is particularly advantageous when the areas expressing the opsins are deep or distributed across the brain. Finally, since BL-OG effects can occur without the use of fiber-optic cables, animals do not need to be tethered or fitted with cranial optical hardware, enabling studies in home cages, social groups, and potentially even seminatural or wild settings.

Similar to the mouse model [10, 14, 16, 17, 23, 24], BL-OG can also be robustly expressed in the brains of NHPs such as marmosets. Figure 13.6a shows expression of the BL-OG molecule LMO7 in area 3b of S1 of a marmoset monkey. In an anesthetized experiment, 300 μL of h-CTZ (6.0 mM) was injected IV (lateral saphenous

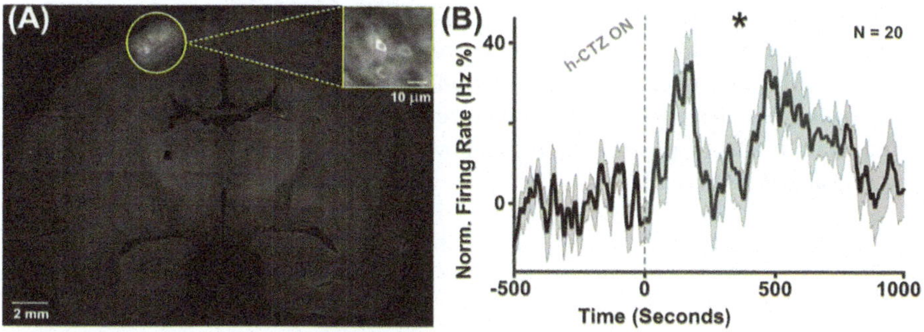

Fig. 13.6 BL-OG effects in a NHP animal model. (**a**) Fluorescent image showing neurons in SI expressing LMO7. The yellow inset is a magnified view of the circular area, showing expression in single neurons. (**b**) Normalized population spiking activity in response to h-CTZ. The shaded box shows the postinjection period with higher spiking relative to baseline. Dashed gray lines in (**b**) indicate the injection onset of h-CTZ. * = $p < 0.01$. Shaded error bars represent SEM

vein), while electrophysiological activity was recorded in S1 using a laminar probe. Figure 13.6b shows averaged activity across 20 neurons in response to h-CTZ injection, with significantly higher spiking activity in the post- (0–750 s) vs. preinjection (−500 to 0 s) period. These data show that systemic h-CTZ induces strong BL-OG effects in neocortical populations of marmosets. Having demonstrated that BL-OG can be expressed in an NHP brain, future studies should use BL-OG to study the neural circuit dynamics of large-scale networks that support complex behavior.

13.7 Conclusion

The development of neuromodulation methods has progressively expanded our ability to causally test how brain circuits support many brain functions, including the following: sensation, action, and cognition. From the early successes of ICMS to the genetic precision of optogenetics and chemogenetics, each advance has addressed limitations of the last while opening new questions. BL-OG continues this trajectory by combining the strengths of both optical and chemogenetic approaches into a single, versatile platform. BL-OG achieves cell class-specific, bidirectional control without the need for implanted hardware, while, at the same time, providing an optical readout of the neuromodulation effects. These features make it possible to move beyond single-site manipulations and begin probing distributed networks that underlie complex behaviors, including those characteristics of NHPs. BL-OG is highly useful because it can interface with many different technologies to achieve better precision, efficacy, and flexibility toward the goal of the experimenter. BL-OG can be combined with FUS to deliver site-specific modulation of different cell classes noninvasively within the same organism, highlighting its potential as a noninvasive circuit dissecting tool in vivo. Importantly, the same features that make BL-OG valuable for basic neuroscience, minimal invasiveness, multiplexing capability, and scalable network-level control also make it attractive for translation to clinical neuromodulation. In general, BL-OG has many advantages over other chemogenetic tools by offering rapid and trackable control of neuromodulation effects, positioning it as a powerful platform for future network-level and translational studies.

References

1. Tabot GA, Dammann JF, Berg JA, Tenore FV, Boback JL, Vogelstein RJ, et al. Restoring the sense of touch with a prosthetic hand through a brain interface. Proc Natl Acad Sci USA. 2013;110(45):18279–84.
2. Romo R, Hernández A, Zainos A, Salinas E. Somatosensory discrimination based on cortical microstimulation. Nature. 1998;392(6674):387–90.
3. Histed MH, Bonin V, Reid RC. Direct activation of sparse, distributed populations of cortical neurons by electrical microstimulation. Neuron. 2009;63(4):508–22.
4. Salzman CD, Britten KH, Newsome WT. Cortical microstimulation influences perceptual judgements of motion direction. Nature. 1990;346(6280):174–7.
5. Shelchkova ND, Downey JE, Greenspon CM, Okorokova EV, Sobinov AR, Verbaarschot C, et al. Microstimulation of human somatosensory cortex evokes task-dependent, spatially patterned responses in motor cortex. Nat Commun. 2023;14(1):7270.
6. Hobbs TG, Greenspon CM, Verbaarschot C, Valle G, Hughes CL, Boninger ML, et al. Biomimetic stimulation patterns drive natural artificial touch percepts using intracortical microstimulation in humans. J Neural Eng. 2025;22(3):036014.
7. Greenspon CM, Valle G, Shelchkova ND, Hobbs TG, Verbaarschot C, Callier T, et al. Evoking stable and precise tactile sensations via multi-electrode intracortical microstimulation of the somatosensory cortex. Nat Biomed Eng. 2025;9(6):935–51.
8. Flesher SN, Collinger JL, Foldes ST, Weiss JM, Downey JE, Tyler-Kabara EC, et al. Intracortical microstimulation of human somatosensory cortex. Sci Transl Med. 2016;8(361):361ra141.
9. Berglund K, Clissold K, Li HE, Wen L, Park SY, Gleixner J, et al. Luminopsins integrate opto- and chemogenetics by using physical and biological light sources for opsin activation. Proc Natl Acad Sci USA. 2016;113(3):E358–67.
10. Gomez-Ramirez M, More AI, Friedman NG, Hochgeschwender U, Moore CI. The BioLuminescent-OptoGenetic in vivo response to coelenterazine is proportional, sensitive and specific in neocortex. J Neurosci Res. 2020;98(3):471–80.
11. Haddock SHD, Choy CA. Life in the midwater: the ecology of deep pelagic animals. Annu Rev Mar Sci. 2024;16:383–416.
12. Haddock SHD, Moline MA, Case JF. Bioluminescence in the sea. Annu Rev Mar Sci. 2010;2:443–93.
13. Prakash M, Murphy J, St Laurent R, Friedman N, Crespo EL, Bjorefeldt A, et al. Selective control of synaptically-connected circuit elements by all-optical synapses. Commun Biol. 2022;5(1):33.
14. Björefeldt A, Murphy J, Crespo EL, Lambert GG, Prakash M, Ikefuama EC, et al. Efficient opto- and chemogenetic control in a single molecule driven by FRET-modified bioluminescence. Neurophotonics. 2024;11(2):21005.
15. Park SY, Song SH, Palmateer B, Pal A, Petersen ED, Shall GP, et al. Novel luciferase–opsin combinations for improved luminopsins. J Neurosci Res. 2020;98(3):410–21.

16. Murphy EF, Means A, Li C, Baez H, Gomez-Ramirez M. Strength of activation and temporal dynamics of bioluminescent-optogenetics in response to systemic injections of the luciferin. Neuroimage. 2024;301:120882.
17. Klein E, Marsh S, Becker J, Andermann M, Lehtinen M, Moore CI. BioLuminescent OptoGenetics in the choroid plexus: integrated opto- and chemogenetic control in vivo. Neurophotonics. 2024;11(2):24210.
18. Tung JK, Berglund K, Gutekunst CA, Hochgeschwender U, Gross RE. Bioluminescence imaging in live cells and animals. Neurophotonics. 2016;3(2):025001.
19. Alexander GM, Rogan SC, Abbas AI, Armbruster BN, Pei Y, Allen JA, et al. Remote control of neuronal activity in transgenic mice expressing evolved G protein-coupled receptors. Neuron. 2009;63(1):27–39.
20. Nagai Y, Miyakawa N, Takuwa H, Hori Y, Oyama K, Ji B, et al. Deschloroclozapine, a potent and selective chemogenetic actuator enables rapid neuronal and behavioral modulations in mice and monkeys. Nat Neurosci. 2020;23(9):1157–67.
21. Whissell PD, Tohyama S, Martin LJ. The use of DREADDs to deconstruct behavior. Front Genet. 2016;7:70.
22. Otto-Duessel M, Khankaldyyan V, Gonzalez-Gomez I, Jensen MC, Laug WE, Rosol M. In vivo testing of Renilla luciferase substrate analogs in an orthotopic murine model of human glioblastoma. Mol Imaging. 2006;5(2):57–64.
23. Crespo EL, Bjorefeldt A, Prakash M, Hochgeschwender U. Bioluminescent optogenetics 2.0: harnessing bioluminescence to activate photosensory proteins in vitro and in vivo. J Vis Exp. 2021;(174). https://doi.org/10.3791/62850.
24. Petersen ED, Sharkey ED, Pal A, Shafau LO, Zenchak-Petersen J, Peña AJ, et al. Restoring function after severe spinal cord injury through BioLuminescent-OptoGenetics. Front Neurol. 2022;12:792643.

Chapter 14
Bioluminescence for Education and Outreach

Justine J. Allen

Abstract Bioluminescence is a phenomenon that is uniquely suited for communicating scientific concepts to diverse audiences. Virtually everyone is captivated by "living light," and simple demonstrations using bioluminescence can help communicate key concepts in biology, chemistry, physics, oceanography, ecology, animal behavior, species diversity and phylogeny, evolutionary biology, cell biology, and more. Through the Bioluminescence Hub collaboration, the BL-OG Team was able to design activities and demonstrations for education and outreach to diverse learners, from elementary school children through senior scientists. These included public and school-based outreach events, a practicum and research internships for undergraduates, and a symposium and workshop for scientists. Based on our experience with education and outreach, I describe the key points educators should communicate and provide descriptions of activities, materials, resources, and tips for success in an effort to inspire other educators to teach scientific concepts using bioluminescence.

Keywords STEM education · Student engagement · Public science · Hands-on activities · Neuroscience

14.1 Introduction

Fireflies in the Garden, Robert Frost [1]

Here come real stars to fill the upper skies,
And here on earth come emulating flies
That, though they never equal stars in size
(And they were never really stars at heart),

J. J. Allen (✉)
Robert J. and Nancy D. Carney Institute for Brain Science, Department of Neuroscience, Brown University, Providence, RI, USA
e-mail: Justine_Allen@brown.edu

U. Hochgeschwender (ed.), *BioLuminescent OptoGenetics*,
https://doi.org/10.1007/978-3-032-18121-3_14

Achieve at times a very starlike start.
Only, of course, they can't sustain the part.

Bioluminescence is a natural phenomenon that has captivated human attention for as long as humans have existed, and certainly for as long as humans have been creating art (e.g. [2, 3]) and recording scientific and naturalist observations [4]. The 1800s saw an increase in written accounts of maritime exploration [5], and Charles Darwin devoted several passages in his *The Voyage of the Beagle* to descriptions of fireflies and their larvae and his encounters with dinoflagellates [6]. Despite being written a century and a half ago, the reader can still feel the awe in Captain Raphael Semmes' words while reading his account of a (putative) milky sea,

> we… passed through a remarkable patch of the sea… there being no moon, but the sky being clear, and the stars shining brightly, we suddenly passed from the deep blue water in which we had been sailing, into a patch of water so white that it startled me… The patch was extensive. We were several hours in running through it. Around the horizon there was a subdued glare, or flush, as though there were a distant illumination going on… Upon drawing a bucket of this water, it appeared to be full of minute luminous particles; the particles being instinct with life, and darting, and playing about in every direction; but upon a deck-lantern being brought, and held over the bucket, the little animals would all disappear, and nothing but a bucket full of *grayish* water would be left…The sun would rise on the morrow, over a sea as blue as usual… this twenty miles square of life would be no longer distinguishable from the surrounding waters. [7]

This milky sea phenomenon continues to puzzle and fascinate humans [5, 8], and, in the age of satellite imaging, has even been examined from space [9]. Importantly, Captain Semmes encountered this milky sea on a moonless night. Bioluminescent light is dim compared with moonlight or sunlight but is bright and spectacular in otherwise dark surroundings, such as the deep sea or inside caves. This contrast, or signal-to-nose, is important when developing research tools that use bioluminescence and when creating experiments and demonstrations to engage students (see the description of our portable ice fishing hut in Sect. 14.4, Resources: The Darker the Better).

Today, the archetypical bioluminescent reaction: luciferase + luciferin = glowing light, is a well-understood chemical reaction (see e.g. [10–12], for overviews of bioluminescent chemical reactions) that continues to capture the attention and awe of sighted[1] learners of all ages. Bioluminescence, and bioluminescence coupled with optogenetics (BL-OG), is an ideal biochemical model for education in biology, ecology, animal behavior, invertebrate biology, oceanography, physics, and

[1] Throughout this chapter, I assume students and outreach participants have functional visual systems. I encourage educators to modify their lesson plans to accommodate learners of all abilities. The invention of sensory substitution devices that can convert visual stimuli into auditory or tactile stimuli is an area of active research (see, for example, Auvray and Myin for a 2009 review [13] and Auvray's recent update in the *Open Encyclopedia of Cognitive Science* [14]). With the dawn of readily available artificial intelligence (AI), creative educators can reach an even more diverse audience. For example, as of this writing, AI can be used to generate music from images via more websites than can be listed here! Although these tools do not remove the challenge, they might provide alternatives that help engage learners of different abilities.

chemistry. Bioluminescence provides a framework for easy, instantly captivating, relatively inexpensive, demonstrations and artwork [2, 15], where scientific principles can literally be seen before one's eyes.

The Bioluminescence Hub

The Bioluminescence Hub is a collaboration between Dr. Christopher Moore's and Dr. Diane Lipscombe's Labs at Brown University, Dr. Ute Hochgeschwender's Lab at Central Michigan University, and Dr. Nathan Shaner's Lab at the University of California, San Diego. The Hub has three Aims: invent new, powerful tools to answer novel research questions, educate learners of all ages ("Vertical Open Science"), and disseminate our tools and findings widely ("Horizontal Open Science," outreach to other scientists) (Fig. 14.1). This leadership team, plus myself, Krystal Literman, MS in science education, and lab members of the partner labs, comprises the BL-OG Team. Hereafter, mentions of "our" or "us" refer to efforts of members of the BL-OG Team.

Here, I describe (I) the key concepts that should be communicated in any lesson that uses bioluminescence (Sect. 14.2), (II) some of the activities, lectures, labs, internships, and other immersive experiences that have helped us use BL-OG to communicate science to learners of all levels (Sect. 14.3), and (III) resources and supplies that can be used to demonstrate bioluminescence, fluorescence, and phosphorescence in a classroom (Sect. 14.4). I hope to inspire you to use some of the features of bioluminescence and bioluminescent life-forms to engage your community.

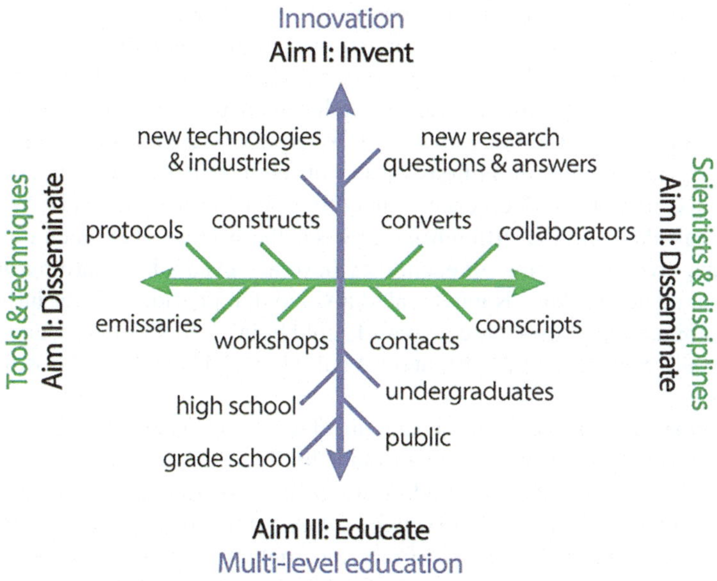

Fig. 14.1 Our approach to education and outreach falls along two axes: Vertical Open Science, reaching students of all ages (purple), and Horizontal Open Science, disseminating our tools and findings to other scientists (green)

14.2 Key Concepts

If your learners understand these key concepts at the end of your lesson or outreach event, you have made great progress in spreading scientific knowledge through bioluminescence.

14.2.1 Bioluminescence Is Different from Fluorescence and Phosphorescence

A common confusion around glowing things comes with terminology, and it doesn't help that historical descriptions of bioluminescent phenomena have used a variety of terms, sometimes interchangeably, or definitions that have since been updated. For example, translators of Aristotle have used "phosphorescent objects," where he was most likely describing bioluminescent phenomena [16]. Charles Darwin wrote about the ubiquity of "phosphorescent" pelagic animals we now recognize as bioluminescent and described "billows of liquid phosphorus" over the bow of a ship and "luminous" tracks of swimming penguins [6] that were almost certainly the glow of bioluminescent dinoflagellates being jostled by the waves, the ship, or swimming animals. In other accounts, mariners have reported a sustained, whitish glow forming eerie "milky seas" that we now attribute to bioluminescent bacteria (*Vibrio harveyi*) [17]. Even Harvey, whose definition of phosphorescence ("the lasting luminescence which results from exposure of a substance to visible or ultraviolet radiation [4].") remains largely unchanged over more than half a century, entitled his comprehensive tome *History of Luminescence* and, in "Part III: Luminescence of Living Organisms. Bioluminescence," included chapters on "Phosphorescence of the Sea" and "Animal Luminescence" [4]. No wonder students are confused!

Teaching the modern definitions of and differences between bioluminescence, fluorescence, and phosphorescence is an entry point for many learners. When your students can describe the differences between these three phenomena and accurately categorize an example of each, they have achieved a key learning objective. One of our students, Raisa Kahn, created a video that provides a brief introduction and disambiguation of these three types of "cold light" [18]. For hands-on activities, consider our "Make It Glow" [19] and "What 'Glows' There?!" [20] downloadable activities.

Bioluminescence is the result of a chemical reaction between molecules made by living (or formerly living) creatures or made by living entities those creatures have eaten or collected from the surrounding water (e.g., the famous symbiosis between the Hawaiian bobtail squid *Euprymna scolopes* collecting the bacterium *Vibrio fischeri* [21]). Usually, this chemical reaction contains two pieces: a luciferase (an enzyme, a type of protein responsible for breaking apart other molecules) and a luciferin (a small molecule). In the presence of oxygen, the luciferase breaks down the luciferin, and a photon (a unit of light) is released. These photons are visible light that we see as bioluminescence. Importantly, luciferase-luciferin pairs are specific. Like a lock-and-key, the molecules must be compatible for the chemical reaction to occur and produce bioluminescence. For an overview of the chemical reactions that result in bioluminescent light, see Haddock et al. 2010 [22] or Henry and Michelson (1974) [10].

Fluorescence occurs when a material capable of fluorescence is excited by a high-energy light source, such as ultraviolet light. Electrons in the material absorb this high-energy light and are excited to a higher energy state. As the electrons fall back to a lower energy state, they emit a photon of light on the way. This photon is seen in a wavelength that is lower energy than the source used to illuminate the material and, when that wavelength is in the visible spectrum, is what we observe as fluorescence. (Visible) fluorescent light can be observed as long as the object is illuminated by the excitatory source but immediately stops when the excitatory source is removed. Noteworthy examples of fluorescent biological materials include: corals [23] and anemones [24], turtle skin and shells [25], shark skin (ibid.), mammal fur [26], and gecko skin [27].

Phosphorescence is similar to fluorescence in that a light source excites electrons in the phosphorescent material and the electrons emit photons as they return to their resting state. However, instead of being instantaneous, these electrons take time to return to their baseline excitation, resulting in a glow (photon emission) that persists after the excitatory light source is removed. Phosphorescent items can glow for minutes or even hours after the excitatory light source is removed. Commercial items that "glow in the dark" (pajama decorations, plastic stars, stickers, and other toys) are phosphorescent. Phosphorescence is rare in living things but has been identified in basswood (putatively *Tilia americana*) [28] and in at least one anecdotal report of the brilliantly fluorescent millipede *Xystocheir dissecta* exhibiting brief phosphorescence [29].

These terms are often used inaccurately (according to modern definitions), imprecisely, and/or interchangeably in the literature, especially in historical accounts [30]. Perhaps worse, reports are sometimes confounded with additional terms including the following: biofluorescence (fluorescence in a biological, often living, material), chemiluminescence (light that is the result of a chemical reaction, regardless of whether a living thing is involved), and photoluminescence (a term sometimes used to describe fluorescence and phosphorescence together). In general, the use of these terms should be avoided, but I mention them here because the reader using these words as search terms might identify resources to bioluminescence that they otherwise would not find.

Although used less frequently and less likely to be confused with bioluminescence, fluorescence, and phosphorescence, the following terms describe additional types of "cold light": thermoluminescence (light that is the result of heating a substance at a relatively low temperature), triboluminescence (light that is the result of rubbing a material), piezoluminescence (light that is the result of pressing on a material), crystalloluminescence (light that appears when solutions crystallize), lyoluminescence (light that appears when crystals go into solution), and radioluminescence (light that appears after exposure to radiation other than light). See Harvey

(1957) for detailed definitions, history of their observation, and discussion of these phenomena.

I encourage the reader researching bioluminescence, fluorescence, and phosphorescence to search broadly using these terms as well as their imprecise or dated counterparts. Then, in the context of the term, consider whether the phenomenon it describes can more accurately fit within the bioluminescence, fluorescence, or phosphorescence matrix. A particular focus should be on "phosphorescence," especially when considering historical documents (e.g. [6, 31–34]). For example, one of the earliest scientific descriptions of bioluminescence in the sea pansy *Renilla* is entitled "The Phosphorescence of *Renilla*" [34].

14.2.2 Bioluminescence Is Ubiquitous Throughout Nature and Serves a Variety of Functions

> But there can be no doubt that very many other pelagic animals, when alive, are phosphorescent. *The Voyage of the Beagle* [6]

Well-known examples of bioluminescent species in terrestrial environments include fireflies, glowworms, beetles, and fungi (e.g. [35, 36]; Table 14.1). To date, only a few bioluminescent freshwater species have been described: the limpet *Latia neritoides* found in New Zealand [37] and snails in the genus *Phuphaina* [39] (although one might also include some insect larvae in this category [22, 67–69]). To date, bioluminescence has not been found in the species most familiar to most students: terrestrial vertebrates (birds, reptiles, amphibians, mammals) and flowering plants [22], resulting in few students being aware of bioluminescence beyond the flash of a firefly (or "lightning bug") or nearshore dinoflagellates (sometimes causing a harmful algal bloom or "red tide" with millions of individuals per liter! [70]). Helping them understand that bioluminescence is not only ubiquitous but, in some environments, the rule rather than the exception, is a core concept.

Bioluminescence is common and widespread in marine environments [22, 30, 54, 71–74]. In one study of video footage of marine water columns from the surface to a depth of nearly 4000 m, around half to three-quarters of all organisms identified were known to be bioluminescent or were likely bioluminescent based on taxonomy, literature, or unpublished observations [74]. Further, bioluminescence is spread throughout the phylogenetic tree. Haddock and colleagues (2010) published a comprehensive review of this topic, including an illustration of the distribution of bioluminescence across taxonomic groups [22], building on the work of earlier efforts to compile a list of known bioluminescent species (e.g. [67]).

Bioluminescence is used for a variety of behaviors, including defense from predators (camouflage, counterillumination, predator distraction, mimicry), prey attraction, and mate attraction [22, 54, 55, 73]. In many species, the specific purpose or use of bioluminescence remains unclear (e.g. [40, 49, 57]). For example, the parchment tube worm *Chaetopterus* sp., an animal that lives buried in marine mud in a

Table 14.1 Examples of species from distinct lineages with a variety of bioluminescence expression patterns and function

Environment	Phylum	Example organism (common name)	Example species	Source of bioluminescence	Possible function of bioluminescence	References
Freshwater	Mollusca	Freshwater limpet	*Latia neritoides*	Secretion of bioluminescent slime	Predator deterrence? (defense)	[37, 38]
Freshwater	Mollusca	Freshwater snail	*Phuphaina* sp.	Steady green light from mantle and foot	Aposematic mimicry? (defense)	[39]
Terrestrial	Annelida	Earthworm	*Eisenia lucens*	Secretes bioluminescent coelomic fluid	Predator deterrence? (defense)	[40]
Terrestrial	Arthropoda	Railroad worm (larvae)	*Phrixothrix* sp.	Red and green luminescent spots, head lantern; steady glow or individual control	Aposematism, Predator deterrence (defense)	[41, 42]
Terrestrial	Arthropoda	Glow worm[a] (beetle larvae)	*Lampyris noctiluca*	Slow pulses or a steady glow combined with pulses	Aposematism, Predator deterrence (defense)	[43]
Terrestrial	Arthropoda	New Zealand Glow worm; Fungus gnats	*Arachnocampa luminosa* (Family Mycetophilidae)	Luminous organ in abdomen, intermittent (pupa) or steady (imago emergence)	Prey attraction (larvae); mate attraction (pupa and adult)	[44]
Terrestrial	Arthropoda	Appalacian fungus gnat (larvae)	*Orfelia fultoni* (Family Keroplatidae)	Constant light from head and tail organs	Prey attraction (larvae)	[45]

(continued)

Table 14.1 (continued)

Environment	Phylum	Example organism (common name)	Example species	Source of bioluminescence	Possible function of bioluminescence	References
Terrestrial	Arthropoda	Cockroaches	*Lucihormetica* sp.	Flashes from lanterns?	Batesian mimicry (defense)	[46], although the validity of this and other reports is disputed by [47]
Terrestrial	Fungi	Fungi	Order Agaricales; 5 lineages	Continuous light emission	Protection against oxidative stress? Prey attraction	[36, 48]
Marine	Annelida	Polychaeta	*Chaetopterus* sp.	Egests bioluminescent mucus when disturbed	Unknown; various potential functions have been hypothesized, including no visual function at all	[49, 50]
Marine	Cnidaria	Crystal jellyfish	*Aequorea aequorea*	Flash along the ring of the bell	Defense?	[51, 52]
Marine	Cnidaria	Angler jellyfish	*Erenna* sp.	Bioluminescent lures on tentacles	Prey attraction	[53]
Marine	Ctenophora	Comb jelly	*Bathocyroe fosteri*	Photocytes; Bioluminescent flickers along comb rows	Communication? Aposematism?	[54–56]
Marine	Dinoflagellata	Dinoflagellates (plankton)	*Pyrodinium. bahamense* var. *bahamense*; *Lingulodinium polyedrum (Gonyaulax polyedra)*	Flashes of light went disturbed; especially evident in high concentrations	Startle display, burglar alarm, aposematic warning (defense)?	[54, 57, 58]
Marine	Echinodermata	Luminous brittle star	*Amphipholis squamata*	Light on arms; steady and intense when autotomized	Predator distraction (autotomized arms) (defense)	[59]

Marine	Echinodermata	Deep-sea sea cucumber	e.g., *Pannychia moseleyi*; other Holothuroidea	Light on body walls or shed skin in response to touch	Unknown	[60]
Marine	Mollusca	Deep-sea finned octopus	*Stauroteuthis syrtensis*	Suckers on arms glow continuously or flicker	Communication, prey attraction	[61]
Marine	Mollusca	Eye-flash squid	*Abralia veranyi*	Photophores, mostly on the ventral surface	Counterillumination (Defense), but see [62]	[63]
Marine	Vertebrata	Flashlight fish	*Anomalops katoptron*	Organs beneath the eyes contain bioluminescent bacteria	Prey attraction; conspecific communication, predator distraction	[64, 65]
Marine	Vertebrata	Anglerfish	Suborder Ceratioidei	Famously, specialized esca that act as lures, in some species, filled with bioluminescent bacteria	Prey attraction	[66]

[a]The term "glow worm" could refer to a female beetle in the Order Coleoptera, Family Lampyridae (family that contains fireflies or lightning bugs; the females look like larvae) or fungus gnats in the Order Diptera, Family Keroplatidae. The reader is encouraged to proceed with caution when using common names as search terms

U-shaped tube of its own construction, secretes bright blue bioluminescent mucus when startled or gently touched near its gills [50, 75]. Although the chemical components of this bioluminescence have been studied [49, 50], the purpose of this bioluminescence in nature (Prey attraction? Secondary defense by illuminating a potential predator? A by-product of the mucus used for tube construction? A side effect of a metabolic process?) remains unknown [49, 50].

An examination of the diversity of bioluminescent species and the ways different species use bioluminescence for a variety of behaviors and functions would be an excellent lesson for undergraduates studying phylogeny and taxonomy, species diversity (especially invertebrate biology), development (in some species, certain life stages are bioluminescent while others are not), and animal behavior. Table 14.1 provides example species from a variety of lineages as inspiration for curricula creation.

14.2.3 Bioluminescence Is Ancient and Evolved Independently Many Times

The origin—and loss—of bioluminescent capabilities in various lineages is a fascinating illustration of the complexity of evolutionary biology and the studies of phylogeny and taxonomy. Recent work estimates that bioluminescent capabilities evolved in Octocorallia (marine soft corals) around 540 million years ago, making this the earliest known emergence of bioluminescence [76]. In contrast, bioluminescence in *Campyloxenus* click beetles emerged only 53 million years ago [77].

Across life-forms, the use of bioluminescent light has independently evolved at least 94 times [78], including at least 27 times in fishes [72]. Nearly every recognized phylum contains a bioluminescent member, and it is sometimes easier to identify lineages *without* bioluminescence [22, 67, 72]. On the other hand, the ability of fungi to generate light most likely has a single origin that has since been lost in many taxa [36]. Within phyla, some clades contain many bioluminescent species while others have only a few (e.g., see Haddock et al.'s comparison of Siphonophores vs. Chaetognaths in Figure 2 of their 2010 review [22]).

These observations suggest that the evolutionary pressures on the ability to make light, or to harness the bioluminescent light made by others (e.g., symbiotic bacteria), are diverse and complex and act on different lineages at different times in their evolutionary history. Despite the excellent and comprehensive work referenced here, many details of these species' chemistry, life history, and behaviors are unknown. Helping students understand that bioluminescence is simultaneously ancient and young, widespread and uncommon, critical and perhaps less important to life processes, and that there remains great opportunity for research in these areas, is key to developing a perspective on evolutionary biology.

14.3 The Bioluminescence Hub: Outreach and Education

The Bioluminescence Hub's education and outreach events and activities were designed to target learners at four levels of engagement: the general public, elementary/grade school, high school, and undergraduate students (this Sect. 14.3). We reached more advanced students (graduate students, postdoctoral scholars) and colleagues through direct contact, workshop and symposium events, our emissary program, and by providing research resources including a constructs database (https://www.bioluminescencehub.org/constructs), Addgene records, and a transgenic mouse (see Sect. 14.3.5).

14.3.1 Outreach at Different Levels: Public

Public engagement is a critical form of science communication. It is important for spreading facts and clarifying understanding, garnering fiscal and cultural support of scientific research, and as a training experience for young scientists and educators. An important challenge is communicating science in fast, approachable ways that engage a diverse, potentially large audience with a broad range of background knowledge, life experience, and interest in the topic. Bioluminescence uniquely captures the attention of virtually all participants, so it is easy to develop demonstrations and activities that illustrate the key concepts described in Sect. 14.2.

A logistically straightforward way to interface with a public audience is by participating as an exhibitor in a larger event. The burden of reserving the space and publicizing the event is handled by the organizers. Further, many organizers can suggest content, swag, and demonstration materials that have worked well during previous iterations of the event. Here, I describe some of our experiences with public outreach events.

During Glow Night in 2017, an annual, public event presented by the Midland, Michigan branch of the American Chemical Society, BL-OG Team members at Central Michigan University presented a lecture and hands-on demonstrations and activities. Approximately 50–100 people interacted with our materials (Fig. 14.2a).

Members of our San Diego contingent manned a booth at the San Diego Festival of Science and Engineering EXPO in 2018 and engaged with approximately 3,000–5,000 attendees. Our booth featured an interactive fluorescence exhibit for children of all ages and provided opportunities for us to talk with parents and teachers about our work in developing research tools from fluorescent and bioluminescent proteins (Fig. 14.2b).

At the public Brown Brain Fair in 2019, we engaged with approximately 200 attendees of all ages (Fig. 14.2c). We presented a video of bioluminescence expressed in a mouse choroid plexus [79], live dinoflagellates in a dish under a microscope, and invited participants to enter our portable ice fishing hut (the large, navy blue cube with red zipper closure in Fig. 14.2c; see also Sect. 14.4, Resources:

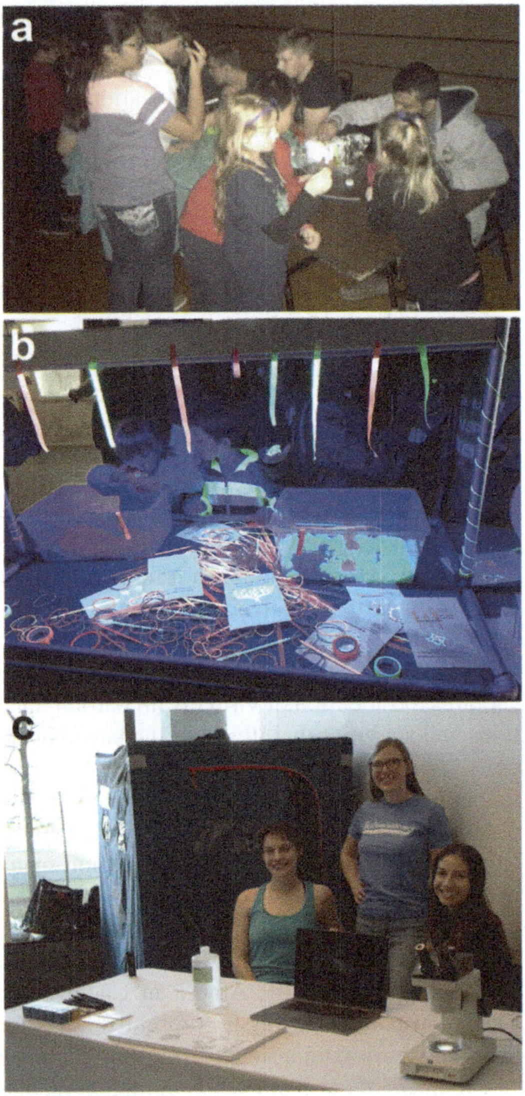

Fig. 14.2 (**a**) Drs. Eric Petersen and Akash Pal teach young students about bioluminescence. (**b**) BL-OG team members hosted a booth about fluorescence at the San Diego EXPO. (**c**) From left, Sophie Marsh, Krystal Literman, and Rebecca Berger-Gutierrez presented hands-on demonstrations at the Brown Brain Fair

The Darker the Better) to view live dinoflagellate and fungal bioluminescence and for a demonstration designed to distinguish bioluminescence from fluorescence and phosphorescence [19] (detailed below). We also distributed our glow-in-the-dark stickers and Activity Placemat [80].

As predicted, it was easy to engage learners of all ages at these events, and we were able to interact with many participants over a short time. Two key challenges were the short attention spans of some of the participants and the physical attributes of our surroundings; people wanted to move through our stations quickly, and it was difficult to create a space that was dark enough to truly appreciate our live bioluminescent specimens. We overcame these challenges by focusing on one key point: the differences between bioluminescence, fluorescence, and phosphorescence, and by using our portable ice fishing hut as a darkroom (see Sect. 14.4, Resources: The Darker the Better). We gave participants time to let their eyes adjust to the darkness by showing them our fluorescent and phosphorescent objects [19] before agitating our live dinoflagellates or directing their attention to our bioluminescent fungi. Finally, we were sure to include items the learners, especially children, could take home with them to continue to explore the concepts after our brief time together.

14.3.2 Outreach at Different Levels: Elementary School Students

Universities often have relationships with nearby schools, whether through formal outreach programs or informally via staff and faculty who are parents. Most classrooms and childcare centers are eager for outside involvement, especially when potential career paths can be modeled by the presenters. My team and I engaged with elementary students through in-person outreach, a real-time virtual event, and freely available downloadable lessons and activities. Table 14.2 lists our activities and provides references to their location online.

The BL-OG Team was invited to present to third grade students at Providence, Rhode Island's Vartan Gregorian Elementary School Science Conference in 2019 (Fig. 14.3). We introduced the terms bioluminescence, fluorescence, and phosphorescence (see Sect. 14.2.1) at an age-appropriate level and set up stations where a total of approximately 50 students over three sessions rotated through shining a UV light on fluorescent and phosphorescent items [19, 20], viewing a luciferase and luciferin chemical reaction on a slide, and examining bioluminescent creatures in our portable ice fishing hut (see also Sect. 14.4, Resources: The Darker the Better).

In 2020, Vartan Gregorian Elementary pivoted, like many of us, to virtual external engagement. My team suggested an "Ask a Scientist" format, where a virtual panel of scientists could field questions from students in grades 3–5 in their area of expertise. I represented neuroscience, and two of my colleagues represented chemistry and earth, environmental, and planetary sciences. We answered questions over Zoom for about an hour, and more than 100 students logged on to participate! The following year, BL-OG Team member Krystal Literman was invited to Virtual Day at Vartan, where she presented a talk on "Glowing Brains" and provided real-time demonstrations using dried ostracods, glow sticks, and live dinoflagellates. More than 60 third and fourth grade students joined her over Zoom. In a time where outreach was especially challenging, we partnered with motivated public educators to find ways to share our expertise with curious children.

We also developed content that can be used asynchronously, and, when used by older children, without an instructor's guidance. Our downloadable and interactive

Table 14.2 Catalogue of our freely available educational activities

Title	Target learner age	Description	References
Make It Glow: A "Cold Light" Kit	8–14, but learners up to 22 (and older!) have enjoyed this activity	Students learn the difference between bioluminescence, fluorescence, and phosphorescence	[19]
What "Glows" There?!	8–14, but learners up to 22 (and older!) have enjoyed this activity	Students use a UV light to discover fluorescent objects in their surroundings	[20]
Who "Glows" There?! Bioluminescent Dinoflagellates	8–14, but learners up to 22 (and older!) have enjoyed this activity	Students are introduced to living dinoflagellates and the basics of bioluminescence. Students will practice making predictions and using their observation and recording skills	[81]
Growin' Glowin' FUNGI: Grow your own bioluminescent fungus	8–14, but learners up to 22 (and older!) have enjoyed this activity	Background information on bioluminescent fungi is provided; students observe and record the growth of a fungus in a dish	[82]
Bioluminescent Dive: Submersible Checklist	8–14	Supplement to our "Bioluminescent Dive, Part 1: Submersibles" online activity (available at bioluminescencehub.org/education). Students design their own submersible considering the challenges of deep-sea exploration	[83]
Bioluminescence Activity Placemat	8–14	Students complete a word search, MadLib, and color drawings of bioluminescent creatures Works well as a handout	[80]

(continued)

Table 14.2 (continued)

Title	Target learner age	Description	References
Bioluminescence Hub Dive: Submersibles and Virtual Dive	8–14	Interactive online	[83–85]
Madlibs: The Bioluminescent Jellyfish The Professor's Glowing Brains	8–14	A mix of science and fun! Includes an "answer key" with appropriate terms. Supports education in parts of speech	[86, 87]
Word Search	8–14	Includes a glossary of terms.	[88]
Bioluminescence Throughout Time	12–18	A timeline illustrating the history, art, and science of human interest in bioluminescence	https://www.bioluminescencehub.org/education; see Bioluminescence Throughout Time section
Junior Trainee Club	5–18	A series of six activities for young scientists	https://www.bioluminescencehub.org/education; see Junior Trainee Club section
Meet our Bioluminescence Hub Scientists	5–18	An introduction to several members of our team, intended to serve as a career exploration resource	[89]

content on our Education page: https://www.bioluminescencehub.org/education (on the website, see the sections: Junior Trainee Club, Bioluminescent Hub Dive, Activities to Try at Home, Fun & Games, Bioluminescence Throughout Time) is targeted to elementary school students but has been appreciated by older learners as well (see Table 14.2).

14.3.3 Outreach at Different Levels: High School Students

In 2019, my team and I were invited to consult the Brown Science Prep group on the design and presentation of their "Bioluminescence: Organisms that Glow and Glitter" lesson for Providence-area high school students (Fig. 14.4). Approximately 20 high school students plus a small team of Brown University undergraduate student mentors attended the event. After a brief overview delivered by the undergraduates, BL-OG Team research assistant Nina Friedman lectured about her work in the brain and how she used BL-OG tools to answer her research questions. After the event, one of the mentors mentioned that Nina's talk made her research accessible and that her presence as a young, female scientist demonstrated the possibility

Fig. 14.3 The Bioluminescence Hub partnered with Vartan Gregorian Elementary School to reach students. (**a**) Krystal Literman used UV lights, fluorescence-enhancing UV-protective safety glasses, and hands-on materials to demonstrate the difference between fluorescence and phosphorescence. (**b**) Dr. Manuel Gomez-Ramirez applied the luciferin coelenterazine to luciferases on a glass slide, and students viewed the chemical reaction through a microscope. (**c**) Nina Friedman used our portable ice fishing hut to show students living, bioluminescent dinoflagellates and fungi. (**d**) The author provided an overview and a short video [18] describing the differences between bioluminescence, fluorescence, and phosphorescence

of a career in research to both the high school student participants and the undergraduate mentors.

During the hands-on portion of the event, the high school students formed small groups and rotated through stations where they viewed bioluminescent creatures and fluorescent or phosphorescent objects in our portable ice fishing hut (see Sect. 14.4, Resources: The Darker the Better and our "Make it Glow: A Cold Light Kit" activity [19]) and hunted for fluorescent objects using UV flashlights (see our "What Glows There?!" activity [20]). At the end of the session, we distributed our glow-in-the-dark stickers and Bioluminescence Activity Placemat [80] for participants to share with younger friends and family members.

In addition to our in-person events and downloadable content, members of our hub created personal introductions, video introductions, and tours of their

Fig. 14.4 Undergraduates in Brown Science Prep teach Providence-area high school students about bioluminescence. We were invited to consult on the curriculum and present our hands-on demonstrations

laboratory space and research projects for our website (see "Meet our Bioluminescence Hub Scientists" in Table 14.2). Targeting a high school student audience, these vignettes were intended to serve both as an introduction to the topic of bioluminescence in neuroscience research and as exposure to the types of careers that are possible in this field.

14.3.4 Outreach at Different Levels: Undergraduates

We served undergraduates through two key activities: an undergraduate practicum in bioluminescence and an internship program.

14.3.4.1 Undergraduate Practicum in Bioluminescence

Our undergraduate practicum was a 1-week, in-person, immersive experience. Students applied for admission via an online form, and a cohort of selected students (13 each year in 2018 and 2019) traveled to the Marine Biological Laboratory (MBL) in Woods Hole, Massachusetts. Held in-person in 2018 and 2019 and virtually in 2021, we trained 42 students (including 29 women and 11 students from historically underrepresented groups) from 38 universities in 19 US states, one US territory, and two foreign countries (Canada and Spain) in the invention and use of bioluminescent tools and technologies for research. Synopses of the practicum, photographs, schedules, and additional testimonials are available on our website at bioluminescencehub.org/mbl.

In-Person Practicum

Each day started with breakfast in the MBL's dining hall, followed by two lectures and lunch. Each afternoon, we visited a nearby attraction to learn more about Woods Hole and its rich history of scientific research (see Sect. 14.5). Back in the lab, students performed hands-on investigation and experimentation with live bioluminescent animals (see Sect. 14.4, Resources: Animals and Supplies Vendors), expressed fluorescence and bioluminescence in cells, and imaged fluorescent and bioluminescent cells using a microscope. In the evenings, students met with their groups to work on their Shark Tank projects, described below.

The retreat-like atmosphere of the practicum allowed for spontaneous idea generation and networking over meals and extra-classroom activities. In the words of one of our students, "It's extraordinary how one (intensive) week can give you new friends and skills to change your trajectory for years to come." Another student noted, "This program forced me to get out of my comfort zone and think like a scientist. It taught me how to work with my peers and come up with great ideas in a short time! I did not know I had that ability and I felt really proud of myself towards the end of this program." For additional testimonials, see Box 14.1.

Box 14.1 Undergraduate Practicum Testimonials

Following each iteration of the practicum, we collected feedback from our students. Below are some of their responses.

> I heard about the Practicum through a Neuroscience announcement and immediately knew that I needed to apply. I had been interested in bioluminescence since freshman year, but had yet to fully indulge in the topic in a formal, classroom setting. This program seemed like the perfect opportunity for me to explore this interest, and it absolutely exceeded my expectations. I loved every component – the lectures, the labs, the projects, the friendships, and the scientific exploration in general. Without a doubt, the MBL Practicum was my favorite week of college because I learned more, and made more positive memories, than I have during any other week in my undergraduate career. I am determined to incorporate bioluminescence into my post-undergraduate and PhD work, and I have a renewed passion for novel Neuroscientific pursuits.—Camille Phaneuf, University of Michigan Ann Arbor, 2018 Student

> The [NeuroNex] Practicum was definitely one of the highlights of my year – I was exposed to an entirely new research field that has so many practical applications, learned so much from some of the most knowledgeable mentors I've ever met, and made some life-long friends along the way! I would recommend anyone to apply—it opens your eyes to real-world research and you make some amazing connections.—Anonymous 2018 Student

> What a great experience! I had so much fun learning about bioluminescence and its applications to modern neuroscience research. It was also super fascinating learning from experts in many different fields of biology including marine biology, neuroscience, and bioluminescence-optogenetics. On top of that, it was great meeting students from other schools who are also interested in pursuing neuroscience research and careers in the medical field. Would highly recommend to anyone interested!—Anonymous 2018 Student

(continued)

Box 14.1 (continued)

Some of my most memorable experiences during the Practicum definitely involved the activities we did (the Nobska Light hike, the aquarium, the rare books collection, the marine resources center), all the meals that we shared as a group, all of [the] lectures, the fluorescent protein lab, the lipofection lab, the lamprey lab, and the technology venture presentation. This whole week was a smashing success, and I immensely enjoyed my time here. It truly was one of the most transformative experiences I have ever had in my life, and I loved how everyone brought something different to the world of bioluminescence, but also to the group as a whole.—Anonymous 2019 Student

I enjoyed all the lectures, and the format of the Shark Tank was extremely fun. I loved being able to think about creatively applying bioluminescence to different aspects of the world and not just science. Also, I really loved the experiments. They were enjoyable despite not having an in-person experience.—Anonymous 2021 Virtual Student

For more testimonials, visit bioluminescencehub.org/mbl/2018-practicum/testimonials (2018) and bioluminescencehub.org/mbl/2019-practicum/testimonials (2019).

Lectures Students received a progressive overview of bioluminescence and its use in scientific research. In both in-person iterations, the following six core lectures were presented. I introduced bioluminescence in nature and why it is important for a variety of life processes, particularly in the ocean, and described how light behaves in water. Dr. Diane Lipscombe provided an overview of how calcium is ubiquitous in living things and is essential for many critical life processes. Dr. Ute Hochgeschwender provided a bit of history of the discovery of luciferase-luciferin pairs and opsins and how they were extracted from their natural sources and modified for use in research. Further, she introduced molecular biology techniques that allow researchers to express bioluminescent, fluorescent, and supporting molecules in cells and mice. Next, Dr. Nathan Shaner spoke on how fluorescent proteins are extracted from natural sources and optimized using molecular evolution, provided an overview of the chemistry of bioluminescence and fluorescence, and introduced fluorescence resonance energy transfer (FRET) and bioluminescence resonance energy transfer (BRET) in the context of creating molecular tools for cellular control. To bridge the gap between chemistry, molecular tools, and animal behavior, systems neuroscientist Dr. Christopher Moore delivered a lecture on the use of BL-OG tools in research (see also Chap. 6). The titles, authors, and references to the digital locations of these recordings are provided in Table 14.3.

In addition to our core lectures, we invited guest speakers to contribute to our practicum. In 2018, practicum teaching assistant (now Dr.) Eric Petersen presented his research using bioluminescence combined with optogenetics as a treatment for spinal cord injury [96, 97]. In 2019, our students were introduced to lamprey as a model system by MBL Senior Scientist and Director of the Bell Center Dr. Jennifer

Table 14.3 Recordings of our undergraduate practicum in bioluminescence lectures

Title	Presenter	References
Light and Living Where It Is Very, Very Dark	Justine J. Allen, PhD	[90]
Calcium—Not Too Little and Not Too Much	Diane Lipscombe, PhD	[91]
From Shrimp to Brain	Ute Hochgeschwender, MD	[92]
Fluorescence and Bioluminescence: Basics and Chemistry	Nathan C. Shaner, PhD	[93]
Bioluminescence: Ideal Natural Engineering for Understanding the Dynamics that Create Behavior	Christopher I. Moore, PhD	[94]
"Brightness" and the Physics of Bioluminescent Light Production and Resonance Energy Transfer	Nathan C. Shaner, PhD	[95]

Morgan, and spent time in her laboratory expressing BL-OG molecules in lamprey to study spinal cord regeneration. Our students also heard from teaching assistants Dr. Jeremy Murphy, a postdoctoral scientist in the Moore Lab, about his work using BL-OG in mouse models to study somatosensation and (now Dr.) Emmanuel Crespo, a graduate student in the Hochgeschwender Laboratory, about his work using bioluminescent optogenetic tools for cellular control. Krystal Literman shared tips and best practices for modern science communication (see Sect. 14.4, Resources: Sharing Your Success and Why we #SciComm [98]). Equally as important as the content of these lectures was our students' exposure to scientists at different career levels working in positions that are different from the standard tenure-track, university faculty these undergraduates had likely already encountered.

Laboratory Investigations To observe bioluminescence in living animals, we purchased specimens from Gulf Specimen Marine Laboratories (see Sect. 14.4, Resources: Animals and Supplies Vendors). In 2018, we acquired a few members of the genera *Mnemiopsis, Renilla,* and *Chaetopterus*; in 2019, we ordered these plus a few *Beroe* sp. These specimens represented diverse phyla (Cnidaria, Ctenophora, and Annelida, respectively; *Beroe* is also a ctenophore but has bioluminescence that is different from *Mnemiopsis*), diverse bioluminescence colors and patterns, and were hardy enough that most individuals survived being shipped from Florida to Massachusetts. After making the experimental room as dark as possible, the students observed bioluminescence in these species, usually by gently simulating them with a cotton-tipped exam probe. A video of one of our *Chaetopterus* spewing bioluminescent mucus can be found in our digital archive [75]; screenshots from the video are shown in Fig. 14.5a. For a detailed review of *Chaetopterus* life history and bioluminescence, see Mirza et al., 2020 [50].

Our students learned a variety of laboratory techniques through hands-on experiments (Fig. 14.6).

With Dr. Jennifer Morgan, students dissected sea lamprey (*Petromyzon marinus*) (Fig. 14.6a) and worked to express BL-OG constructs (luminescent opsin LMO3) in lamprey brain (reticulospinal) neurons [100] (Fig. 14.7).

Drs. Shaner and Hochgeschwender led the students through isolating and cloning the expressed genes from bioluminescent species to purifying proteins and

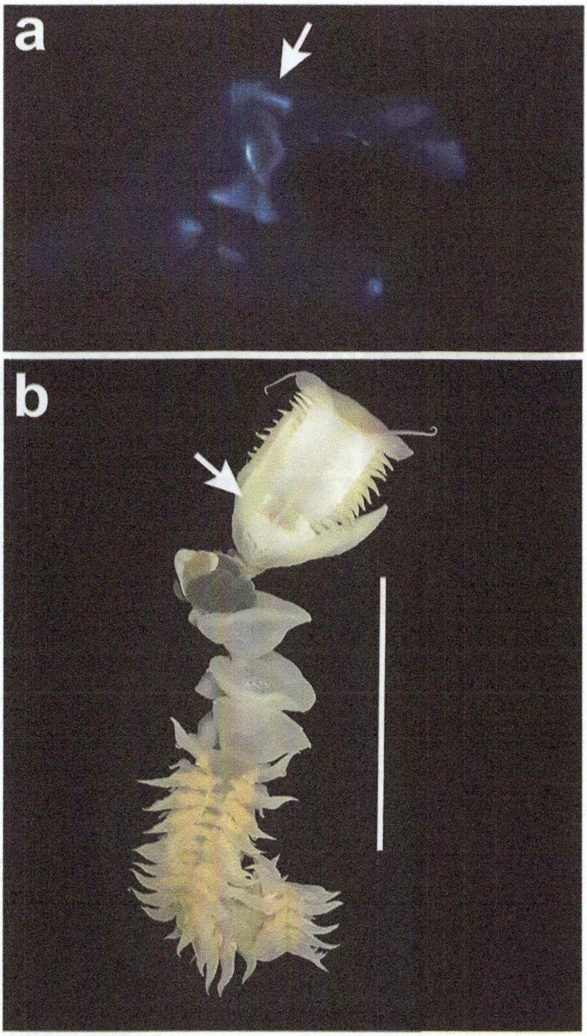

Fig. 14.5 (**a**) Bioluminescent mucus secreted near the anterior end of a parchment tube worm, *Chaetopterus* sp. Screenshots were taken from a movie recorded during the Practicum. (Full movie available in the Brown Digital Repository [75]). (**b**) Brightfield image of a *Chaetopterus* sp. parchment tube worm outside of its tube from Britayev et al. (2017) [99]. (Used under Creative Commons license CC-BY-4.0). Scale bar is 5 cm; images are approximately to scale. To show orientation, *white arrows* point to the same area just anterior to the winglike parapodia on both animals

transforming mammalian cells to make them glow. The students extracted RNA from our *Renilla* and *Mnemiopsis* specimens, ran gels to check the quality of their preparations, reverse transcribed the material into DNA, inserted it into cloning vectors, and transformed bacteria. The next day, students sprayed bacterial plates with luciferin to look for bioluminescent light, the output of a luciferase-luciferin interaction, and thereby identify bacterial colonies that contained cloned luciferases.

With Dr. Shaner, students also purified protein from the bioluminescent species (Fig. 14.6b). With Dr. Hochgeschwender, students lipofected HEK293 cells with different plasmids encoding for BL-OG constructs, examined them with fluorescence microscopy to check expression, added the appropriate luciferin (coelenterazine, CTZ), and imaged fluorescence and bioluminescence (Fig. 14.6c–g). For

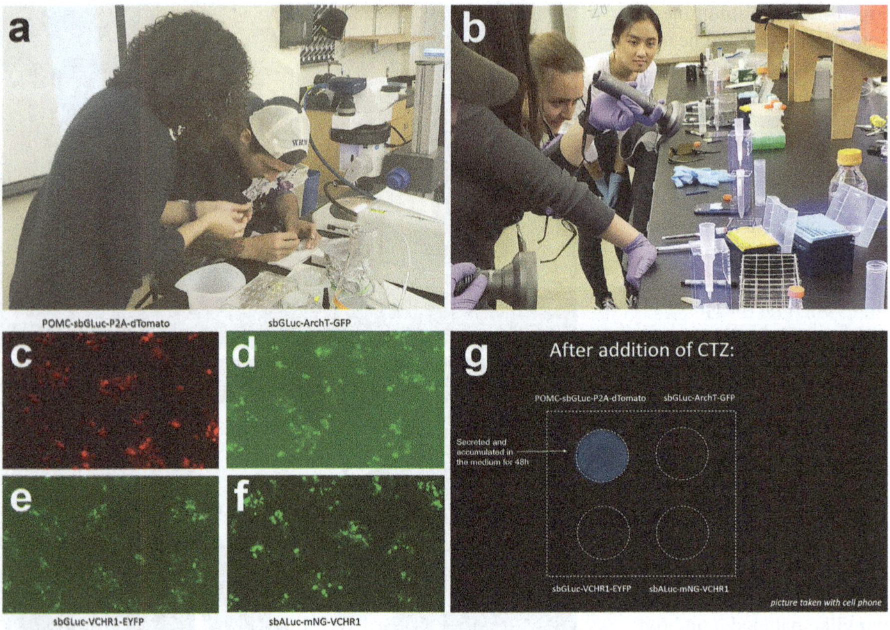

Fig. 14.6 (**a**) Under the direction of Dr. Jennifer Morgan, students used lamprey as a model system to test the expression of the BL-OG tool luminopsin LMO3 [100]. (**b**) Students purifying luciferase proteins from sea pansy *Renilla* preparations. (**c–f**) HEK293 cells were lipofected with four different plasmids containing luciferases and opsins. Expression was tested 24–48 h later by fluorescence microscopy. (**g**) The luciferin CTZ was added to each well for bioluminescence emission

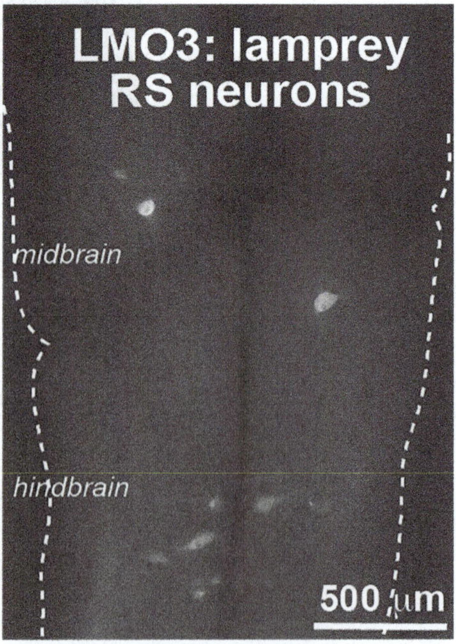

Fig. 14.7 The LMO3 construct has successfully been expressed in lamprey brain (reticulospinal) neurons, expanding the list of potential models for BL-OG tools in neuroscience research. (Image kindly provided by Dr. Jennifer Morgan)

Fig. 14.8 Students celebrate after winning the inaugural NeuroNex Undergraduate Practicum Shark Tank competition

additional details on experimental methods, view Dr. Hochgeschwender and Shaner's recorded lectures [92, 93].

Technology Venture Presentations, or "Shark Tank" Projects On the first evening of the practicum, students were assigned to small teams and given the following challenge: In 5 days, design and pitch a product that uses bioluminescence for health, entertainment, societal improvement, scientific research, or some other function. After dinner on the last day, inspired by the TV show *Shark Tank*, each team presented their product(s) as a pitch to mock investors (the faculty and other students). "Investors" voted for their favorite pitch (as defined by the investor, but could include the most creative, most feasible, best presentation, most persuasive, most pressing need for the product, etc.), and a winner was crowned (Fig. 14.8).

Our students presented many creative ideas. For example, their products included tattoos where bioluminescent components could be triggered for special events, municipal lighting driven by bioluminescent dinoflagellates in a network of canals along streets, a smart wristband for patients with HIV that is calibrated to decrease the risk of the disease progressing to AIDS, and engineered oligodendrocytes that grow myelin to treat dementia and other neurodegenerative diseases. While challenging, this project and event was very popular with the students during all three iterations of the practicum. The students had the opportunity to practice their research skills and drive an idea from inception to pitch in a short amount of time, and also to hone their soft skills of working with a small group of relative strangers with diverse backgrounds and expertise, collecting and organizing their thoughts into a short presentation, and effectively delivering a persuasive pitch.

Virtual Practicum

Like many groups, we were forced to pivot our program from in-person to a virtual format during the COVID-19 pandemic. Although we faced new challenges (competing schedules, diverse locations, multiple time zones, loss of lab space, loss of educational excursions), we were able to meet those challenges in a variety of ways that resulted in much of our content becoming freely available online. First, we shipped materials (e.g., live dinoflagellates, UV lights, program t-shirts) to students' homes so they could explore bioluminescence and fluorescence with hands-on materials. Next, we recorded our lectures and made them available online (Table 14.3) so students could watch them at their own convenience. We followed each lecture with unstructured virtual office hours where students could ask questions about the content of the lectures while getting to know and learn about the research interests and unique career trajectories of each faculty member. We assigned "Shark Tank" groups with an awareness of each student's location to ease the burden of scheduling across time zones. Finally, we added a new component to our practicum, our "interviews with experts."

Interviews with Experts

During the virtual iteration of our practicum, we conducted a series of five "interviews with experts." We invited senior scientists working on a variety of facets of bioluminescence and/or optogenetics to join us for interviews about their research, career path, and thoughts on scientific discovery; our interview questions are provided in Box 14.2. Students were given the opportunity to ask questions and were keen to learn how simultaneously exciting, disappointing, surprising, tedious, and generally nonlinear a typical scientist's career path can be. These interviews were recorded and are freely available on our website; Table 14.4 lists the names of our experts, links to their laboratory websites, and references to the recordings.

Box 14.2 Questions for Interviews with Experts

1. Please tell us a little about your research and path to your current work. Is there anything unusual about it that you recommend doing (or not!)?
2. What is your favorite discovery/innovation or proudest moment in your research?
3. What is your favorite discovery/innovation made by someone else?
4. Every career has hard moments: Talk about a failure, or moment your career was in peril—how did you get through?
5. What's a moment when your data led you in a completely unexpected direction?
6. In 25 years, what do you think will be the dominant topic in your field?
7. When/where do you get your best ideas? Walking, reading science, daydreaming...or?
8. What's most important in your environment for your scientific progress?

Table 14.4 Names of our experts, links to their laboratory websites, and references to the freely available recordings of their interviews [101]

Interviews with experts	Laboratory website	References
Henry Colecraft, PhD	https://www.physiology.columbia.edu/colecraft.html	[102]
Steven Haddock, PhD	https://www.mbari.org/person/steven-haddock/	[103]
Stefan Herlitze, PhD	https://dev3.imp10.ruhr-uni-bochum.de/neurobiol/forschungsinteressen/index.html.en	[104]
Jennifer Prescher, PhD	https://www.chem.uci.edu/~jpresche/	[105]
Peter Hegemann, PhD	https://www.biologie.hu-berlin.de/de/gruppen/expbp_portal	[106]
Elena Goun, PhD	https://chemistry.missouri.edu/people/goun	[107]

14.3.4.2 Undergraduate Internships

In-Person Internships

To date, we have had the pleasure of training two high school students and 31 undergraduate or recently graduated interns (18 women, nine students from historically underrepresented groups). These research experiences were held in Bioluminescence Hub labs at Brown University, Central Michigan University, the Scintillon Institute, and the University of California, San Diego, over approximately ten weeks in the summer. Students worked on real research questions—projects for which we didn't have a known result—and whose data were published in scientific journal articles (e.g. [108–111]). This allowed our students to experience the thrill of discovery and, in turn, the frustration of a failed experiment or the puzzle of an unexpected result. At the end of each summer, our interns presented a brief synopsis of their projects and findings to the other members of the collaborating Bioluminescence Hub labs. Table 14.5 shows the variety of research questions pursued by our undergraduate interns.

We received overwhelmingly positive feedback from our students. These immersive internships exposed our students to real research and focused mentorship that is generally not possible in undergraduate laboratory courses. In all cases, our internships helped our students decide whether to pursue additional training or a career in research in a relatively low-stakes, fixed-term setting that is otherwise difficult to simulate in employment or a degree-granting training program.

Virtual Internships

During the COVID-19 pandemic, eight students over two summers studied as virtual interns with our group. These students were charged with identifying and locating all published references to bioluminescent molecules and extracting as much descriptive or measured attributes of those molecules as possible. They then complied and organized those data in a format that was understandable to other scientists and published it in a new resource, BLBase.org. This database was published

Table 14.5 Our interns pursued a variety of research projects during their time in Bioluminescence Hub laboratories

Student	Location	Year	Topic and reference(s)
Tariq Brown	Scintillon	2018	Directed evolution of fluorescent and bioluminescent proteins, improving the brightness of a green fluorescent protein [112]
Hira Choudhry	Scintillon	2018	Optimizing the spectral qualities of monomeric red fluorescent protein from a coral *Acropora millepora* (mAmil) [112]
Ryann Felton	Scintillon	2018	Investigating the spectral qualities and increasing the fluorescence of a photoswitching fluorescent protein from an unknown species of yellow-orange coral (mYOC) [112]
Raisa Kahn	Brown	2019	Created education and outreach materials including: A card to succinctly communicate the properties and function of bioluminescent-optogenetic (BL-OG) constructs [113], a video to disambiguate bioluminescence, fluorescence, and phosphorescence [18], and a gif created to disambiguate bioluminescence, fluorescence, and phosphorescence [114]
Patrick Friend	UCSD	2019	Directed evolution of mAbb, a high quantum yield GFP [115]
Dylan Gregory	UCSD	2019	Photoswitching fluorescent proteins [115]
Sabrina Tran	UCSD	2019	Red fluorescent protein [115]
Brazil Bartholomew Priscilla Louis Cameron MacIntyre Mia Roberts Rochelle van der Merwe	Brown (Virtual)	2020	This cohort of five students performed an extensive literature search and collected the attributes of all formally studied bioluminescent molecules. The result of their study is the database of bioluminescent molecules, BLBase.org [116, 117]
Brazil Bartholomew Reid Funderburk Virginia Groce Emmanuel Buckman Rochelle van der Merwe	Brown (Virtual)	2021	This cohort of five students built on the work of the previous cohort and added to our collection of the attributes of all formally studied bioluminescent molecules. The result of their study is the database of bioluminescent molecules, BLBase.org [116, 118]
Grant Folkert	CMU	2021	Machine learning; DeepLabCut to track and label videos of mouse behavior [119]
Macy Knoblock	CMU	2021	Use of different chemogenetic technologies impacted neuronal electrophysiology, morphology, and the transcriptome in vitro [119]
Jacob Simkins	CMU	2021	Use of molecular cloning techniques to design new luminopsins (LMOs) [119]

(continued)

Table 14.5 (continued)

Student	Location	Year	Topic and reference(s)
Nicholas Coon	CMU	2021	Viral injections, electrophysiology of the substantia nigra pars reticula, optogenetic alteration of brain activity [119]
Stephanie Venn	CMU	2021	Quantification of calcium levels in cortical neurons [119]
Om Naphade	Brown	2022	Blood-brain barrier permeability using BL-OG [120]
Leila Paul	Brown	2022	Effects of activating VTA dopaminergic neurons on the blood-brain permeability using BL-OG
Lisa Kraisriwatana	UCSD	2022	Evaluation of the efficacy and brightness of novel bioluminescent substrates; evolving a FRET linker to produce red-shifted emission [121]
Selena Luong	UCSD	2022	Improving the brightness and increasing the calcium affinity of a calcium sensor [121]
Frederic Nguyen	UCSD	2022	Photomaturable fluorescent proteins [121]
Matthew Tate	UCSD	2022	Engineering various binding affinities between complementary pieces of a split GFP [121]
Camille Phares	UCSD	2022	Evolving a FRET linker to produce red-shifted emission and developing a fluorescent protein that is sensitive to low pH
Nipun Gorantla	CMU	2022	Recording and gauging real-time cortical neuron response to more than 20 novel luminopsin constructs [122]
Shashwat Maharjan	CMU	2022	Creating scripts for automating pipelines for single- and multiunit activity; machine learning algorithms using electrophysiology and behavior for predictive metrics to distinguish healthy vs aberrant brain functions [122]
Lily Mclean	CMU	2022	Automated data analysis at the cellular level (neuron-muscle communication) and behavior level (driven turning behavior in mice) [122]
Jeel Patel	CMU	2022	Chemogenetic technologies and their effects on the electrophysiology and morphology of cortical rat neurons in vitro [122]
Anusha Uprety	CMU	2022	Systematically enhancing neocortical pyramidal activity levels in early development using noninvasive bioluminescence optogenetics (BL-OG) to examine the effects of hyper-excitation of layer V neocortical pyramidal neurons during early postnatal development on adult behavior [122]

in the spring of 2023 and, to date, includes entries for 84 published luciferase-luciferin combinations.

Freely available worldwide, this resource provides an overview of the breadth of species, colors, and luciferase-luciferin pairs that have been described in detail. Our hope is that BLBase.org is a starting place for researchers looking for a natural source of bioluminescence with particular characteristics or references to historical accounts of the discovery of bioluminescent species and their chemical attributes, and that it serves as a source of inspiration for future bioluminescent tool development and research projects.

14.3.5 Horizontal Open Science: Reaching Other Scientists

One of the aims of the Bioluminescence Hub is the broad dissemination of our methods, tools, and resources. We have accomplished this goal through in-person experiences, online content, and archival of our materials in digital data repositories and with companies whose mission is to distribute research materials (Addgene, Jackson Laboratories). This multipronged approach has helped us build a community of researchers aware of and using BL-OG tools and has expanded our collective ability to address pressing research questions.

In 2018, we hosted a 2-day symposium, "Next Generation Technologies for Neuroscience," where approximately 150 guests with diverse levels of experience, from undergraduate researchers to senior scientists, visited Brown University (Fig. 14.9a). Each senior scientist was encouraged to bring at least one trainee, at our expense, to present a poster during the event. Visitors were treated to a discussion with a panel of experts in academic research, industry, and scientific publishing about how discoveries can be translated into tools that can help other researchers and how knowledge can be broadly shared. Next, nine subject matter experts spoke about a variety of tools and technologies used to image and investigate the nervous system. Graduate students and postdoctoral trainees were able to network and receive feedback on their research projects during a poster session and sponsored lunch (Fig. 14.9b).

During our "Workshop for Scientists on Bioluminescence Methods," scientists ranging in experience from research assistant to assistant professor visited Brown University for a lecture on the utility of bioluminescent tools for neuroscience research and to participate in hands-on demonstrations. During the in vitro session, participants learned how to handle the constituents of a bioluminescent chemical reaction, set up an experiment in cells, initiate the reaction, and image or record the bioluminescence (Fig. 14.9c). During the in vivo session, participants worked with a mouse that expressed the LMO3 BL-OG construct in its cortex (see Resources). They learned how to excite the construct with an LED (optogenetic stimulation) and record the neuronal response. They next prepared the luciferin (coelenterazine) for injection and added it to the mouse's cortex to initiate the bioluminescent reaction and excite the LMO3 construct with bioluminescent light. Finally, they imaged the resultant bioluminescence and recorded the neuronal response (Fig. 14.9d).

One of the faculty participants of our Workshop invited one of our a graduate student emissaries to visit her lab and help her group get started with bioluminescent calcium imaging. (Now Dr.) Akash Pal spent a week at Loyola University Chicago, working with the lab and helping them get started with various techniques. He also had the opportunity to deliver a lecture on his thesis work and network with many potential future colleagues and collaborators. While time-intensive, this type of exchange can be mutually beneficial and is often used by industrial purveyors of equipment and supplies.

Fig. 14.9 (**a**) Experts in tool development traveled to Brown University to present their research at our Next Generation Technologies for Neuroscience Symposium. (**b**) Trainee attendees shared their research and made connections during the poster session component of our symposium. (**c**) Dr. Ute Hochgeschwender led workshop participants through the use of BL-OG tools in vitro. (**d**) Dr. Manuel Goez-Ramirez led workshop participants through the use of BL-OG tools in vivo

In-person, retreat-like events can be expensive but are well worth the effort. By investing money and time in travel and preparing a poster or talk, participants will arrive engaged and ready to interact. Some costs can be defrayed by charging a registration fee (although this requires additional administrative management), requiring that guests secure their own transportation, lodging, and most meals, and hosting the event in a space that does not charge a facility use fee. If a group meal or two is possible, it can be a way to encourage informal discussion between scientists, particularly trainees and potential future mentors, or among future collaborators. We offered a grab-and-go bag lunch and a buffet-style pizza and salad dinner, both with unstructured seating, to great effect.

To foster collaboration after the event, we have had success encouraging participants to convert their posters and talks into a format that can be shared digitally [123]. Presented with an abstract and contact information, such an archive allows researchers to build relationships and asynchronously interact with each other. This extends what might otherwise be an ephemeral event and increases the likelihood of researchers forming collaborations.

In addition to our in-person events, we developed freely available online content to reach our scientific colleagues worldwide. These resources include a recorded lecture on "Bioluminescent Optogenetics—Molecular Control with Biological Light" presented by Dr. Ute Hochgeschwender at the Max Planck University of Toronto Centre for Neural Science and Technology, a journal publication that includes a video demonstration of how to get started with bioluminescent constructs [124] and an "E-Z Guide" with links to materials (also distributed in hard copy at the Society for Neuroscience Annual Meeting). We also created a database of the attributes of our bioluminescent molecular constructs, sent our plasmids to Addgene for easy distribution, and ensured our LMO3 mouse is readily available for purchase from The Jackson Laboratory. See Table 14.6 for links to additional information about these activities and resources.

As described here and in Sect. 14.3, creating online content can be a viable way to reach a broad audience with a limited budget, especially as quality personal audio/video equipment and software become increasingly accessible. Likewise, tapping existing outreach and dissemination mechanisms, including commercial industry partners such as Addgene and Jackson Laboratory, can be effective.

14.4 Resources

The Darker the Better

One of the challenges of viewing natural[2] bioluminescence in the classroom or during public outreach events is finding or creating a space that is dark enough to

---

[2] I distinguish natural from engineered bioluminescence because the latter is much brighter. For example, luciferases sbGLuc or NanoLuc have been molecularly engineered for higher light emission; their at least tenfold higher brightness allows them to be seen in ambient light.

Table 14.6 Resources and additional information for other scientists

Activity/resource	Link(s) to more information
Symposium in Next Generation Technologies for Neuroscience	https://www.bioluminescencehub.org/symposium
Workshop for Scientists on Bioluminescence Methods	https://www.bioluminescencehub.org/workshops/august-2018
(now Dr.) Akash Pal's Emissary visit to Loyola University Chicago	https://www.bioluminescencehub.org/research/akash-loyola
JoVE paper and video	[124]
E-Z Guide	Web page: https://www.bioluminescencehub.org/LMO3 Printer-friendly version: https://www.bl-hub.s3.amazonaws.com/E-Z_Guide_to_LMO3.pdf
Dr. Ute Hochgeschwender's lecture	https://www.bioluminescencehub.org/education/Hochgeschwender_MPUoT_Lecture
Constructs database	https://www.bioluminescencehub.org/constructs
Hochgeschwender Lab Plasmids	https://www.addgene.org/Ute_Hochgeschwender/
Lipscombe Lab Plasmids	https://www.addgene.org/Diane_Lipscombe/
Shaner Lab Plasmids	https://www.addgene.org/Nathan_Shaner/
LMO3 mouse at the Jackson Laboratory	https://www.jax.org/strain/034853

appreciate the bioluminescent glow. Lowering the shades in a standard classroom is generally not dark enough and should only be relied on as a last resort. Closing the doors of an interior closet or stairwell can work, but these spaces often allow large gaps of light around doors or have illuminated emergency exit signs. In comparatively dark surroundings, this light pollution is enough to dull the brilliance of the bioluminescent light, so be prepared to temporarily cover these openings and lights with dark paper, tape, or cloth.

If budget allows, a collapsible, portable ice fishing hut has worked well for our demonstrations using natural bioluminescence (shown in Figs. 14.2c and 14.3c). These tools come in a variety of shapes and sizes with features that could influence portability, setup, the footprint within a classroom or exhibit hall, ease of use, number of occupants, and durability. Our ice fishing hut was relatively inexpensive, rectangular, and dark blue and can be quickly assembled and broken down, transported in a passenger vehicle, and carried by two adults. Ours has flaps that cover the built-in windows via a hook and loop tape attachment and two doors with zipper closures. To exclude as much light as possible, we covered the edges of the window covers with duct tape. Very little light penetrates the zipper closures, especially when the ice fishing hut is used in an otherwise dark room (such as an interior closet), but these closures could also be covered with tape. Our ice fishing hut can be comfortably occupied by two adults or one adult and three young children, depending on how many educational materials are also present. The space becomes warm quickly, so it is best to have your demonstration ready to show as soon as your audience's eyes have adjusted to the darkness.

Next, it is important to strike a balance between waiting long enough for your participants' eyes to adjust to view an impressive bioluminescent display and exhausting their patience in a small, dark, warm space. Older students, high school and undergraduate, of course have more patience than younger students. Once participants have entered the hut, settled, and closed the door, the facilitator starts with a demonstration of fluorescence and phosphorescence. The facilitator directs an ultraviolet (UV) light-emitting diode (LED) flashlight at a variety of objects and discusses the difference between fluorescence and phosphorescence (the former is the instantaneous return of an electron to a lower energy state; the latter is a slower return of the electron and thus the glow persists over time). Table 14.7 lists suggestions for fluorescent and phosphorescent objects that work well for illustrating these

Table 14.7 Readily available objects that demonstrate fluorescence or phosphorescence

Object/creature	Fluorescent or phosphorescent	Vendor, notes
Scorpion in resin	Fluorescent	https://theevolutionstore.com/chinese-black-scorpion-in-resin
Laundry detergent	Fluorescent	Readily available. Choose a variety that includes "optical brighteners"
Canola oil, olive oil	Fluorescent	Readily available at grocery stores, large retail stores
Irish Spring Original bar soap	Fluorescent	Readily available at drug stores, large retail stores
Shortwave fluorescent minerals collection	Fluorescent	www.carolina.com/earth-science-rocks-minerals/long%2D%2Dand-short-wave-fluorescent-minerals-collection/GEO2317.pr
Shortwave UV fluorescent rocks minerals—fluorescent willemite & calcite + franklinite, andradite	Fluorescent	https://www.etsy.com/listing/647227125/1-to-20-pound-lots-franklin-new-jersey
Lichens	Fluorescent	Look at rocks and trees outside; easiest to find at night in areas with little light pollution
Glow-in-the-dark toys, stickers, clothing	Phosphorescent	Readily available; e.g., https://www.amazon.com/Glow-Dark-Insects-Halloween-Skeleteen/dp/B07FTTF596/ https://www.amazon.com/UNIME-Pebbles-Walkways-Decorative-Luminous/dp/B073VFFPGC/ https://www.amazon.com/AOWDIAO-Glow-Dark-Stickers-Sizes/dp/B0F7WKPYGL/ https://www.carters.com/o/glow-in-the-dark-clothes-pajamas
Mineral apatite	Phosphorescent	https://www.rocksinahardplace.com/minerals/p/apatite

concepts. Visualizing fluorescence and phosphorescence can be enhanced by wearing yellow-tinted, UV-protective safety glasses.

During this demonstration, the participants' eyes adjust to the darkness of the space (their photoreceptors regenerate rhodopsin), and they become more sensitive to the bioluminescence they are next shown from dried ostracods, live dinoflagellates, and/or bioluminescent fungi. In my experience, it is important to direct the participants' gaze to a particular location in the hut, perhaps with audible cues, instructions when they enter the hut, or by being consistent with the location of the fluorescent objects, before gently jostling live dinoflagellates. The dinoflagellate bioluminescence is reliable and beautiful but fast enough that it could be missed if participants are looking in the wrong direction. Table 14.8 lists examples of bioluminescent specimens that we have used for demonstrations and their relative brightness and ease of use. Notes and links to suppliers are included.

We have successfully presented these demonstrations at the Brown Brain Fair (public event), the annual Science Conference at Vartan Gregorian Elementary (elementary school), for Brown Science Prep (high school), and as part of our NeuroNex Practicum for Undergraduates (Sect. 14.3). Unsurprisingly, attention span increased with student age and individual interest. Some of our undergraduates were happy to crowd into our hut and wait many minutes for their eyes to adjust to see the glow of our bioluminescent fungi!

We have created several downloadable activities for educators and learners, freely available on our website, bioluminescencehub.org. Printed versions of these instructions are excellent handouts at public-facing events.

Sharing Your Success

One of the key outcomes of these education and outreach efforts is the creation of a community, both local and online, of supporters and collaborators. As we planned, advertised, and held our in-person and virtual events, we communicated our success via weekly team meetings and our Instagram and X (Twitter) accounts with the handle @biolumhub. This became an excellent way to keep our alumni engaged, and we continue to highlight the achievements of our alumni through our "Where Are They Now?" series on Instagram. We also used this communication effort as a training opportunity for our students, and BL-OG Team member Krystal Literman created a presentation for our Practicum students and a freely available, downloadable resource for our website [98]. In short, we encouraged the sharing of factual scientific information in a positive, constructive way and continue to suggest this as a means of sharing knowledge and building community.

Animals and Supplies Vendors

Given an adequate seawater supply, several species can be acquired and at least temporarily maintained for classroom or laboratory observation. I had specimens shipped to and held by the Marine Biological Laboratory, but products such as Instant Ocean can also be used to create temporary classroom aquaria. We selected the sea walnut *Mnemiopsis macrydi* (a Ctenophore), the sea pansy *Renilla mulleri* (a Cnidarian), and the parchment tube worm *Chaetopterus variopedatus* (an Annelid) (Table 14.9). We purchased these animals from Gulf Specimen Marine

Table 14.8 Comparison of natural bioluminescence demonstrations (ostracods vs. dinoflagellates vs. fungi): brightness, ease of use, repeatability, maintenance (light cycling, fungal contamination, humidity)

Bioluminescent creature	Brightness	Cost	Ease of use or maintenance	Advantages	Disadvantages	Notes and URL
Dried ostracods ("sea firefly" *Cypridina hilgendorfii* or *Vargula hilgendorfii*)	[a]	[b]	[a]	Easy to keep in storage	Demonstration of bioluminescence is not repeatable	Dried; nonliving Need a means to crush the ostracods, e.g., mortar and pestle, just before adding water https://www.carolina.com/cellular-physiology-enzymes/sea-firefly-cypridina-hilgendorfii-500-mg/203430.pr?question=bioluminescence
Dinoflagellates (*Pyrocystis* sp.)	[c]	[a]	[c]	Impressive "sparkle" of bioluminescence Easy for a student to gently jostle to produce bioluminescence	Fatigue quickly; need time to "recharge" before the next demonstration	Living Usually arrive light cycled for use during daytime classroom activities; requires maintenance https://www.carolina.com/algae/bioluminescent-dinoflagellates-living/153305.pr?question=bioluminescent
Fungi (*Panellus stipticus*)	[b]	[b]	[c]	Greenish bioluminescence, distinct from blue of other demonstrations	Relatively easy to contaminate with pathogenic mold	Living Must be patient to allow eyes to adjust to see dim bioluminescence https://www.carolina.com/fungi/panellus-stipticus-living-plate/156135.pr?question=bioluminescent

[a] Is best/easiest
[b] Is moderate
[c] Is least good/most challenging

Table 14.9 Name and attributes of specimens we have used for undergraduate education

Animal (species)	Phylum	Bioluminescence color (wavelength in nm)	Bioluminescence expression	Hypothesized function	Relevant biology topics (examples)
Mnemiopsis macrydi (sea walnut)	Ctenophora	Blue	Brief bursts of full-body flashing	Startle response?	Anatomy (oral/aboral, transparency, clila); ecology (pelagic coastal ocean)
Beroe ovata (pink comb jelly)	Ctenophora	Blue-green	Series of flashes	Startle response?	Anatomy (oral/aboral, transparency, clila); ecology (pelagic coastal ocean)
Renilla mulleri (sea pansy)	Cnidaria	Green	Appears as green waves moving across the flat surface of the animal	Startle response?	Anatomy (polyps); ecology (benthic reef)
Chaetopterus variopedatus (parchment tube worm)	Annelida	Blue	Bright blue mucus secreted from the animal near its head	Startle response? Prey attraction? Secondary defense by illuminating a potential predator? A by-product of the mucus used for tube construction?	Ecology (benthic mud flats, crab symbiote, builds tube)

Laboratories, but there are other excellent sources of marine specimens for education (Table 14.10). I strongly recommend asking the vendors about bioluminescence before ordering. Not only will this help them collect individuals from bioluminescent species instead of closely related non-bioluminescent species, but the collectors are often able to recommend other species that are hardy and exhibit specific behaviors of interest.

As with all animal behavior experiments and observations, the specimens must be healthy and happy to exhibit naturalistic bioluminescence. After a brief acclimation period, we were able to observe bioluminescence in our specimens by gently jostling their holding tanks, disturbing their surrounding water, or *very* gently

Table 14.10 Sources of bioluminescent specimens for laboratory examination

Organization	URL to order specimens	Notes and other available resources
Gulf Specimen Marine Laboratories	https://gulfspecimen.org/purchase-live-marine-specimens-new/	Experiment Kits https://gulfspecimen.org/product-category/experiment-kits/
Marine Biological Laboratory's Marine Resources Center	https://www.mbl.edu/research/resources-research-facilities/marine-resources-center/mrc-services	Catalog of Specimens and Services https://www.mbl.edu/sites/default/files/2025-05/MRC_Catalog_2025.pdf
Carolina Biological Supply	https://www.carolina.com/living-organisms/classroom-animals/marine-and-saltwater-animals/10534.ct?utm_campaign=home_nav_10476&utm_source=internal&utm_medium=link	Inquire about the availability of bioluminescent species.
Sea Dwelling Creatures	https://www.seadwelling.com/collections	Inquire about the availability of bioluminescent species. Aquarium Supplies https://www.seadwelling.com/collections/dry-goods

touching the edge of their bodies with a cotton-tipped exam probe or the tip of a disposable plastic pipette. In my experience, unless they have taken Invertebrate Biology or spent a lot of time exploring the sea, very few undergraduates are familiar with these animals. It is thus worthwhile to provide an overview of gross anatomy, ecology, and phylogeny to illustrate how these diverse and distantly related animals have converged on using bioluminescence despite their substantially different behaviors and habitats.

Quality Sources of Information
In addition to our website, bioluminescencehub.org, there are several online sources of high-quality information about bioluminescence, bioluminescent creatures, fluorescence, and phosphorescence. I list a few of these websites in Table 14.11, in no particular order.

14.5 Woods Hole

I include this section to both inspire the reader to consider taking part in one of the many opportunities for scientific engagement in Woods Hole (Table 14.12 lists opportunities currently available at the MBL) and to encourage educators to explore their local resources (e.g., natural attractions, libraries, academic institutions, community centers and special interest groups) to find creative ways to incorporate field trips and community engagement when developing curricula.

Table 14.11 Additional sources of reliable information about bioluminescence, bioluminescent species, fluorescence, and phosphorescence

Title	Institution	Description and URL
Illuminating the facts of deep-sea bioluminescence [125]	Monterey Bay Aquarium	Definition and description of bioluminescence, distribution and diversity in nature, colors, use in life processes Excellent image and video content https://www.montereybayaquarium.org/stories/bioluminescence
The Bioluminescence Web Page	The University of California, Santa Barbara Monterey Bay Aquarium Research Institute (MBARI)	Comprehensive website on bioluminescent life, distribution, life processes, discoveries Excellent image and video content https://biolum.eemb.ucsb.edu/
Bioluminescence	Monterey Bay Aquarium Research Institute	Collection of News articles from research performed at MBARI Excellent image and video content https://www.mbari.org/know-your-ocean/bioluminescence/
Latz Laboratory: Bioluminescence Questions and Answers	University of California, San Diego Scripps Institution of Oceanography	Answers to ten common questions; wide range of information https://latzlab.ucsd.edu/bioluminescence/bioluminescence-questions-and-answers/
Bioluminescence	Smithsonian National Museum of Natural History	Introductory overview and descriptions of how bioluminescence is used in nature https://ocean.si.edu/ocean-life/fish/bioluminescence
Glow Rocks	Field Museum	Demonstration showing fluorescent and phosphorescent minerals https://www.fieldmuseum.org/blog/glow-rocks

Woods Hole Attractions Enhanced Our Undergraduate Practicum

During our in-person undergraduate practicum, we took time to explore some of the attractions and learn the scientific history of the village of Woods Hole, Massachusetts [126]. Among the various research institutions in Woods Hole, the Marine Biological Laboratory (MBL), the Woods Hole Oceanographic Institute (WHOI), and the National Oceanic and Atmospheric Administration (NOAA) Woods Hole Laboratory of the Northeast Fisheries Science Center are the oldest and most famous. We were fortunate to interact with all three of these institutions during our stay.

The MBL was founded in 1888 and has since been a hub of scientific research [127]. We received guided tours of the MBL's Marine Resources Center, the MBLWHOI Library's rare books and Nobel Prize collection, WHOI's Ocean Science Discovery Center, and NOAA's Woods Hole Science Aquarium (Fig. 14.10).

In the Marine Resources Center, Scott Bennett, Marine Resources Services and Outreach Coordinator, led us through a guided tour of the state-of-the-art facility that collects and maintains local and exotic research animals (Fig. 14.10a). Students

Table 14.12 The MBL provides a range of opportunities for engagement among scientists and with the public

MBL opportunity	URL	Participant rank
NSF REU Biological Discovery in Woods Hole	https://www.mbl.edu/education/internships/nsf-reu-biological-discovery-woods-hole	Current undergraduates
Advanced Research Training Courses	https://www.mbl.edu/education/advanced-research-training-courses/course-offerings	Graduate students and above
Grass Foundation Fellowships	https://www.mbl.edu/research/funding-opportunities/grass-foundation https://grassfoundation.org/	Advanced graduate students, postdoctoral researchers
Faculty Opportunities in MBL Courses	https://www.mbl.edu/education/advanced-research-training-courses/faculty-opportunities-mbl-courses	Faculty
Stay a Season at the MBL	https://www.mbl.edu/stay	Faculty
Friday Evening Lectures	https://www.mbl.edu/events/friday-evening-lectures	Everyone

learned about the ocean currents that bring cold water to the north side of Cape Cod and warm water to the south side of Cape Cod, resulting in rich local biodiversity and making Woods Hole an ideal place for marine science research.

In the MBLWHOI Library, Librarian Matthew Person treated us to a tour of the rare books collection (Fig. 14.10b). There, we paged through albums created during ocean exploration cruises more than a century ago, filled with drawings of real and imagined sea creatures (some with likenesses similar to political figures of the time) and hand-painted plates showing the anatomical details of collected specimens. Students viewed an inscribed sixth edition *Origin of Species* gifted from Charles Darwin himself to Alpheus Hyatt, Massachusetts Institute of Technology professor, curator of the Boston Society for Natural History, founder of the MBL antecedent the Annisquam Summer School, and first President of the MBL.

A gifted storyteller, Matt told us a moving story about a note, now mounted on the wall of the MBLWHOI library, left by Japanese researchers at the University of Tokyo Oceanographic Institute during World War II, asking allied forces to spare their scientific research materials [128].

> This is a marine biological station with her history of over sixty years. If you are from the Eastern Coast, some of you might know Woods Hole or Mt Desert or Tortagus [sic]. If you are from the West Coast, you may know Pacific Grove or Puget Sound Biological Station. This place is a place like one of these. Take care of this place and protect the possibility for the continuation of our peaceful research. You can destroy the weapons and war instruments But save the civil equipments [sic] for Japanese students When you are through with your job here Notify to the University and let us come back to our scientific home [signed] The last one to go

The MBLWHOI Library is home to Thomas Hunt Morgan, PhD's 1933 Nobel Prize in Physiology or Medicine. In addition to viewing his Nobel diploma, students

Fig. 14.10 Students enjoyed viewing (**a**) local specimens in one of the tanks at the MBL's Marine Resources Center, (**b**) the MBLWHOI Library's rare books collection, (**c**) exhibits in WHOI's Ocean Science Discovery Center, and (**d**) residents of NOAA's Woods Hole Science Aquarium

were allowed to handle Morgan's gold Nobel Prize medal, a treat for any young scientist (Fig. 14.11).

The WHOI Ocean Science Discovery Center [129] has a variety of exhibits with a focus on deep-sea exploration (Fig. 14.10c). Highlights include a timeline of how

Fig. 14.11 Emmanuel Crespo, PhD, Practicum teaching assistant, standing in front of Thomas Hunt Morgan's 1933 Nobel Diploma and holding Morgan's Nobel Prize medal

WHOI researchers found the wreck of the *Titanic*, videos of deep-sea hydrothermal vents captured by WHOI submersibles and remotely operated vehicles (ROVs), and a replica of the submersible *Alvin* that visitors can touch and imagine boarding for a dive.

Opened in 1875, the Woods Hole Science Aquarium is the oldest aquarium in the United States (Fig. 14.10d). The facility maintains separate cold and warm water systems and features local species that thrive in either condition, some collected from the cold waters of Cape Cod Bay and points northward and some collected in the summer and early fall from warm water in Nantucket Sound and points southward. Free and open to the public, our students especially enjoyed the touch tank exhibit, where they could gently handle and examine some hardy specimens.

Opportunities at the Marine Biological Laboratory

The Marine Biological Laboratory, now part of the University of Chicago, remains a rich source of training opportunities for junior scientists and teaching or sabbatical opportunities for faculty (Table 14.12). According to the MBL's website (https://www.mbl.edu/), around 1,500 scientists and students visit the MBL each year, and 63 Nobel Prize winners have been linked to the institution. This includes Osamu Shimomura, PhD (1928–2018), co-recipient of the 2008 Nobel Prize in Chemistry for his work discovering green fluorescent protein in the bioluminescent jellyfish *Aequorea* [130]. Shimomura spent decades studying bioluminescence and fluorescence and was the first person to isolate aequorin, a calcium-sensitive photoprotein that produces bioluminescence in *Aequorea*. The combination of readily available research and educational resources, the history of discovery, and the scientific community affiliated with Woods Hole makes the village one of the best possible places to introduce students to bioluminescence. We were fortunate to be able to visit and work there and encourage the interested reader to explore their own path to Woods Hole.

14.6 Summary

It is easy to captivate learners with bioluminescence, and quick, relatively inexpensive demonstrations can be used to convey scientific information from a variety of disciplines. From my experience working with diverse audiences, I suggest that educators designing their own curricula incorporate hands-on activities, supply items participants can take home with them, and organize visits to significant places related to the topic of choice whenever possible. Even in a virtual or limited resources setting, a creative approach with a focus on key learning concepts can be effective in communicating scientific information. Finally, I encourage educators to share their success and experiences to build and engage a community interested in spreading understanding through awe and curiosity.

Acknowledgments The activities described here were supported by NSF NeuroNex 1707352. Our education and outreach efforts were designed and conducted under the leadership of the principal investigators of that award, Drs. Christopher Moore, Diane Lipscombe, Ute Hochgeschwender, and Nathan Shaner. Dedicated staff and trainees, including Krystal Literman, Nina Friedman, and (alphabetically) Drs. Dmitrijs Celinskis, Emmanuel Crespo, Manuel Gomez-Ramirez, Jeremy Murphy, Akash Pal, Eric Petersen, and Mansi Prakash, developed educational content and worked with participants. Brown University Librarian Andrew Creamer built our NeuroNex Collection in the Brown Digital Repository and continues to help us archive research and educational materials. MBLWHOI Librarian Matthew Person and MBL Marine Resources Services and Outreach Coordinator Scott Bennett provided excellent and memorable tours of their MBL facilities. Senior Scientist and Director of the Bell Center at the Marine Biological Laboratory, Dr. Jennifer Morgan, lectured and led a laboratory section during our NeuroNex Practicum for Undergraduates and tested our LMO3 construct in her lamprey model system. Finally, the BL-OG Team is grateful to all of our educational and outreach participants, particularly our undergraduate interns and practicum alumni, for your enthusiasm, inquisitiveness, and persistence.

References

1. Frost R, Lathem EC. The poetry of Robert Frost. London: Vintage; 2001. 607 p.
2. Latz MI. The artistry of dinoflagellate bioluminescence. Mater Today Proc. 2017;4(4):4959–68.
3. Prischmann-Voldseth DA. Fireflies in art: emphasis on Japanese Woodblock Prints from the Edo, Meiji, and Taishō Periods. Insects. 2022;13(9):775.
4. Harvey EN. A history of luminescence from the earliest times until 1900, vol. 44. Philadelphia: American Philosophical Society, Memoirs of the American Philosophical Society; 1957.

5. Hudson J, Miller SD. From sailors to satellites: a curated database of bioluminescent milky seas spanning 1600-present. Earth Space Sci. 2025;12(4):e2024EA004082.
6. Darwin C. The voyage of the beagle. 1st ed. Auckland: The Floating Press; 2008.
7. Semmes R. Memoirs of service afloat: during the war between the states [Internet]. Baltimore: Kelly, Piet & Co.; 1869 [cited 2025 Aug 4]. Available from: https://link.gale.com/apps/doc/FZ0100095163/SAS?u=prov98893&sid=bookmark-SAS&xid=2fb0995e&pg=737.
8. Nick S. What are milky seas? [Internet]. SOURCE. 2025 [cited 2025 Sept 25]. Available from: https://source.colostate.edu/what-are-milky-seas-centuries-old-ship-logs-diaries-may-help-solve-this-maritime-mystery/.
9. Miller SD, Haddock SHD, Elvidge CD, Lee TF. Detection of a bioluminescent milky sea from space. Proc Natl Acad Sci USA. 2005;102(40):14181–4.
10. Henry JP, Michelson AM. Bioluminescence: physiological control and regulation at the molecular level. Photochem Photobiol. 1978;28(2):293–310.
11. Viviani V. Introduction to the themed issue on bioluminescence. Photochem Photobiol Sci. 2008;7(2):145.
12. Fleiss A, Sarkisyan KS. A brief review of bioluminescent systems (2019). Curr Genet. 2019;65(4):877–82.
13. Auvray M, Myin E. Perception with compensatory devices: from sensory substitution to sensorimotor extension. Cogn Sci. 2009;33(6):1036–58.
14. Auvray M. Sensory substitution. Open encyclopedia of cognitive science [Internet]. 2025 [cited 2025 May 21]. Available from: https://oecs.mit.edu/pub/lmiweq0s.
15. Latz MI. The making of *infinity cube*, a bioluminescence art exhibit. Limnol Oceanogr Bull. 2019;28(4):130–4.
16. Aristotle. In: Hicks RD, editor. De anima. New York: Arno Press; 1976. (History of ideas in ancient Greece).
17. Lapota D, Galt C, Losee JR, Huddell HD, Orzech JK, Nealson KH. Observations and measurements of planktonic bioluminescence in and around a milky sea. J Exp Mar Biol Ecol. 1988;119(1):55–81.
18. Khan R. Cold light [Internet]. Brown University; 2019 [cited 2025 Sept 9]. Available from: https://repository.library.brown.edu/studio/item/bdr:847081/.
19. Literman K. Make it glow: a "cold light" kit. 2020 [cited 2025 Sept 9]. Available from: https://repository.library.brown.edu/studio/item/bdr:1111550/.
20. Literman K. WHAT "GLOWS" THERE?! Fluorescent discoveries around the home. 2019 [cited 2025 Sept 9]. Available from: https://repository.library.brown.edu/studio/item/bdr:880615/.
21. Nyholm SV, McFall-Ngai MJ. A lasting symbiosis: how the Hawaiian bobtail squid finds and keeps its bioluminescent bacterial partner. Nat Rev Microbiol. 2021;19(10):666–79.
22. Haddock SHD, Moline MA, Case JF. Bioluminescence in the sea. Annu Rev Mar Sci. 2010;2(1):443–93.
23. Ben-Zvi O, Lindemann Y, Eyal G, Loya Y. Coral fluorescence: a prey-lure in deep habitats. Commun Biol. 2022;5(1):537.
24. Clarke DN, Rose NH, De Meulenaere E, Rosental B, Pearse JS, Pearse VB, et al. Fluorescent proteins generate a genetic color polymorphism and counteract oxidative stress in intertidal sea anemones. Proc Natl Acad Sci USA. 2024;121(11):e2317017121.
25. Gruber DF, Sparks JS. First observation of fluorescence in marine turtles [Internet], vol. 3845. New York: American Museum of Natural History; 2015. 1 p. Available from: https://www.biodiversitylibrary.org/page/58570410.
26. Reinhold LM. Widespread fluorescence in the fur of mammals of the Australian Wet Tropics. Proc R Soc Queensl. 2023;132:3–40.
27. Prötzel D, Heß M, Schwager M, Glaw F, Scherz MD. Neon-green fluorescence in the desert gecko *Pachydactylus rangei* caused by iridophores. Sci Rep. 2021;11(1):297.
28. Yuan J, Zhai Y, Wan K, Liu S, Li J, Li S, et al. Sustainable afterglow materials from lignin inspired by wood phosphorescence. Cell Rep Phys Sci. 2021;2(9):100542.

29. Nature's night lights: learn about the glow-in-the-dark organisms of the Santa Cruz Mountains | Midpeninsula Regional Open Space District [Internet]. [cited 2025 Aug 4]. Available from: https://www.openspace.org/stories/natures-night-lights-learn-about-glow-dark-organisms-santa-cruz-mountains.
30. Widder EA. Bioluminescence. In: Archer SN, Djamgoz MBA, Loew ER, Partridge JC, Vallerga S, editors. Adaptive mechanisms in the ecology of vision [Internet]. Dordrecht: Springer Netherlands; 1999 [cited 2025 Sept 10]. p. 555–81. Available from: http://link.springer.com/10.1007/978-94-017-0619-3_19.
31. Jousset de Bellesme GLMF. Experimental researches on the phosphorescence of the glow-worm. Ann Mag Nat Hist Zool Bot Geol. 1880;5:345–7.
32. McAlpine D. Phosphorescent fungi in Australia. Proc Linnean Soc NSW. 1901;25:548–58.
33. Lloyd RE. Notes on phosphorescence in marine animals. Rec Indian Mus. 1907;1:257–61.
34. Parker GH. The phosphorescence of *Renilla*. Proc Am Philos Soc. 1920;59(220):171–5.
35. Oba Y, Schultz DT. Firefly genomes illuminate the evolution of beetle bioluminescent systems. Curr Opin Insect Sci. 2022;50:100879.
36. Perry BA, Desjardin DE, Stevani CV. Diversity, distribution, and evolution of bioluminescent fungi. J Fungi (Basel). 2024;11(1):19.
37. Meyer-Rochow VB, Moore S. Biology of *Latia neritoides* Gray 1850 (Gastropoda, Pulmonata, Basommatophora): the only light-producing freshwater snail in the world. Int Rev Hydrobiol. 1988;73(1):21–42.
38. Shimomura O, Johnson FH, Kohama Y. Reactions involved in bioluminescence systems of limpet (*Latia neritoides*) and luminous bacteria. Proc Natl Acad Sci USA. 1972;69(8):2086–9.
39. Pholyotha A, Yano D, Mizuno G, Sutcharit C, Tongkerd P, Oba Y, et al. A new discovery of the bioluminescent terrestrial snail genus *Phuphania* (Gastropoda: Dyakiidae). Sci Rep. 2023;13(1):15137.
40. Pes O, Midlik A, Schlaghamersky J, Zitnan M, Taborsky P. A study on bioluminescence and photoluminescence in the earthworm *Eisenia lucens*. Photochem Photobiol Sci. 2016;15(2):175–80.
41. Harvey EN. The nature of the red and green luminescence of the South American "railroad worm," *Phryxothrix*. J Cell Comp Physiol. 1944;23(2):31–8.
42. Viviani VR, Bechara EJH. Bioluminescence and biological aspects of Brazilian railroad-worms (Coleoptera: Phengodidae). Ann Entomol Soc Am. 1997;90(3):389–98.
43. Cock RD, Matthysen E. Aposematism and bioluminescence: experimental evidence from glow-worm larvae(Coleoptera: Lampyridae). Evol Ecol. 1999;13(7):619–39.
44. Richards AM. Observations on the New Zealand glow-worm (Skuse) 1890. Trans R Soc N Z. 1960;88:559–74.
45. Sivinski J. Prey attraction by luminous larvae of the fungus gnat *Orfelia fultoni*. Ecol Entomol. 1982;7(4):443–6.
46. Vršanský P, Chorvát D, Fritzsche I, Hain M, Ševčík R. Light-mimicking cockroaches indicate Tertiary origin of recent terrestrial luminescence. Naturwissenschaften. 2012;99(9):739–49.
47. Oba Y, Meyer-Rochow VB. A supposedly bioluminescent species of cockroach that would not glow: a case of scientific fraud, erroneous observation or bacterial infection? Arthropod Struct Dev. 2025;86:101438.
48. Sivinski J. Arthropods attracted to luminous fungi. Psyche J Entomol. 1981;88(3–4):383–90.
49. De Meulenaere E, Bailey JB, Tezcan FA, Deheyn DD. First biochemical and crystallographic characterization of a fast-performing ferritin from a marine invertebrate. Biochem J. 2017;474(24):4193–206.
50. Mirza JD, Migotto ÁE, Yampolsky IV, De Moraes GV, Tsarkova AS, Oliveira AG. *Chaetopterus variopedatus* bioluminescence: a review of light emission within a species complex. Photochem Photobiol. 2020;96(4):768–78.
51. Mills C. Bioluminescence and other factoids about Aequorea, a hydromedusa. 2009 [cited 2025 Sept 10]. Bioluminescence of Aequorea. Available from: https://faculty.washington.edu/cemills/Aequorea.html.

52. Shimomura O. The jellyfish *Aequorea* and other luminous coelenterates. Singapore: World Scientific Publishing Company; 2012. p. 91–165.
53. Haddock SHD, Dunn CW, Pugh PR, Schnitzler CE. Bioluminescent and red-fluorescent lures in a deep-sea siphonophore. Science. 2005;309(5732):263.
54. Widder E. Bioluminescence and the pelagic visual environment. Mar Freshw Behav Physiol. 2002;35(1–2):1–26.
55. Mashukova O, Silakov M, Temnykh A. Ecological role of bioluminescence of Black sea ctenophores. Biophys Rev. 2023;15(5):947–54.
56. Santhanam R. Bioluminescent ctenophores. Singapore: Bentham Science Publishers; 2022.
57. Hanley KA, Widder EA. Bioluminescence in dinoflagellates: evidence that the adaptive value of bioluminescence in dinoflagellates is concentration dependent. Photochem Photobiol. 2017;93(2):519–30.
58. Valiadi M, Iglesias-Rodriguez D. Understanding Bioluminescence in dinoflagellates—how far have we come? Microorganisms. 2013;1(1):3–25.
59. Deheyn D, Mallefet J, Jangoux M. Expression of bioluminescence in *Amphipholis squamata* (Ophiuroidea: Echinodermata) in presence of various organisms: a laboratory study. J Mar Biol Assoc. 2000;80(1):179–80.
60. Bessho-Uehara M, Mallefet J, Haddock SHD. Chapter 22 – Glowing sea cucumbers: bioluminescence in the Holothuroidea. In: Mercier A, Hamel JF, Suhrbier AD, Pearce CM, editors. The world of sea cucumbers [Internet]. Academic Press; 2024 [cited 2025 Sept 9]. p. 361–75. Available from: https://www.sciencedirect.com/science/article/pii/B9780323953771000035.
61. Johnsen S, Balser EJ, Widder EA. Light-emitting suckers in an octopus. Nature. 1999;398(6723):113–4.
62. Johnsen S, Widder EA, Mobley CD. Propagation and perception of bioluminescence: factors affecting counterillumination as a cryptic strategy. Biol Bull. 2004;207(1):1–16.
63. Herring PJ, Widder EA, Haddock SHD. Correlation of bioluminescence emissions with ventral photophores in the mesopelagic squid *Abralia veranyi* (Cephalopoda: Enoploteuthidae). Mar Biol. 1992;112(2):293–8.
64. Gruber DF, Phillips BT, O'Brien R, Boominathan V, Veeraraghavan A, Vasan G, et al. Bioluminescent flashes drive nighttime schooling behavior and synchronized swimming dynamics in flashlight fish. Radford CA, editor. PLoS ONE. 2019;14(8):e0219852.
65. Hellinger J, Jägers P, Donner M, Sutt F, Mark MD, Senen B, et al. The flashlight fish *Anomalops katoptron* uses bioluminescent light to detect prey in the dark. De A, editor. PLoS ONE. 2017;12(2):e0170489.
66. Pietsch TW. Oceanic anglerfishes: extraordinary diversity in the deep sea. 1st ed. Berkeley: University of California Press; 2009.
67. Herring PJ. Systematic distribution of bioluminescence in living organisms. J Biolumin Chemilumin. 1987;1(3):147–63.
68. Buschman LL. Biology of the firefly *Pyractomena lucifera* (Coleoptera: Lampyridae). Fla Entomol. 1984;67:529–42.
69. Blair KG. An aquatic lampyrid larva from S. Celebes. Trans R Entomol Soc Lond. 1927;75(1):43–5.
70. Zheng B, Lucas AJ, Franks PJS, Schlosser TL, Anderson CR, Send U, et al. Dinoflagellate vertical migration fuels an intense red tide. Proc Natl Acad Sci USA. 2023;120(36):e2304590120.
71. Claes JM, Haddock SHD, Coubris C, Mallefet J. Systematic distribution of bioluminescence in marine animals: a species-level inventory. Life. 2024;14(4):432.
72. Davis MP, Sparks JS, Smith WL. Repeated and widespread evolution of bioluminescence in marine fishes. Thuesen EV, editor. PLoS ONE. 2016;11(6):e0155154.
73. Letendre F, Twardowski M, Blackburn A, Poulin C, Latz MI. A review of mechanically stimulated bioluminescence of marine plankton and its applications. Front Mar Sci. 2024;10:1299602.
74. Martini S, Haddock SHD. Quantification of bioluminescence from the surface to the deep sea demonstrates its predominance as an ecological trait. Sci Rep. 2017;7(1):45750.

75. Shaner NC, 2019 NeuroNex Practicum Class. Bioluminescent parchment tube worm [Internet]. Brown University; 2019 [cited 2025 Sept 10]. Available from: https://repository.library.brown.edu/studio/item/bdr:1147336/.
76. DeLeo DM, Bessho-Uehara M, Haddock SHD, McFadden CS, Quattrini AM. Evolution of bioluminescence in Anthozoa with emphasis on Octocorallia. Proc R Soc B. 2024;291(2021):20232626.
77. Motyka M, Kusy D, Arias-Bohart ET, Bybee SM, Bocak L. Enigmatic *Campyloxenus*: shedding light on the delayed origin of bioluminescence in ancient Gondwanan click beetles. iScience. 2023;26(12):108440.
78. Lau ES, Oakley TH. Multi-level convergence of complex traits and the evolution of bioluminescence. Biol Rev. 2021;96(2):673–91.
79. Klein E. Movie: bioluminescence in the choroid plexus [Internet]. Brown University; 2017 [cited 2025 Sept 18]. Available from: https://repository.library.brown.edu/studio/item/bdr:788093/.
80. Berger-Gutierrez R. Bioluminescence activity placemat [Internet]. Brown University; 2019 [cited 2025 Sept 18]. Available from: https://repository.library.brown.edu/studio/item/bdr:847428/.
81. Allen JJ. WHO "GLOWS" THERE?! Bioluminescent dinoflagellates. 2019 [cited 2025 Oct 2]. Available from: https://repository.library.brown.edu/studio/item/bdr:919150/.
82. Literman K. Growin' Glowin' FUNGI: grow your own bioluminescent fungus. 2019 [cited 2025 Oct 2]. Available from: https://repository.library.brown.edu/studio/item/bdr:929626/.
83. Literman K. Submersible checklist. 2020 [cited 2025 Oct 2]. Available from: https://repository.library.brown.edu/studio/item/bdr:1144736/.
84. Literman K. The bioluminescence hub. 2020 [cited 2025 Oct 2]. Bioluminescent dive: part 1: submersibles. Available from: https://www.bioluminescencehub.org/education/dive-1.
85. Literman K. The bioluminescence hub. 2020 [cited 2025 Oct 2]. Bioluminescent dive: part 2: the dive. Available from: https://www.bioluminescencehub.org/education/dive-2.
86. Berger-Gutierrez R. The bioluminescence hub. 2019 [cited 2025 Oct 2]. The bioluminescent jellyfish. Available from: https://www.bioluminescencehub.org/education/madlib-bioluminescent-jellyfish.
87. Berger-Gutierrez R. The bioluminescence hub. 2019 [cited 2025 Oct 2]. The professor's glowing brains. Available from: https://www.bioluminescencehub.org/education/madlib-professors-brains.
88. Salem H, Salem Z, Allen J. The bioluminescence hub. 2024 [cited 2025 Oct 2]. Word Search. Available from: https://www.bioluminescencehub.org/education/wordsearch.
89. Literman K, Friedman N, Crespo E, van der Merwe R, Celinskis D, Murphy J, et al. The bioluminescence hub. 2020 [cited 2025 Oct 2]. Junior trainee club part 1: welcome to the team! Available from: https://www.bioluminescencehub.org/JTC-welcome.
90. Allen JJ. Light and living where it is very, very dark [Internet]. Brown University; 2021 [cited 2025 Sept 18]. Available from: https://repository.library.brown.edu/studio/item/bdr:1167673.
91. Lipscombe D. Calcium — not too little and not too much [Internet]. Brown University; 2021 [cited 2025 Sept 18]. Available from: https://repository.library.brown.edu/studio/item/bdr:8pjntyf5/.
92. Hochgeschwender U. From shrimp to brain [Internet]. Brown University; 2021 [cited 2025 Sept 18]. Available from: https://repository.library.brown.edu/studio/item/bdr:1169225/.
93. Shaner N. Fluorescence and bioluminescence: basics and chemistry [Internet]. Brown University; 2021 [cited 2025 Sept 18]. Available from: https://repository.library.brown.edu/studio/item/bdr:1169212/.
94. Moore CI. Bioluminescence: ideal natural engineering for understanding the dynamics that create behavior [Internet]. Brown University; 2021 [cited 2025 Sept 18]. Available from: https://repository.library.brown.edu/studio/item/bdr:1170548/.

95. Shaner N. "Brightness" and the physics of bioluminescent light production and resonance energy transfer [internet]. Brown University; 2021 [cited 2025 Sept 18]. Available from: https://repository.library.brown.edu/studio/item/bdr:56u8rvus/.
96. Ikefuama EC, Kendziorski GE, Anderson K, Shafau L, Prakash M, Hochgeschwender U, et al. Improved locomotor recovery in a rat model of spinal cord injury by BioLuminescent-OptoGenetic (BL-OG) stimulation with an enhanced luminopsin. Int J Mol Sci. 2022;23(21):12994.
97. Petersen ED, Sharkey ED, Pal A, Shafau LO, Zenchak-Petersen J, Peña AJ, et al. Restoring function after severe spinal cord injury through BioLuminescent-OptoGenetics. Front Neurol. 2022;12:792643.
98. Literman K. #Why we SciComm. 2021 [cited 2025 Sept 10]. Available from: https://repository.library.brown.edu/studio/item/bdr:85dak8qf/.
99. Britayev TA, Mekhova E, Deart Y, Martin D. Do syntopic host species harbour similar symbiotic communities? The case of *Chaetopterus* spp. (Annelida: Chaetopteridae). PeerJ. 2017;5:e2930.
100. Morgan J. Use of lamprey as a model for studying the neural control of locomotion. 2019 [cited 2025 Sept 18]. Available from: https://repository.library.brown.edu/studio/item/bdr:kaejcxzn/.
101. Allen JJ. The bioluminescence hub. 2021 [cited 2025 Sept 18]. Interview with experts. Available from: https://brown.hosted.panopto.com/Panopto/Pages/Viewer.aspx?id=a5cc658f-da53-473f-ac40-ad2f01517ccf.
102. Colecraft H, Moore CI, Lipscombe D. Interview with experts: Henry Colecraft, PhD [Internet]. Brown University; 2021 [cited 2025 Sept 19]. Available from: https://repository.library.brown.edu/studio/item/bdr:eefr6skt/.
103. Haddock S, Moore CI, Shaner NC. Interview with experts: Steven Haddock, PhD [Internet]. Brown University; 2021 [cited 2025 Sept 19]. Available from: https://repository.library.brown.edu/studio/item/bdr:23yje4k9.
104. Herlitze S, Moore CI, Hochgeschwender U. Interview with experts: Stefan Herlitze, PhD [Internet]. Brown University; 2021 [cited 2025 Sept 19]. Available from: https://repository.library.brown.edu/studio/item/bdr:nwpsrhtw/.
105. Prescher J, Moore CI, Shaner NC. Interview with experts: Jennifer Prescher, PhD [Internet]. Brown University; 2021 [cited 2025 Sept 19]. Available from: https://repository.library.brown.edu/studio/item/bdr:4f23ydub/.
106. Hegemann P, Moore CI, Hochgeschwender U. Interview with experts: Peter Hegemann, PhD [Internet]. Brown University; 2021 [cited 2025 Sept 19]. Available from: https://repository.library.brown.edu/studio/item/bdr:8s8ypt6w/.
107. Goun E, Moore CI, Lipscombe D. Interview with experts: Elena Goun, PhD [Internet]. Brown University; 2021 [cited 2025 Sept 19]. Available from: https://repository.library.brown.edu/studio/item/bdr:zbamwm4e/.
108. Björefeldt A, Murphy J, Crespo EL, Lambert GG, Prakash M, Ikefuama EC, et al. Efficient opto- and chemogenetic control in a single molecule driven by FRET-modified bioluminescence. Neurophotonics. 2024;11(2):021005.
109. Lambert, G.G., Crespo, E.L., Murphy, J. et al. CaBLAM: a high-contrast bioluminescent Ca2+ indicator derived from an engineered *Oplophorus gracilirostris* luciferase. Nat Methods. 2026;23:205–15. https://doi.org/10.1038/s41592-025-02972-0
110. Slaviero AN, Gorantla N, Simkins J, Crespo EL, Ikefuama EC, Tree MO, et al. Engineering luminopsins with improved coupling efficiencies. Neurophotonics. 2024;11(2):024208.
111. Zenchak JR, Palmateer B, Dorka N, Brown TM, Wagner L, Medendorp WE, et al. Bioluminescence-driven optogenetic activation of transplanted neural precursor cells improves motor deficits in a Parkinson's disease mouse model. J Neurosci Res. 2020;98(3):458–68.
112. Allen JJ, Brown T, Choudhry H, Felton R. The bioluminescence hub. 2018 [cited 2025 Sept 19]. Summer 2018 at the Scintillon Institute. Available from: https://www.bioluminescence-hub.org/education/scintillon-students.

113. Khan R. BL-OG: bridging optogenetics and chemogenetics through bioluminescence [Internet]. Brown University; 2019 [cited 2025 Sept 18]. Available from: https://repository.library.brown.edu/studio/item/bdr:847090/.
114. Khan R. Flourescence, phosphorescence, and bioluminescence [Internet]. Brown University; 2019 [cited 2025 Sept 18]. Available from: https://repository.library.brown.edu/studio/item/bdr:847089/.
115. Allen JJ, Friend P, Gregory D, Tran S. The bioluminescence hub. 2019 [cited 2025 Sept 18]. Summer 2019 at the University of California, San Diego. Available from: https://www.bioluminescencehub.org/education/UCSD-interns.
116. Allen JJ, Friedman N, van der Merwe R, Bartholomew B, Buckman E, Funderbunk R, et al. BLBase. 2023 [cited 2025 Sept 24]. BLBase. Available from: https://www.blbase.org/.
117. Allen JJ, Bartholomew B, Louis P, Roberts M, van der Merwe R. The bioluminescence hub. 2020 [cited 2025 Sept 24]. Summer 2020 Virtual Internship at Brown. Available from: https://www.bioluminescencehub.org/education/BLbase-interns.
118. Allen JJ, van der Merwe R, Buckman E. The bioluminescence hub. 2021 [cited 2025 Sept 24]. Summer 2021 Virtual Internship at Brown. Available from: https://www.bioluminescencehub.org/education/BLbase-interns-2021.
119. Allen JJ, Folkert G, Knoblock M, Simkins J, Coon N, Venn S. The bioluminescence hub. 2021 [cited 2025 Sept 24]. Summer 2021 at Central Michigan University. Available from: https://www.bioluminescencehub.org/education/CMU-interns-2021.
120. Allen JJ, Naphade O. The bioluminescence hub. 2022 [cited 2025 Sept 24]. Summer 2022 at Brown University. Available from: https://www.bioluminescencehub.org/education/Brown-interns-2022.
121. Allen JJ, Kraisriwatana L, Luong S, Nguyen F, Tate M. The bioluminescence hub. 2022 [cited 2025 Sept 24]. Summer 2022 at UCSD. Available from: https://www.bioluminescencehub.org/education/UCSD-interns-2022.
122. Allen JJ, Gorantla N, Maharjan S, McLean L, Patel J, Uprety A. The bioluminescence hub. 2022 [cited 2025 Sept 24]. Summer 2022 at Central Michigan University. Available from: https://www.bioluminescencehub.org/education/CMU-interns-2022.
123. Abhat D. NeuroNex. 2022 [cited 2025 Sept 19]. NeuroNex community presents research at 4th NeuroNex Investigator Meeting. Available from: https://neuronex.org/conference/92.
124. Crespo EL, Bjorefeldt A, Prakash M, Hochgeschwender U. Bioluminescent optogenetics 2.0: harnessing bioluminescence to activate photosensory proteins in vitro and in vivo. J Vis Exp. 2021;(174):62850.
125. Cooper H. Monterey Bay Aquarium. 2021 [cited 2025 Sept 26]. Illuminating the facts of deep-sea bioluminescence | Stories | Monterey Bay Aquarium. Available from: https://www.montereybayaquarium.org/stories/bioluminescence.
126. History of Woods Hole [Internet]. [cited 2025 Oct 1]. Available from: https://woodshole.com/about/history-of-woods-hole/.
127. History & Archives | Marine Biological Laboratory [Internet]. [cited 2025 Sept 30]. Available from: https://www.mbl.edu/about/history-archives.
128. Steinbach Ulbrich M. The last one to go. Spritsail J [Internet]. 2016 [cited 2025 Oct 2];30(2). Available from: https://woodsholemuseum.org/oldpages/sprtsl/v30n2-lastone.pdf.
129. Ocean Science Discovery Center – Woods Hole Oceanographic Institution [Internet]. https://www.whoi.edu/. [cited 2025 Oct 1]. Available from: https://www.whoi.edu/who-we-are/visit-whoi/ocean-science-discovery-center/.
130. Shimomura O | Marine Biological Laboratory [Internet]. 2018 [cited 2025 Sept 30]. Available from: https://www.mbl.edu/news/obituaries/osamu-shimomura.

Chapter 15
Outlook

Eric D. Petersen and Ute Hochgeschwender

Abstract While the current arsenal of BioLuminescent OptoGenetics (BL-OG) tools for controlling and integrating biological processes within and across cells is powerful, it is also continuously evolving and expanding. With respect to light emission, further developments will provide luciferases with brighter photon production and more diverse spectral output that, together with the synthesis of unique luciferins without cross-reactivity, will increasingly enable multiplexing of BL-OG tools. With respect to light sensing, refined and new photoactivatable proteins with improved properties and color-shifted activation spectra will substantially widen the partnerships with the improved light emitters. These parallel developments will allow higher flexibility in spatial placement and tool pairings, increasing creativity in experimental designs and expanding the applications of BL-OG in modulation, integration, and communication in biological systems. New areas of application are anticipated across cells, tissues, and organ systems, including intercellular communication over longer distances across the body, research in freely behaving experimental animals in natural habitats, interfacing with AI for real-time monitoring and adjustment of cellular behavior, and clinical applications in the human body.

Keywords Bioluminescence · Optogenetics · Luciferase · Luciferin · Photoreceptor · Biological systems

The chapters in this primer on BioLuminescent OptoGenetics (BL-OG) introduced the essential components of this tool platform, light-emitting luciferases and light-sensing photoreceptors, and their combination for controlling and integrating biological processes within and across cells. This chapter reflects briefly on where developments and where improvements might go and on novel possibilities this technology platform may enable.

E. D. Petersen · U. Hochgeschwender (✉)
College of Medicine, Central Michigan University, Mt. Pleasant, MI, USA
e-mail: peter1ed@cmich.edu; hochg1u@cmich.edu

U. Hochgeschwender (ed.), *BioLuminescent OptoGenetics*,
https://doi.org/10.1007/978-3-032-18121-3_15

With respect to luciferases, improvements will encompass brightness and kinetics, together with improved functionality of luciferins. Higher light output will decrease the concentration of substrate needed for light emission and expand the variety of light-sensing target molecules that can be used in conjunction with a luciferase. This requirement can be met by engineering luciferase-fluorescent protein pairs, taking advantage of intramolecular resonance energy transfer. Partnering luciferases with fluorescent proteins can greatly increase brightness and benefit from the rapid and ongoing expansion in the diversity of fluorescent molecules. Resonance energy coupling between luciferase and a high quantum yield fluorophore will enable selection for luciferases that have much higher substrate turnover rates, extending the versatility of light emitters. Another way higher photon emission at various wavelengths can be achieved is by creating new luciferase systems that use orthogonal luciferins, which could be achieved by traditional molecular evolution or AI-guided de novo design.

In addition to making bioluminescent photon production brighter, another important goal is to make it more diverse in spectral output, which is crucial to the growing utility of this powerful modality. The majority of BL-OG applications reviewed here utilized blue light-emitting luciferases with blue light-sensing photoreceptors, reflecting the prevalence of the most commonly used optogenetic elements. Red-light responsive optogenetic actuators are becoming increasingly available, enabling the use of red-light-emitting luciferases for BL-OG applications. While there are some red-light emitters provided by nature [1, 2], partnering with fluorescent proteins will aid in shifting color spectra, in addition to providing brighter light emitters [3–5]. Complementing these efforts in broadening the color spectrum of luciferases are the development of a variety of luciferin analogues and/or evolving or creating de novo luciferases to create different color outputs [6, 7]. In addition to allowing color tuning, unique luciferins can be synthesized that do not cross-react with other luciferases [8]. Both lines of development, unique color spectra and unique specificities, will ultimately enable multiplexing BL-OG tools. One application is the simultaneous use of Interluminescence by expressing two sets of non-cross-reacting presynaptic luciferases and postsynaptic optogenetic channels in a single synapse for sophisticated manipulation of synaptic transmission. Another example is the simultaneous sensing and resulting modulation of cellular states by expressing multiple sets of non-cross-reacting integrators (sensor–actuator pairs) within a single cell for multi-dimensional self-regulation of intracellular fluctuations. Yet, another application is the simultaneous sensing of different kinases with distinct effects through activating diverse optogenetic actuators within a single cell.

Matching improvements and extensions of light emitters are continuing developments of refined and new light-sensing molecules. Channelrhodopsins with altered photocurrent properties and color-shifted activation spectra are becoming available regularly [9, 10]. New photoactivatable proteins are being discovered, and the photophysical properties of existing photoactivatable proteins are being continuously improved [11, 12], opening many opportunities for combining them with luciferases and bioluminescent sensors in BL-OG tools. In addition, engineered tools that add photosensing to a variety of molecules not inherently photoactivatable [13] will substantially widen the partnerships with bioluminescent tools and the applications of BL-OG in modulation, integration, and communication in biological systems.

There are two major avenues to obtain both light-emitting and light-sensing molecules with new properties: bioprospecting and evolution or creation in the laboratory. Natural diversity is the oldest and richest source of new tool options, and exploration and discovery is arguably the most potent, and fun, pathway for achieving these goals [14, 15]. At the same time, it is important to recognize the role of molecular evolution and protein engineering. Initial combinations of natural luciferases and photoreceptors hardly worked for BL-OG applications when initially testing this approach, and engineering ultimately enabled this powerful system. There is still a lot of room in the random mutagenesis and rational design space to achieve substantial advances. More recently, computational de novo enzyme design using deep-learning-based approaches has aided in generating luciferases with higher substrate specificity [16]. A machine learning-guided engineering approach led to the generation of high-performance channelrhodopsins designed with high photocurrent and high light sensitivity [17]. Lastly, current BL-OG tools have all had light emitter and acceptor evolved separately and logically paired together. There might be benefits that could offer significant improvements in the co-evolution of light emitters and sensors for BL-OG tools, as optimizing the pairing of the optogenetic sensor with a photon/energy donor is critical to generating the most efficient partners.

Another challenge is the delivery of luciferin to experimental systems. Considerable progress has been made in synthesizing compounds with much improved tissue access [18–20], increasing the efficacy of optogenetic actuator activation after systemic luciferin administration, including in the brain. Another way to enable better crossing of membranes and/or targeting delivery in specific spaces and situations is the caging of luciferins. Caged luciferins are inactive, modified luciferin molecules designed to release active luciferin only when a specific trigger, such as an enzyme, or a chemical reaction, is applied. Caged luciferins can be designed to cross membranes more efficiently, be protected from degradation, and allow for cell-specific uncaging and light emission [21–24]. Another emerging solution might be genetically encodable and autonomous luciferin synthesis within the organism, an option especially attractive for applications with BL-OG integrators where light emission is dependent on fluctuations of intracellular factors. However, luciferin synthetic pathways are currently only known for the bacterial and fungal systems that will have to be adapted for efficient use in mammalian organisms [25–28].

In addition to the question of how to make BL-OG tools better is the question of what the next frontiers for applying BL-OG tools might be. Combining bioluminescence with opsins, transcription factors, intracellular signaling molecules, etc., for controlling processes within cells will permit even more creativity in experimental designs as improved light emission/sensing pairs allow higher flexibility in spatial placement (non-tethered partnering) and tool pairings. The widening development of bioluminescent sensors for ions, kinase activity, mechanical tension, or blood components will directly benefit their application in BL-OG tools ([29–34] and

Chap. 10). The progress made in developing split luciferase scaffolds [30] opens the door for taking advantage of many more sensor domains for integrating conditional bioluminescence emission with optogenetic actuators for numerous opportunities for self-regulation of cellular states. By placing the light emitter and light acceptor in separate cells, BL-OG enables cell-to-cell communication and control of cellular processes across cells. While this powerful method has been demonstrated for synaptic transmission in neurons, communication between cells can be extended from neuron-neuron to neuron-glia or vice versa [35, 36] or from neurons to tumor cells [37, 38], etc. Further, there are countless possibilities in combining bioluminescent sender and photoreceptor receiver cells for controlling many biological events, including membrane potential, transcription, and intracellular signaling, and across various types of cells, e.g., of the immune system. So far, bioluminescence-mediated cell-to-cell communication has been accomplished over short distances (pre-synapse to post-synapse). Intercellular communication over longer distances could potentially be achieved by targeting luciferases to extracellular vesicles, such as microvesicles or exosomes [39, 40], that are released by cells of origin to be taken up by specific recipient cells across the body expressing an optogenetic actuator. Utilizing this cell-cell crosstalk system would allow researchers to modulate, via BL-OG, inflammatory responses or responses to tumor growth, potentially influencing disease progression and/or alleviation.

A number of BL-OG applications have been demonstrated in neurons, and BL-OG tools will continue to be used to control neuronal membrane potential and thus neuron activity, synaptic transmission between neurons, and integration of intracellular fluctuations, e.g., calcium or pH, with neural activity states. It is anticipated that BL-OG applications will spread across cell, tissue, and organ systems as researchers outside of neuroscience become more aware of the potential of bioluminescence for activating optogenetic elements for a wide variety of actions that can be achieved in addition to regulating ion flow.

Finally, further developments and improvements of BL-OG tools are prone to catalyze progress in two important, parallel areas of biomedical research. The first is research in awake, unencumbered, freely behaving experimental animals, potentially in "natural" habitats rather than laboratory settings. With the continuing development of highly sensitive cameras, BL-OG enables monitoring of intracellular events and activation of photoreceptors simultaneously with animal behavior. At the other end of the spectrum are applications in 2- and 3-dimensional cell systems in cultures, where bioluminescence enables tracking changes in cells experiencing BL-OG integration and control within and across cells that can be interfaced with AI for real-time monitoring and adjustment of cellular behavior [41].

Looking out further and anticipating advances in clinical use of viral vectors, many of the BL-OG tools that were developed for basic and preclinical research could have broad clinical applications, taking advantage of the orthogonality of luciferase–luciferin–photoreceptor interactions to endogenous signaling in the human body. For example, tools such as PhAST and Interluminescence that enable cell-to-cell communication or NT-LMOs that modulate the actions of endogenous neurotransmitters could be applied to virtually any neurological or neurodegenerative disorder by using the BL-OG tools to correct an underlying circuit pathology, e.g., rescuing the actions of a circuit that has degenerated or dampening the actions of a hyperactive circuit. Another example is applying the bioluminescent kinase sensors that can be adapted to respond to most cell signaling pathways and can act on optogenetic actuators. Here, the pathologic signaling events trigger activation of the optogenetic components to correct the aberrant signaling either through direct action or transcription of a desired protein.

References

1. Viviani VR, Bevilaqua VR, de Souza DR, Pelentir GF, Kakiuchi M, Hirano T. A very bright far-red bioluminescence emitting combination based on engineered railroad worm luciferase and 6'-amino-analogs for bioimaging purposes. Int J Mol Sci. 2020;22(1):303.
2. Branchini BR, Fontaine DM, Kohrt D, Huta BP, Racela AR, Fort BR, et al. Systematic comparison of beetle luciferase-luciferin pairs as sources of near-infrared light for in vitro and in vivo applications. Int J Mol Sci. 2022;23(5):2451.
3. Torrey ZR, Halbers LP, Scipioni L, Tedeschi G, Digman MA, Prescher JA. A versatile bioluminescent probe with tunable color. RSC Chem Biol. 2024;5(11):1097–103.
4. Hattori M, Wazawa T, Orioka M, Hiruta Y, Nagai T. Creating coveted bioluminescence colors for simultaneous multi-color bioimaging. Sci Adv. 2025;11(4):eadp4750.
5. Tian X, Zhang Y, Du H, Huang W, Bizimana LA, Nishimura N, et al. An enhanced red bioluminescent indicator for responsive detection of physiological calcium dynamics in cells and mice. ACS Sens. 2025;10(8):5826–33.
6. Yeh HW, Xiong Y, Wu T, Chen M, Ji A, Li X, et al. ATP-independent bioluminescent reporter variants to improve in vivo imaging. ACS Chem Biol. 2019;14(5):959–65.
7. Tamaki S, Kitada N, Kiyama M, Fujii R, Hirano T, Kim SB, et al. Color-tunable bioluminescence imaging portfolio for cell imaging. Sci Rep. 2021;11(1):2219.
8. Williams SJ, Hwang CS, Prescher JA. Orthogonal bioluminescent probes from disubstituted luciferins. Biochemistry. 2021;60(8):563–72.
9. Alekseev A, Hunniford V, Zerche M, Jeschke M, El May F, Vavakou A, et al. Efficient and sustained optogenetic control of sensory and cardiac systems. Nat Biomed Eng. 2025. https://doi.org/10.1038/s41551-025-01461-1.
10. Govorunova EG, Sineshchekov OA, Li H, Gou Y, Chen H, Yang S, et al. Blue-shifted ancyromonad channelrhodopsins for multiplex optogenetics. elife. 2025;14:RP106508.
11. Xu Q, Vogt A, Frechen F, Yi C, Küçükerden M, Ngum N, et al. Engineering bacteriophytochrome-coupled photoactivated adenylyl cyclases for enhanced optogenetic cAMP modulation. J Mol Biol. 2024;436(5):168257.
12. Hallmann A. Sensing a rainbow of colors: algal photoreceptors. Front Plant Sci. 2025;16:1684559.
13. McCue AC, Kuhlman B. Design and engineering of light-sensitive protein switches. Curr Opin Struct Biol. 2022;74:102377.
14. Hellinger J, Jägers P, Donner M, Sutt F, Mark MD, Senen B, et al. The flashlight fish Anomalops katoptron uses bioluminescent light to detect prey in the dark. PLoS One. 2017;12(2):e0170489.
15. Lambert GG, Depernet H, Gotthard G, Schultz DT, Navizet I, Lambert T, et al. Aequorea's secrets revealed: new fluorescent proteins with unique properties for bioimaging and biosensing. PLoS Biol. 2020;18(11):e3000936.

16. Yeh AHW, Norn C, Kipnis Y, Tischer D, Pellock SJ, Evans D, et al. De novo design of luciferases using deep learning. Nature. 2023;614(7949):774–80.
17. Bedbrook CN, Yang KK, Robinson JE, Mackey ED, Gradinaru V, Arnold FH. Machine learning-guided channelrhodopsin engineering enables minimally invasive optogenetics. Nat Methods. 2019;16(11):1176–84.
18. Su Y, Walker JR, Park Y, Smith TP, Liu LX, Hall MP, et al. Novel NanoLuc substrates enable bright two-population bioluminescence imaging in animals. Nat Methods. 2020;17(8):852–60.
19. Su Y, Walker JR, Hall MP, Klein MA, Wu X, Encell LP, et al. An optimized bioluminescent substrate for non-invasive imaging in the brain. Nat Chem Biol. 2023;19:731–9.
20. Gao C, Wu Y, Fitzgerald C, Wang H, Ugo T, Uyeda T, et al. An optimized luciferin formulation for NanoLuc-based in vivo bioluminescence imaging. Sci Rep. 2025;15(1):12884.
21. Tian X, Zhang Y, Li X, Xiong Y, Wu T, Ai HW. A luciferase prosubstrate and a red bioluminescent calcium indicator for imaging neuronal activity in mice. Nat Commun. 2022;13(1):3967.
22. Orioka M, Eguchi M, Mizui Y, Ikeda Y, Sakama A, Li Q, et al. A series of furimazine derivatives for sustained live-cell bioluminescence imaging and application to the monitoring of myogenesis at the single-cell level. Bioconjug Chem. 2022;33(3):496–504.
23. Navarro MX, Brennan CK, Love AC, Prescher JA. Caged luciferins enable rapid multicomponent bioluminescence imaging. Photochem Photobiol. 2023;100(1):67–74.
24. Tian X, Zhang Y, Ai HW. PEGylated ATP-independent luciferins for noninvasive high-sensitivity high-speed bioluminescence imaging. ACS Chem Biol. 2025;20(1):128–36.
25. Kusuma SH, Kakizuka T, Hattori M, Nagai T. Autonomous multicolor bioluminescence imaging in bacteria, mammalian, and plant hosts. Proc Natl Acad Sci USA. 2024;121(41):e2406358121.
26. Kiszka KA, Dullin C, Steffens H, Koenen T, Rothermel E, Alves F, et al. Autonomous bioluminescence emission from transgenic mice. Sci Adv. 2025;11(28):eads0463.
27. Shakhova ES, Karataeva TA, Markina NM, Mitiouchkina T, Palkina KA, Perfilov MM, et al. An improved pathway for autonomous bioluminescence imaging in eukaryotes. Nat Methods. 2024;21(3):406–10.
28. Gleneadie HJ, Veland N, Warren EC, Sardini A, Barroso C, Fadeeva AA, et al. A step towards animal models with self-sustained fungal bioluminescence. bioRxiv. 2025;2025.08.04.668132.
29. Zhao S, Xiong Y, Sunnapu R, Zhang Y, Tian X, Ai HW. Bioluminescence imaging of potassium ion using a sensory luciferin and an engineered luciferase. J Am Chem Soc. 2024;146(19):13406–16.
30. Lambert GG, Crespo EL, Murphy J, Boassa D, Luong S, Celinskis D, et al. CaBLAM – a high-contrast bioluminescent Ca(2+) indicator derived from an engineered Oplophorus gracilirostris luciferase. Nat Methods. 2026;23(1):205–15.
31. Wu Y, Walker JR, Westberg M, Ning L, Monje M, Kirkland TA, et al. Kinase-modulated bioluminescent indicators enable noninvasive imaging of drug activity in the brain. ACS Cent Sci. 2023;9(4):719–32.
32. Zhong BL, Elliot JM, Wang P, Li H, Hall RN, Wang B, et al. Split luciferase molecular tension sensors for bioluminescent readout of mechanical forces in biological systems. ACS Sens. 2024;9(7):3489–95.
33. Tanaka R, Sugiura K, Osabe K, Hattori M, Nagai T. Genetically encoded bioluminescent glucose indicator for biological research. Biochem Biophys Res Commun. 2025;742:151092.
34. Chen JYH, Peng X, Xi C, Lee GR, Baker D, Yeh AHW. De novo design of high-performance cortisol luminescent biosensors. J Am Chem Soc. 2025;147(31):27494–505.
35. Gundersen V, Storm-Mathisen J, Bergersen LH. Neuroglial transmission. Physiol Rev. 2015;95(3):695–726.
36. Gu N, Makashova O, Laporte C, Chen CQ, Li B, Chevillard PM, et al. Microglia regulate neuronal activity via structural remodeling of astrocytes. Neuron. 2025;113(20):3408–3423.e5.
37. Drexler R, Drinnenberg A, Gavish A, Yalçin B, Shamardani K, Rogers AE, et al. Cholinergic neuronal activity promotes diffuse midline glioma growth through muscarinic signaling. Cell. 2025;188(17):4640–4657.e30.

38. Savchuk S, Gentry KM, Wang W, Carleton E, Biagi-Junior CAO, Luthria K, et al. Neuronal activity-dependent mechanisms of small cell lung cancer pathogenesis. Nature. 2025;646(8087):1232–42.
39. Pitt JM, Kroemer G, Zitvogel L. Extracellular vesicles: masters of intercellular communication and potential clinical interventions. J Clin Invest. 2016;126(4):1139–43.
40. Liu YJ, Wang C. A review of the regulatory mechanisms of extracellular vesicles-mediated intercellular communication. Cell Commun Signal. 2023;21(1):77.
41. Ahmed SA, Saikia A, Devi SG, Borah JC. AI-assisted cell culture system. Methods Mol Biol. 2025;2952:149–67.

Index

U. Hochgeschwender (ed.), *BioLuminescent OptoGenetics*,
https://doi.org/10.1007/978-3-032-18121-3

The manufacturer's authorised representative in the EU is Springer Nature Customer Service Centre GmbH, Europaplatz 3, 69115 Heidelberg, Germany. If you have any concerns regarding our products, please contact ProductSafety@springernature.com

Printed and bound by CPI Group (UK) Ltd, Croydon, CR0 4YY

05/10/2026

02228057-0004